THE FORCE OF THE VIRTUAL

The Force of the Virtual

. . . .

Deleuze, Science, and Philosophy

Peter Gaffney, Editor

University of Minnesota Press
Minneapolis
London

Portions of chapter 4 were previously published in Steven Shaviro, *Without Criteria: Kant, Whitehead, Deleuze, and Aesthetics* (Cambridge, Mass.: The MIT Press, 2009).

Published by the University of Minnesota Press
111 Third Avenue South, Suite 290
Minneapolis, MN 55401–2520
http://www.upress.umn.edu

Library of Congress Cataloging-in-Publication Data
The force of the virtual : Deleuze, science, and philosophy / Peter Gaffney, editor.
p. cm.
Includes bibliographical references and index.
ISBN 978-0-8166-6597-6 (hc : alk. paper) — ISBN 978-0-8166-6598-3 (pb : alk. paper)
1. Science—Philosophy.   2. Deleuze, Gilles, 1925–1995.   I. Gaffney, Peter.
Q175.F7126 2010
501—dc22
2009026453

Printed in the United States of America on acid-free paper

The University of Minnesota is an equal-opportunity educator and employer.

17 16 15 14 13 12 11 10   10 9 8 7 6 5 4 3 2 1

Contents

Preface

THE IDEA TO MAKE A BOOK OF THIS KIND began at the annual meeting of the American Comparative Literature Association in 2006 at a panel discussion organized by Catherine Liu, "Individuals, Groups, Multiplicities: Humans and Others." The question was raised whether something on the order of definitions is disturbed, if somebody somewhere ought to take offense, whenever people in the social sciences—often with the best intentions—talk about science as if it were a purely creative enterprise. It is evident that *something* is disturbed, that *someone* is offended. Simply to ask this question calls attention to a certain divide, not only between one discipline and another, one methodology and another, one truth-system and another. More critically, it points to a set of privileges, perhaps even an ethics, that is necessarily invoked in the production of scientific knowledge but is not typically invoked with the notion of creativity (creativity has its own "ethics"). How have we, scientists and nonscientists alike, arrived at such a distinction? With what aim have we molded our language around the creation of phenomena, on the one hand, and the observation or description of phenomena on the other? To follow this line of inquiry and yet maintain the legitimacy of science *as a creative enterprise* is the challenge we face when we try to understand the philosophy—and "scienticity"—of a thinker like Gilles Deleuze.

The need to engage this dimension of Deleuze's work becomes evident when we consider a trend in the social sciences that Bruno Latour calls "antifetishism."[1] The critical gesture, he claims, has two moments. In the first moment, it reveals that process by which we (naïve believers) transform an object into a screen for our own wishes; the material entity "does nothing at all by itself." In the second moment, the critical gesture shows how these same wishes are the direct effect of our genes, interests, drives, etc.—in other words, that we are acted on by some other "causal" entity. When it comes to the objects of science, however, we find that they are

> much too strong to be treated as fetishes and much too weak to be treated as indisputable causal explanations of some unconscious action. And this is not true of scientific states of affairs only. . . . Once you realize that scientific objects cannot be explained, then you realize too that the so-called weak objects, those that appear to be candidates for the accusations of antifetishism, were never mere projections on an empty screen either. They too act, they too do things, they too *make you do* things. . . . To accuse something of being a fetish is the ultimate gratuitous, disrespectful, inane, and barbarous gesture.[2]

Latour's attempt to reorient critique around what things *do* rather than what things *are* is anticipated by nearly every work in Deleuze's corpus, from *Subjectivity and Empiricism* to *Bergsonism*, and from *A Thousand Plateaus* to *What Is Philosophy?* Throughout the development of his philosophy, Deleuze does not entertain an "antifetishism" that would undermine the objects of science. Indeed, he is always more critical of the cherished objects of psychoanalysis or of a general "decalcomania" that is at work behind the scenes of the critical gesture.

To be sure, there are other challenges as well. It will be nearly impossible at times to examine the difference between science and philosophy and yet think critically about what we mean when we make that distinction. As Manuel DeLanda points out in his contribution to this volume, we make something of a misstep simply by speaking of "science in general."[3] By the same token, it will be difficult to focus our attention on the convenient (but often misleading) metaphors we use to describe the world, without losing the very means to describe it. As Steven Shaviro writes, "It still remains the case that biologists can only explain an organism by speaking *as if* its features (eyes, or reproductive behaviors, or whatever) were purposive."[4] There will be many other difficulties, some of which, no doubt, this volume neither addresses nor avoids. It is our hope that the various approaches presented here will lead to other questions and provide the grounds on which to give such questions legitimate expression.

Acknowledgments

THIS PROJECT HAS COME A LONG WAY since our initial conversations on the topic, and its final format is more extensive than I had originally planned. It has involved a number of people, both scientists and nonscientists, to whom I owe a tremendous debt of gratitude. First, I give my most sincere thanks to the contributors, who have taught me so much and have given me something even greater in the way of their trust, patience, and encouragement at all stages.

For my part, this book began in long conversations with Tom Kelso and Julie Kruidenier, and my ongoing admiration for them, more than anything, has been the guiding motivation for this project—as it has been, I suspect, for many other projects I have undertaken in the past few years. I thank Jean-Michel Rabaté, Meta Mazaj, and Gerry Prince for the support and inspiration (much, much inspiration) they gave along the way. I am indebted to Jean-Jacques Lecercle and Gregory Flaxman for their vital role at the beginning of this project; Ilinca Iurascu (who helped with German translations), Ian Buchannan, and David Holdsworth for their help in crucial last stages; and Doug Armato and Danielle Kasprzak at the University of Minnesota Press for invaluable assistance. I also thank Cris Neil, Greg Hurlock, Shanti Oram, Rolf Lakaemper, Michael Schupp, and Petar Mamula, the scientists and doctors who added presence and accountability to my understanding of science as a discipline.

Finally, I thank Esther Alarcon Arana, Nicola Gentili, Michael Koshinskie and Ginny Chimel, David Humphries, Dana Grozdanic, Anna Frangiosa, Cedric Tolliver, Robert Barta, Michael and Sonia Agnew, Rob Goenen, Gregg Greene, Adam Cornell, Lucie Hawa Goldin, Beni Shwartz, and Michael Oristian, as well as my parents and sister, for their friendship, love, and support.

Science in the Gap

Peter Gaffney

There are notions that are exact in nature, quantitative, defined by equations, and whose very meaning lies in their exactness: a philosopher or writer can use these only metaphorically, and that's quite wrong, because they belong to exact science. But there are also essentially inexact yet completely rigorous notions that scientists cannot do without, which belong equally to scientists, philosophers, and artists. They have to be made rigorous in a way that's not directly scientific, so that when a scientist manages to do this he becomes a philosopher, an artist, too. This sort of concept's not unspecific because something's missing but because of its nature and content.

—Gilles Deleuze, *Negotiations*

I N *WHAT IS PHILOSOPHY?* Deleuze and Guattari introduce a paradoxical gap in the order of becoming that rules out any straightforward reading of the opposition between the actual and the virtual: "The actual is not what we are but, rather, what we become, what we are in the process of becoming—that is to say, the Other, our becoming-other."[1] They draw on various constructions of history, by Péguy, Nietzsche, and Foucault (to whom Deleuze and Guattari attribute this definition), to show how thought emerges not from the actual *or* the virtual, but from a process of actualization that joins the two together in a productive tension, and that supersedes any sense in which these terms represent ontologically distinct modes of Being. The relevant opposition is not between the actual and the virtual, but between the continuous actual-virtual system and an already-vanishing present that Deleuze and Guattari describe as "what we are and, thereby, what we are ceasing to be."[2] Similarly, in the opening pages of *What Is Philosophy?* we find a formulation of the "Other Person" as that which "makes the world go by" with respect to an "I"; the latter, in turn, can only designate a world already past (18). If philosophy only reaches maturity when it turns away from this world already past (the self), that is because thought itself takes place in a gap where everything is already becoming other, already anticipating (and participating in) the emergence of a new world. To think means not only to occupy this gap in the order of becoming, but to take part in and sustain the force of the virtual that is the basic condition for qualitative change "in the world"—that is, in a world that is both fundamentally plural and never fully actualized.

Science in the Gap

But what about science? If philosophical thought sustains the force of the virtual, if it takes place in the gap between worlds, can we confer a similar status to scientific thought? Is science, too, a becoming-other? What would this tell us about the virtual? In what ways, conversely, do scientific practices already presuppose the force of the virtual, acting on the history of science "behind the scenes," as it were? Deleuze and Guattari's final analyses of the actual-virtual system give us a tantalizing glimpse of the creative energy (and agency) proper to the plane of immanence; yet they also challenge the classical notion of scientific praxis as a domain perpetually unfolding in relation to an already subjectivized body of knowledge—i.e., *physical realism:* a representation of, rather than an idea in, the real world. If the scientific community is engaged in organizing knowledge around an existing state of affairs—around a vanishing present as "unity of science"—it is no less oriented toward the horizon of what it is capable of knowing, where it never ceases to take part in the actualization of the world. This is as good as saying that what is actualized is never fully actual, that the set of relations that emerges in scientific knowledge is never fully determinate, much less unified: "Philosophy proceeds with a plane of immanence or consistency; science with a plane of reference. In the case of science it is like a freeze-frame. It is a fantastic *slowing down,* and it is by slowing down that matter, as well as the scientific thought able to penetrate it with propositions, is actualized. A function is a Slow-motion."[3] This concept will be essential to our understanding of science and the various strategies it deploys in face of a perpetually receding actuality. To say that a plane of reference slows down the virtual is not the same as saying that it arrests or isolates it in a discrete state of affairs. Science, like matter, does not escape the logic of becoming: its object is not a static world, but one that remains always in the process of (qualitative) change. Nor, in these terms, do scientific practices disengage a world from the plane of immanence, so much as establish a set of coordinates that allow thought and matter to coincide, to create a block of becoming—not a unity of science, but a "unity of composition" as regards a particular historical stratum.[4]

The first problem to consider in a discussion of Deleuze and science is this new paradigm, which matches thought to a world that has no essences, only singularities, flows, processes of *individuation* and *singularization*. Whereas the first term refers to patterns of change that effectively negate time (organizing phenomena around static variables), the second refers to dynamic physical processes in which the variables are themselves subject to sudden reversals, dissolutions, and emergences (in the form of "strange attractors," for example). The question here is: How will science reorient its production of knowledge in such a way that it may grasp these singularities, without thereby plunging into a virtual chaos that defies all efforts to comprehend it? What are the adequate terms in which to evaluate scientific knowledge, when it cannot simply be verified with respect to a point of reference outside the process of qualitative change it seeks to describe? "Science is haunted not by its own unity but by the plane of reference constituted by all the limits or borders through which it confronts chaos."[5] The new scientific paradigm presented by Deleuze and Guattari involves principles and practices that do not simply observe, but directly confront and, in the final analysis, *construct* a world. The plane of reference is a tool for engaging chaos, but it is not a readymade "site" of knowledge (Neoplatonic conception of the *nous*). It belongs instead to a transcendental empiricism, actively deploying processes that exceed both the subject and object of knowledge, and which comprise the basis of all lived experience in the form of affect, intensity, difference—a sense of movement in all things that Alfred North Whitehead called *appetition*.[6]

In other words, science cannot occupy a position of neutrality vis-à-vis its object, nor, strictly speaking, can it presume that this object already exists. At most we can say that a particular (historically specific) body of scientific thought has a *reciprocal* relationship with the object it determines, each one participating in the actualization of the other and simultaneously traversing a diversity of social, intellectual, and material processes. We may think, for instance, of the field of actualization occupied by technology as it enters into one alliance after another with the political and social paradigms that regard these technologies as instrumental. At the same time that a scientific breakthrough influences nonscientific spheres of thought and action (i.e., social and political movements), we see how these other historically specific "becomings" have already had a large part in orienting material and theoretical resources in the scientific community toward a particular set of problems to be solved (as in the rapid growth of nuclear physics during the Cold War). To this extent, a Deleuzian philosophy of science can be said to coincide with Jean Piaget's notion of "genetic epistemology," the process by which a body of knowledge organizes an experiential world by organizing itself.[7] In its confrontation with chaos, scientific thought leads to accelerations and decelerations in the technological, social, and political strata that comprise its milieu. This network of connections has an important reciprocal effect on scientific thought, so that we cannot speak of first causes, only co-occurrences in thought and matter that emerge simultaneously from a field of immanence.

The constructivist attitude toward science has profound consequences for the way we understand the status of thought with respect to its object. In *Intensive Science and Virtual Philosophy*, Manuel DeLanda explains how epistemological problems that emerge from the relationship between thought and the world are always already determined by the ontological relationship between the actual and the virtual:

> In Deleuze's approach the relation between the well-posed explanatory problems and their true or false solutions is the epistemological counterpart of the ontological relation between the virtual and the actual. Explanatory problems would be the counterpart of virtual multiplicities since, as he says, "the virtual possesses the reality of the task to be performed or a problem to be solved." Individual solutions, on the other hand, would be the counterpart of actual individual beings: "An organism is nothing if not the solution to a problem, as are each of its differentiated organs, such as the eye which solves a light problem."[8]

Deleuze himself identifies this move from epistemology to ontology with the Bergsonian conception of the problem—

> It is true that, in Bergson, the very notion of the problem has its roots beyond history, in life itself or in the *élan vital*: Life is essentially determined in the act of avoiding obstacles, stating and

solving a problem. The construction of the organism is both the stating of a problem and a solution.[9]

—and more emphatically, with any attempt to derive meaning from matter:

Neither the problem nor the question is a subjective determination marking a moment of insufficiency of knowledge. Problematic structure is part of objects themselves, allowing them to be grasped as signs.[10]

We find yet another formula for the mutual determination of thought and being in *What Is Philosophy?*

Infinite movement is double, and there is only a fold from one to the other. It is in this sense that thinking and being are said to be one and the same. Or rather, movement is not the image of thought without being also the substance of being. When Thales's thought leaps out, it comes back as water. When Heraclitus's thought becomes *polemos,* it is fire that retorts. It is a single speed on both sides . . .[11]

How then does thought—scientific or otherwise—work? Any given body of knowledge sustains an essential ambiguity or movement that perpetually exceeds it, resulting in a zone of "ontological difference" that is inseparable from the same gap in matter—what Foucault calls the "outside" of thought.[12] This means that, even as it turns toward matter and away from thought, science is confronted with the problem of difference. Science, just like philosophy, finds itself "in the gap." It, too, is situated within an indeterminate "zone of immanence," and its capacity to interrogate matter, to render signs from experimental data, is only possible by virtue of this smallest distance between a continuously renewed problem—the force of the virtual as such—and the imminent solution. Science thus makes what Deleuze calls a "leap into ontology" simply by bringing its own laws and principles into contact with the problem—the "chaos that haunts it"—thereby facilitating and allowing itself to be "swept away" by the movement of becoming. This is in spite of our customary image of the scientist as an indifferent observer, who makes sense of disparate phenomena by assuming a purely objective view vis-à-vis the observable world (Locke's empiricism, for instance, in which observation is limited to measurable or "primary" qualities).

Inspired in large part by Michel Serres's conception of atomism as a science of flows (*The Birth of Physics*), Deleuze and Guattari will seek out those areas of contemporary science where the act of observation is taken up by the logic of ontological difference—by "vortical movement," "smooth space," and other phenomena that take place on a plane of reference that is dissolved and recreated in an accelerated manner.[13] The moment science turns from the imminent solution to the continuous renewal of the problem, its production of scientific "truth" is suspended and ceases to act as the validating mechanism for political and social enunciations (its role as "royal" science).[14] In our view, this is what constitutes science's maturity: a process of "nomadization" in the role of the scientific observer that allows a field of knowledge to "singularize" alongside its object. We have to be careful here not to think of "nomadology" in teleological terms, or as a new set of criteria for evaluating the progress of science. The maturity of science, the distinction between nomadic thought and its sedentary or royal counterpart, refers only to a differential in the rhythm and scope of the actual-virtual system. It means speeding up or slowing down the designation of new and divergent worlds, raising or lowering the threshold of tolerance for disjunctions on the plane of reference, or between one plane of reference and another. From our own historically specific point of view, some terms and concepts will necessarily appear more adequate to the task than others. The classical notion of statistical variance, for example, is less productive than that of singularities, strange attractors, and turbulence. Lacan's theory of science as a symbolic chain that seeks (and fails) to "suture" the subject[15] must be qualified by Thomas Kuhn's concept of scientific breakthrough along the lines of a "paradigm-induced gestalt switch,"[16] and by Guattari's own investigation of scientific breakthrough in his essay "Machine and Structure."[17] In these cases, if scientific thought has begun to resemble a creative process, converging to some degree with the disciplines of art and philosophy, it is because thought is inseparable from a process of actualization with which all forms of creativity are coextensive. Science does not describe an objective state of affairs so much as inscribe a more or less mobile point of view within things themselves, causing a plurality of worlds to emerge from the virtual state of flux.

Is it too early to declare that science, too, has reached its maturity? Has it not already begun, in some cases, to consider the knowledge it produces as the result of an ongoing engagement with social and technological imperatives, not merely observable phenomena? And would this not necessarily imply—paradoxically, problematically—that it might proceed as an affirmation of simulacra, interest and creativity, rather than acting as mere instrument of technology or authentication of positivistic reason? The aim of the present volume is to consider the possibilities of such a science as it engages (or is engaged by) the force of the virtual, and how this influences our understanding of Deleuze's process-oriented ontology. The question is not only whether science withstands, confirms, or refutes the role that Deleuze attributes to it, but how this challenges contemporary literary and social criticism to "play by new rules." What are these rules, and what consequences will they have for the philosopher as producer? What mutations can we expect in our perceptions of time and space? How does Deleuze's theory of the virtual provide the basis for new alliances ("interbreedings" or "interferences") among diverse fields of production, from mathematics to digital technology, architecture to social theory—what may we expect them to produce?

The "Scienticity" Polemic

Questions like these will be especially pertinent in light of a polemic that has recently emerged from discussions regarding what we might call Deleuze's "scienticity."[18] Throughout its development, Deleuze's thinking does not intersect with science so much as intercept it, in the space between one discourse and another, refusing to maintain a critical distance where it would be possible to make the traditional evaluations of good or bad, true or false. This makes it all the more tempting to see his metaphysics in terms of a scienticity that would put it on equal footing with empirically based knowledge claims—i.e., physics proper—or to consider, in any case, what such a move on the part of Deleuzian studies might imply. In recent years, important scholarship has emerged that calls attention to the way works by Deleuze (and Guattari) point to a fundamental shift in the relationship between Continental philosophy and science. Jean-Clet Martin describes a Deleuzian "microphysics" of qualities,[19] and Rosi Braidotti refers to his work as a "materialist, high-tech brand of vitalism."[20] John Mullarkey also speaks of Deleuze's tendency to formulate internal, affective experience in physicalist terms, such as *conatus* and *spatium*.[21] But he sees many of these categories as reinforcing, rather than dissolving, the distinction between virtual and actual, as they tend to give only derivative status to the latter. There are still others who see this tendency as a result of Deleuze's effort to reformulate science according to the rules of metaphysics, as an act of resistance to the unifying logic of scientific thought with respect to a single historically specific point of view.

For our purposes, we can reduce these various positions to two divergent approaches in Deleuzian studies. The first compares Deleuze's work to theories and discoveries advanced by contemporary "radical" science (the notion of complexity in various branches of theoretical science: symbiogenesis in biology; topology in physics and mathematics, etc.). The second approach tries to limit this comparison, using terms that point nonetheless to radical developments in the philosophy of science (intuitionism, constructivism), developments that run parallel to Deleuze's nonepistemic methods and aims. Both approaches render valuable insight on Deleuze's work, but they also represent diverging paths for the future of Deleuzian studies. In the first instance, writers like Manuel DeLanda and Keith Ansell Pearson (as well as Timothy Murphy, Mark Hansen, and many of the contributors to this volume) have reconstructed key Deleuzian concepts using counterparts in science and mathematics: manifolds, complex dissipative systems, vector fields, topological space, embryogenesis, etc.[22] The advantage of such an approach is that it bridges the gap between Deleuze's philosophical inheritance properly so-called—Spinoza's expressionism, Nietzsche's eternal return, Bergson's *élan vital,* etc.—and texts which tend to be more obscure (to nonscientists) by the likes of Weismann, Varela, Uexküll, and Whitehead, to which Deleuze *also* owes a significant debt of influence. More controversially, the scientistic approach has the edifying effect of a "second-order viability," a term that describes the scientific imperative to corroborate findings by way of independent research. As DeLanda writes in his introduction to *Intensive Science and Virtual Philosophy,* a reconstruction of Deleuze's metaphysics in scientific terms would

> show that his conclusions do not depend on his particular choice
> of resources, or the particular lines of argument he uses, but that
> they are *robust to changes* in theoretical assumptions and strategies.
> Clearly, if the same conclusions can be reached from entirely differ-
> ent points of departure and following entirely different paths, the
> validity of those conclusions is thereby strengthened.[23]

The upshot of this approach, beyond its immediate implications for Deleuzian studies, is a more general merger of metaphysics and theoretical science, a strategy that is all the more controversial as it satisfies a general demand (in popular culture, but also in the academe) for a more rigorous approach to the social sciences.

The second, critical approach proceeds by drawing a line between physics and metaphysics, highlighting the epistemic truth claims of one as they contrast with the "philosophical openness" of the other. James Williams describes Deleuze as "a vital metaphysician [who] opens life on to a realm of conditions and potentials that resist modern day naturalism (the reduction of life to the findings of the natural sciences defined in a very limited way)."[24] Deleuze's most important contribution to contemporary thought, in these terms, is not his reconciliation of philosophy with science, but a strategy of resistance that aims primarily at disengaging thought from the logic of second-order viability, extending well beyond science to the positivistic tendencies of "modern day naturalism." Williams does not discredit DeLanda's interpretation per se, and he will even praise its explanatory power as regards Deleuze's more oblique theories of natural and social systems. What bothers Williams, rather, is this tendency to draw Deleuze's thinking into the validating light (and deceptive clarity) of scientific objectivity: "This leads to a devaluing of the transcendental move in Deleuze's work and to a science-based definition of terms such as multiplicity and the virtual, when they cannot be simply determined in this way without inviting grave problems with respect to historical situation. How can we know that any given science is definitive or even on the way to being definitive?"[25] It is not this or that scientific theory that is at issue here, but a tension that emerges between the virtual consistency of philosophical concepts, on the one hand, and the necessarily symbolic, plural/disjunctive character of scientific knowledge, on the other, through which the actual opposes this continuity. Williams goes further here, characterizing modern scientific practices in terms of a "dialectical position between theory and fact," a claim that points to the nomadization of science at the level of discourse:

Modern science demands a dialectical position between theory and fact, indeed, between theories, facts and further theories. It is not that facts contradict theories, neither is it that theories straightforwardly contradict one another. Rather, the nature of the debate is a two-directional one. Facts only appear thanks to theories, notably on how to simplify reality so that it may reveal facts. Theories make claims about reality that are undermined by the complexity revealed by scientific discovery. (102)

Contemporary scientists tend to regard their achievements with a more sophisticated metascientific attitude, and they acknowledge the epistemological limits of the exercise as a whole. We have seen this recently in the writings of theoretical physicist Wolfram Schommers, for example, who argues that "Even in the case of extreme objectivity it is not possible to evade metaphysics . . . even such aspects of life, which are describable by mathematics are unavoidably metaphysically interspersed, and this is valid to an increased extent for those aspects which cannot be expressed mathematically."[26] As a consequence, science as a whole can no longer pretend to arrive at a privileged frame of reference with respect to real physical processes. This necessarily precludes the use of any one scientific frame of reference (complexity theory or topology, for example) to explain philosophical thought, however compelling it may be from our historically specific point of view.

The scientistic approach, in these terms, fails to account for a pronounced difference between philosophical and scientific strategies for "inhabiting" the virtual. Williams's notion of philosophical openness indicates the capacity for metaphysical thought to coincide with difference, rather than subjecting it to a process of selection and simplification. The philosophical concept does not have the means to establish a state of affairs; according to Deleuze and Guattari, it may only extract "a consistent event from a state of affairs."[27] Similarly, metaphysics does not arrive at a fully articulated image of the world; rather, it tends to move in the opposite direction, following "a process of reaction internal to language" that spans the interval between an event in the present moment and a time that Deleuze calls "infinitive."[28] In other words, philosophical thought is coextensive with processes of both actualization and de-actualization, thus occupying the entire process of change and not simply its results (this or that scientific theory, model, or "image"). Deleuze and Guattari describe this as "a grandiose time of coexistence that does not exclude the before and after but *superimposes* them in a stratigraphic order . . . an infinite becoming of philosophy that crosscuts its history without being confused with it."[29] Philosophy thus spans all the disjunctions that separate one theory from another in a series of paradigmatic shifts, not only as regards the history of science (the theoretical divide between classical physics and complexity theory, for example) but also between one philosophical concept and another. If philosophical thought is able to "crosscut" history, it is because it remains distinct from its concepts, or indeed from any fully actualized world or state of affairs. "For Deleuze," Williams writes, "actual worlds are discontinuous, but presuppose a virtual continuity" (112). The movement of science in history, as Kuhn has already shown, is one of ruptures or "revolutions," a veritable tectonics of thought rather than an indivisible flow. Deleuze and Guattari account for this by distinguishing between the time of philosophical thought and that of science:

> Science is not confined to a linear temporal progression any more than philosophy is. But, instead of a stratigraphic time, which expresses before and after in an order of superimpositions, science displays a peculiarly serial, ramified time, in which the before (the previous) always designates bifurcations and ruptures to come, and the after designates retroactive reconnections.[30]

In this regard, philosophy has a marked advantage over science, inhabiting interstitial processes of becoming within the social, economic, technological, and political spheres that are necessarily excluded from the scientific image of physical processes.

Science thus has an active role in the process of actualization; yet the tendency that Williams calls "modern day naturalism" is unable to extricate the production of knowledge from the moment in which its theoretical object (a theory or model) assumes determinate form. Indeed, this is all we mean by the term "objectivity": a coalescing of relations within thought and matter at any given point in time. To produce this determinacy is the singular goal of normative physical science, even if it means engaging only the narrowest section of qualitative change—even if it means that each set of relations drawn from the plane of reference is already in danger of becoming obsolete the moment it emerges. The possibility of making objective claims about reality seems to come only when the process of actualization has reached completion and in reference to an object that has been fully actualized. (Yet we have already seen how such designations are problematic in view of an actual that is the "now of our becoming.") The most characteristic Deleuzian categories, on the other hand, are based on a notion of change that supposes, as its basic condition, a dimension that is always in excess of the actual, and is therefore incompatible with the basic requirements of scientific truth claims (i.e., repeatability, second-order viability, and so on). In other words, the scientistic approach (as Williams writes) "conflates virtual and actual in ways that conflict with Deleuze's transcendental deductions" (113). If science begins by constructing a frame of reference, if it can only "see" a world in terms of what is imminently useful to a particular historical stratum, this means that the virtual will always be elsewhere, always take place in a different time, and always elude the scientific gaze. The force of the virtual, meanwhile, will continue to haunt the scientific observer from behind the scenes.

Ontological versus Epistemological Realism

It is not so much that science lacks access to questions concerning its own process, but that it actively renounces the conditions under which (and by virtue of which) it would come to confront this process *as an ontological question,* i.e., in terms of the experience of phenomena that act directly on the senses and that positivism designates as secondary qualities. The science of early modernity emerges at the same moment in which thought arrives at a methodology for expurgating these secondary qualities from experience, in order to focus exclusively on those that can be measured, verified and tested—a tendency that can be found not only in Locke's empiricism but in any approach that imposes the rules of "good sense" on the production of knowledge.[31] The orthodoxy of philosophical "common sense" begins, as Gregory Flaxman explains in chapter 7, with the advent of "the principle that distributes its universal organization universally, over time and throughout space." The relation between philosophy and science thus takes place according to their respective doxologies, the means with which each thought system "circumscribes and regulates chaos out of existence." This will have profound consequences for philosophy in its confrontation with thought as "infinite movement or the movement of the infinite"[32]—that is, *as chaos.* The history of early modern philosophy is traversed by tensions, ruptures, and ambiguities that arise from its investment in scientific or "Cartesian" realism[33]—the Baconian philosophy of inductive reasoning and natural laws, for example, but also Pascal's religious anti-empiricism, and Geulincx and Malebranche's occasionalism ("vision in God" and the redundancy of material substance), and so on. By virtue of this investment, philosophy shares with science the assumption of a mind-independent reality, but also a deeply rooted suspicion that any mental image by which we relate to this reality will be: (1) limited in scope by the time and place of the observer, (2) structured according to the modalities of sense, and (3) ontologically secondary to the thing it represents. The initial phenomena that are assumed to cause the mental image and which must be said for that reason to precede it are simultaneously obscured by, and often impossible to distinguish from, the internal and subjective conditions under which this image is finally given to the understanding. Classical science, therefore, prepares a special kind of image in which observable phenomena are given idealized, diagrammatic form, providing the necessary criteria in advance of any real encounter with the world, for selecting those characteristics that have immediate and incontrovertible bearing on the problem at hand—i.e., those that render the conditions of observable phenomena in subjective-independent terms. This goes beyond the formulation of scientific theories properly speaking, which provide only a set of universal, analytical propositions.[34] The role of the scientific image is to use these propositions to synthesize or interpret the anticipated data of a specific phenomenon, in this way bridging the gap between theory and experience.

As the philosophical offspring of these assumptions, Cartesian realism relies on the scientific image to edify its own production of knowledge. More significantly, it considers, not without some apprehension, that the possibility of a denotative but purely theoretical language is dependent on the ongoing validity of a set of practices that bridge the gap between necessary truths and contingent ones—i.e., those that link thought to a real encounter with the world. Philosophy, on one side, will occupy itself wholly with necessary truths, but it will build its *conception* of knowledge on the ever-growing foundation provided by scientific facts and methods. We see this most clearly in the positivism of Auguste Comte, who categorically rejects the notion of "internal" fact, proposing that philosophy proceed instead by overseeing the ongoing expansion of external, scientifically determinable factuality to all branches of the social sciences.[35] Sociological theories will thus be written side by side with those of biology (Comte's "social physics"), or they will simply extend the biologists' findings to events in the sociopolitical domain. More recently, the skeptical attitude toward metaphysics that finds recourse in the scientific image is sustained by pragmatism and logical positivism, which seek (more modestly) to limit philosophical disputes to those which may be verified by reference to an external, scientifically determinable reality. As A. J. Ayers writes, "We are able to eliminate the possibility of perpetual, undecidable rational disagreements only in those areas where unquestioned links to external reality provide a common ground for the disputants."[36] In opposition to this tendency, the philosophers that inform Deleuze's conceptual apparatus (Spinoza, Nietzsche, Bergson) belong to another lineage, one that provides the basis for a rejection of positivism as inconsistent with the primary goals of philosophy. Bergson, reacting to Comte's positivism, is particularly interested in the question of the production of knowledge: "The metaphysic or the critique that the philosopher has reserved for himself he has to receive ready-made . . . it being already contained in the description and analyses, the whole care of which he left to the scientists."[37] It is not only that Cartesian realism reduces philosophy to simple reformulations of scientific designations; these designations themselves have a constructed, synthetic quality. Philosophy is not something that can be made to circulate freely, independent of experience, but it is contained in the description and analysis of a real encounter with the world *as the movement of thought.* Bergson sees science as a set of strategies for further quantifying, manipulating, and instrumentalizing the material world synthesized by concrete perception. By seeking to represent life in this way, science cannot help but misrepresent it, in the sense that its objects and goals are *only* practical, abstracted from the nonrepresentational "internal" conditions of life that unfold at a distance from the sphere of action. To an even greater extent, with technology, science does not simply "think matter as mechanism" but actively intervenes, with the result that both reality and our intuited knowledge of it are actualized according to the immediate conditions of practical human interests. Technology is no more than the concrete form in which thought, at one extreme, becomes alienated from its ontological root in the virtual. Deleuze (and Guattari) reverse this process by insisting on the status of subjectivity *as process,* that is, as a "machinic" production of thought and matter that takes place behind the scenes of the actual.

To some, this move will represent no more than a regression to the naïve and unproblematic unity between thought and world, as if one might escape the epistemological problem by insisting on the ascendant role of belief; however, in Deleuze's analysis, Cartesian realism escapes a far more difficult problem when it takes for granted the process of actualization by which scientific objectivity arrives at its image of the world (indeed,

through which science, as technology, actualizes real, material objectivities). We simply cannot know "what is," not because it is distorted by our
senses, but because sensing and knowing take place alongside the process
of actualization which *produces* "what is." This is the basis for Deleuze's
characteristically *ontological* realism: to think, in this case, is to participate
in the quasi-material process that produces, and never ceases to produce,
the real. ("Ontology is the dice throw, the chaosmos from which the
cosmos emerges.")[38] As Eric Alliez writes in *The Signature of the World,*

> What is the interest of *What Is Philosophy?* if it does not lie in the
> attempt to sketch the programme for a physical ontology up to
> the task of superseding the opposition between "physicalism"
> and "phenomenology" by integrating the physico-mathematical
> phenomonology of scientific thought into a superior material-
> ism founded on a general dynamics? . . . The detournement of
> Bergsonian (immediate) intuition is prepared both before and
> beyond the realism-idealism opposition, when the act of knowing
> tends to coincide with the act that generates the real.[39]

We might say then that the ambiguity confronted by epistemological realism has less to do with the creation of true propositions than with the act
of creation itself, arising from a gap not between representation and reality
but between the problem and the solution, between the inventor and his
imminent invention, between the artist standing before a blank canvas and
the same artist before a finished painting. In his article, "Digital Ontology
and Example," Aden Evens describes how a virtual but objectless plenitude, by way of problemata, gives rise to an actual object:

> Two dancers, strangers to each other, meet on the dance floor;
> how will they relate, where will they touch, where will they move?
> There is not enough energy in the batteries to support the typical
> demands of life-support systems and engines in the earth orbiter—
> what to switch off, what trajectory to follow, what to jettison? These
> problems have no solution but must generate solutions in them-
> selves, determining their own boundaries through a play of tension
> and release, conservation and expenditure, despair and elation.
> Thus, the problem is desiring, but it is a desire without object, a
> desire that eventually gives rise to its own object. (chapter 5)

This gap between the urgency of desire and the object to which it gives rise is the same gap we find between intuition and intelligence in the Bergsonian conception—not a gap at all, but a tension that gives rise to form in the process of release. This suggests that we can cure the "slackness" of science (where everything exists in a state of postrelease) by coupling it with the "tension" of ontology, a science of problems. It also means we must affirm the fundamentally creative nature of thought, knowledge, truth, etc., these designations referring, in the final analysis, to an imaginative process that has no unified will or consciousness at its helm, which takes place in a virtual domain crisscrossed by desiring machines, the fastidious stewards of Whiteheadian appetition.[40]

Mullarkey raises a pertinent question here: "if we are each of us a desiring-machine running blindly under the illusion of its own self-conscious volition, and if our language does not *represent* the real but orders, shapes and creates it (as Deleuze would have us think), then according to which values should we prefer the thoughts of Deleuze over those of, say, Gottlöb Frege? *How* did *he* get it right?"[41] Deleuze's conceptual system can have no basis in fact, only in active forces, creativity, simulacra. It is not a body of knowledge per se—philosophy does not aim to constitute such a body—but a practice in thought, like Nietzsche's "*gaya scienza*," or the *scientia* of God in the Spinozan conception: "not a science of the possible, but the knowledge God has of himself and of his own nature."[42] We will have a hard time, in these terms, finding any conception of science in Deleuze's works that proceeds by way of verification, justification, corroboration, indeed by any method that takes science's positivism at face value. "The epistemic category of truth by correspondence is obviously inappropriate to machinism on account of its position of transcendence."[43] Mullarkey finds an alternative to the epistemic approach in Deleuze and Guattari's notion of "consistency," the basic formal condition for the synthesis of a concept (in *What Is Philosophy?*). Consistency does not apply to the correspondence between two conceptual systems; it measures only the extent to which a single system is able to adequately pose a problem. Philosophy, as the well-posed problem, engages the constant renewal of experience as motor of the concept, providing a purely internal justification within a "machinic" system of values. Though Mullarkey starts out by describing Deleuze's "biologism" as an example of the more general post-Continental trend in philosophy (an anti-anti-scientism), he will stop short of calling him an ontological realist (as Timothy Murphy, for example, has suggested)[44] because machinism requires no presupposition of a reality beyond that which it deploys.

In these terms, of course, classical science will prove no more adequate to the task. Its ability to work directly with matter is exemplary, but its tools are only designed to identify quantifiable differences in time and space—which is to say that it does not concern itself with difference as such, only with questions or problems that arise during the individuation of a particular difference, from the more or less stable diagram of its outcome. And modern science? We may argue, with Prigogine and Stengers, for example, that the introduction of revolutionary scientific theories at the beginning of the twentieth century has led to radical changes in the way science is "done."[45] In the first place, scientists have turned their attention away from systems in a state of equilibrium, from "optimization models" that provide discrete representations of matter that have been stripped of contingent secondary phenomena and tend for that reason to confirm the classical notions of causality, determinism, universality, etc. With Einstein's law of special relativity, and later with quantum mechanics, time and space can no longer designate the continuous, linear axes against which phenomena are plotted or their outcomes interpolated. Not only does this mean the end of Newtonian universality (the claim that physical laws are applicable throughout time and space), challenging claims and assumptions based on the unity of science, but also it means the end of a mechanistic worldview in which matter passively fills out a set of determinate spatio-temporal relations. In other words, what modern science indicates is the decisive reversal of a Cartesian project to reduce matter to the simple principle of extension.[46] This allows for an increasingly complex picture of nature, characterized by far-from-equilibrium models of change and a capacity for spontaneous activity that even a rationalist like Descartes might have been compelled to describe as "intelligent."

If it is true, in this sense, that the virtual pervades the universe, and if it is intrinsic to matter in the same way that it is intrinsic to thought, we too shall be compelled to rethink the way science is done, beyond the immediate implications of the latest scientific revolutions, but also beyond a philosophy of science that sustains many of the classical assumptions concerning the relationship between thought and its "outside." Above all, we shall have to resist the lingering tendency to approach the problem from a purely epistemological point of view, a "dialogue" between thought and nature.[47] This essentializing dualism can no longer be sustained as such.

It has been supplanted by another, more immediate opposition, another gap in the order of being and in the production of knowledge. It is of little consequence now to ask how (or whether) thought makes itself the measure of worldly phenomena. This approach follows the classical mode of thought as a "striating of mental space" which "operates with two 'universals,' the Whole as the final ground of being or all-encompassing horizon, and the Subject as the principle that converts being into being-for-us."[48] But it will be of no greater help to presume, with philosophical anti-realism (a preclassical Protagorean relativism, for example), that thought is destined to fall back perpetually on an indeterminacy immanent to the modalities of the senses, or to a universalized individual consciousness. The question is more difficult than that. Indeterminacy is rooted elsewhere: not in a dialogue between thought and nature, but in a movement that engages the conditions for the continuous renewal of *qualitative* change and that bridges the gap between the determinate and indeterminate states of being. It is an ontological leap into the subjective-independent conditions of being-in-itself, not the process by which the for-itself of being is converted into universal laws (or the failure to carry out this process with adequate results). In order to make such a leap, we will have to invent new tools for registering and putting into motion strictly *internal* differences, which is to say, for engaging the endless capacity for creative transformation that occurs simultaneously in thought and nature.

We might begin then by searching for a possible "cohabitation" of science and philosophy that takes place within the process of actualization. In what ways does the openness of one join with the objectivity of the other? In what region of the mind do we find an intersection between planes of consistency (transversal alignment of thought with the movement of becoming) and planes of reference (the actual, historically specific and pluralistic set of relations that constitute the scientific image)? Deleuze and Guattari provide one answer in the figure of the brain as the "*junction*—not the unity—*of the three planes.*"[49] This junction is most succinctly expressed in terms not of the brain as such, but of sensation, which preserves the movement of the virtual like a string that resonates with its surroundings—a vibration or harmonic. Whereas we see that sensation, in these terms, is necessarily passive, it is also productive. This "passive creativity" (the passive synthesis) is the manner in which the virtual constitutes an inhabitable time, from which emerges the subject as a simple "I feel" on the side of perception, or an "I think" on the side of thought.

But this is still only a "larval" subjectivity and not yet a thinking subject. Sensation as a process of actualization or "contraction" accounts only for "pure contemplation without knowledge," or what Deleuze and Guattari identify as "a pure internal Awareness": a vitalistic force that exists in thought and matter but does not correspond to a unified subject. Before there is scientific knowledge, before there can be an encounter between thought and the world, there must therefore occur this vibration of awareness and sensation, which is common to both science and philosophy. At first glance, any cohabitation of philosophy and science ends there. The latter is characterized neither by the "I think" of philosophy nor the "I feel" of art, but by an "I *function*" that is intrinsic to a subjective-independent and materially bound block of becoming.[50] It does not describe or explain the real so much as *generates* it.

Does this mean that Deleuze, who proceeds by way of affect and intuition rather than any formal logic, rejects a priori the aims and methods of the exact sciences? And would this not leave the sciences with only the narrowest pathway for producing new knowledge, and even then only of a speculative kind, with no claim to empirical proof or, for that matter, theoretical certainty? At times, conclusions like these will be inevitable. On the other hand, they will be difficult to square with Deleuze's vast interest in, and debt of influence to, nearly all branches of the exact sciences. What is so compelling about the early collaborations with Guattari (who began his university studies as a pharmacist)[51] is that they create a theoretical language that draws equally from all fields of knowledge, culminating in an entirely new way of "doing" philosophy. The body without organs begins with Artaud but extends, in the final analysis, to Riemannian space and Mandelbrot's fractals.[52] Abstract machines merge with organic chemistry in the molecular properties of silicon, and Hjelmslev's linguistics with nucleotides and nucleic acids.[53] Bergsonian intuition finds its complement in the mathematics of Brouwer, Heytig, Griss and Bouligand.[54] The objective of this collection is to provide a set of terms in which to examine this debt of influence, but also to question, with Williams, the commensurability of philosophy with science.

Modern Science Has Not Found Its Metaphysics

Rather than viewing this debt of influence in terms of a metaphysics that lacks a rigorous method for testing its concepts and must therefore turn to science, we might consider the opposite scenario: a science that lacks the metaphysics it needs. In 1981, when Arnaud Villani suggested to Deleuze that the *physis* played a large part in his work and that the conclusion of *A Thousand Plateaus* undertakes a "transposition" of metaphysics into biological and mathematical categories, the latter responded:

> It's true. I think I'm getting at a certain idea of Nature, but I haven't yet managed to consider this notion directly.... You ask if there is a possible mathematical and biological transposition. It is without a doubt the other way around. I consider myself to be Bergsonian, in the sense that Bergson says that modern science has not found its metaphysics, the metaphysics it needs. It is this metaphysics that interests me.[55]

In light of this "statement of interest," *The Force of the Virtual* sets out to examine the specific ways in which Deleuze proceeds to invent a metaphysics proper to science, at the same time opening scientific thought to a production of ideas that far exceed its aims and means. In this sense, the new metaphysics does not directly compensate for what science lacks so much as communicates with a matrix of scientific determinations from beyond its plane of reference, that is, from the "essentially inexact yet completely rigorous" dimension of thought that science subtracts in order to arrive at empirical truth.

In the first section, "The Virtual in Time and Space," we trace the effects of this mutual "haunting," with the aim of grasping the different ways in which thought (philosophical and scientific) sustains and embodies the process of actualization. The question here is two-fold: Where and when does the event of thought take place, and how does this coincide with—or disengage itself from—the time and space of the physical events it reproduces? For Villani, the answer lies in a notion of lived experience as an embodiment of the virtual. The virtual only emerges when the indivisible time of lived experience ("I feel that I am") comes to displace purely symbolic designations ("I say that I am") as the first principle of thought. It is for this reason, Villani argues, that Deleuze's concept of the virtual can be reduced neither to the enumeration of possible worlds that constitute the "modal real" nor to the phenomenological assumption of a broken totality. To insist on the virtual as transcendental condition of becoming is rather to proceed from the ontological root of both thought and matter—to let the "sensualist" cogito have its word again: "we are not speaking of an abstract and measurable time, but of real time, wearing away at bodies, acting on bodies" (chapter 1). In "The Intense Space(s) of Gilles Deleuze," Thomas Kelso adds an essential theoretical problem to the question of embodiment. The issue here is not time (duration, memory) but virtual space (a *spatium* or depth) in which the body exceeds its actual, extensive coordinates: "we retrace routes that exist virtually in our bodies, but not abstractly in a map or a diagram in our minds" (chapter 3). For the dancer, the archer, the pilot, etc., the virtual is not simply a logical necessity; nor, at the other extreme, is it restricted to a basic proprioceptive faculty that acts in lieu of conscious mental processes. Rather, the virtual is that space in which all movements in a series have *already* been played out—these movements, by definition, exceeding the actual limits of the individuated body. This means that the virtual is everywhere: "our perception is riddled with traces . . . as if the virtual and the actual were interlaced or superimposed on one another. This superimposition is what Deleuze means by the intensive spatium: there is an originary depth to all experience, which we cannot directly sense, though we are lost without it" (chapter 3). In contrast, the scientific method acts as a kind of filter on the superimposition of actual and virtual. It proceeds by subtracting the depth of experience (the virtual) so as to superpose all data in a single diagrammatic model or image (the actual). "Everything is actual in science," writes Villani; "time never begins. There is at most a fourth dimension, never a 'worldly' fabric, the flesh of things, the essence of being. If the virtual does not belong to science, that is because it escapes it entirely" (chapter 1).

However, to say that everything is actual in science, that time never begins, will create as many problems as it resolves. From the point of view of the scientist, there are at least two ways that time—or something like time—enters the production of scientific knowledge, the most obvious being the way a scientific model is only ever complete when it has accurately described the entire succession of states that comprise a particular physical process *in time.* (Erwin Schrödinger refers to this as the "fourth dimension" of the scientific image.)[56] Indeed, to speak of the "truth" of a scientific theory, or of science generally, is to acknowledge *empirically* how the outcome of a process may be accurately and consistently derived from given knowledge of its initial state. Yet this notion of time also presumes that a particular theory or model, given *complete* knowledge of the initial state, should be able to predict every outcome with total accuracy. Or in any case, it is under this assumption that science proceeds, through repeated observation and experimentation, to develop a more and more complex picture of the world. Here we see the second sense in which time *appears* to enter the production of scientific knowledge. After each "breakthrough" or gestaltic shift in the dominant paradigm, a body of scientific knowledge undergoes incremental changes *in historical time,* adapting to findings in the new field. Yet science cannot even proceed without first considering that the natural principles under investigation *transcend* time, that all possible outcomes of a given process lie hidden (and fully determined) in the infinite complexity of its initial state. Even the more speculative theoretical approaches tend to treat physical processes in this way. Complexity theory, for example, is based in large part on the principle of *sensitive dependence on initial conditions,* also known as the "butterfly effect."[57]

In the general attitude described above (royal science), there is a marked negation of the notion of time as process of *differentiation*—that is, as a process of qualitative change, as opposed to *individuation* or "differenciation," which involves the actualization of a single set of variables.[58] Science means the latter when it speaks of the "historical adaptation" of a particular theory or model; or, as Flaxman writes, "Inasmuch as good sense rests upon an irreversible synthesis of time, its arrow always shoots from the differences of the past to the homogenized diversity ('differenciation') of the future, from 'the particular to the general' . . . from the improbable to the probabilistic" (chapter 7). In contrast, we find various passages throughout *A Thousand Plateaus* (based, once more, on the works of Michel Serres) in which it is not simply the case that a physical process undergoes a reconfiguration of its elements as it advances from moment to moment. Rather, time—"real time" in the Bergsonian sense—effects change in the very laws under which the process advances. In other words, the so-called constants that guide the development of a phenomenon are themselves in flux.[59] The counterpart of royal science is thus the ambulant or nomadic sciences, which "follow" the emergence and development of novelty, engaging "a continuous variation of variables" rather than submitting them to a "legal model," or reterritorializing them around a single point of view.[60] The result is not scientific truth as such but the invention of problems:

> The ambulant sciences confine themselves to *inventing problems* whose solution is tied to a whole set of collective, nonscientific activities but whose *scientific solution* depends, on the contrary,

on royal science and the way it has transformed the problem by introducing it into its theoremetic apparatus and its organization of work.[61]

We first find the theoretical groundwork for this "science of problems" in *Bergsonism*, where Deleuze specifies the first rule of Bergsonian intuition as "the stating and creating of problems; the second, the discovery of genuine differences of kind; the third, the apprehension of real time"; [62] and in *Difference and Repetition*, where he develops an aphorism for the constant variation of variables on a cosmic scale: "God makes the world by calculating, but his calculations never work out exactly [*juste*], and this inexactitude or injustice in the result, this irreducible inequality, forms the condition of the world. The world 'happens' while God calculates; if the calculation were exact, there would be no world."[63]

Yet this is precisely the problem we face when trying to understand the Deleuzian conception of science. Given God's fundamental "fallibility," one would expect science itself to repeatedly miss its mark, lost in a Heraclitean flux with no reliable basis on which to "go beyond" the given; however, it is apparent—and to a remarkable degree—that science *really does work*. Even complexity theory, in theory incompatible with any standard based on reproducibility, has been effectively applied in software that helps meteorologists make more accurate predictions.[64] How is this possible? One very simple answer is that science is goal oriented, working within parameters determined by an implicit (and sometimes obfuscated) pragmatism, rather than an abstract and holistic truth. Setting out from hypotheses that bracket in advance any ostensibly unrelated aspects of a phenomenon, the scientific observer is able to avoid "problems" in the Deleuzian sense and to assume a point of view from which qualitative change appears as no more than statistical variance. This leads to a rather sclerotic production of thought, one that is nonetheless punctuated now and then by the cumulative effect of real variation, i.e., scientific revolution in the Kuhnian sense.

But Deleuze's answer to this question will be more radical. To speak of science in terms of truth and accuracy, to evaluate the reproducibility of a certain outcome according to the so-called *predictive success* of a scientific theory, is already to fall back on epistemological concerns that do not fully address the problem of thought as a creative process. Science creates what it predicts—or, more succinctly, the position of the scientific observer is created in the same moment, and by virtue of the same process of differentiation, in which the real creates itself. "We are not in the world, we become the world; we become by contemplating it. Everything is vision, becoming. We become universes."[65] What Deleuze introduces to the philosophy of science, effectively bringing it beyond an investigation of the mere *status* of knowledge, is a metaphysics that outlines the specific ways in which thought and matter are simultaneously and mutually organized. This is to say that thought, like the material processes it reduplicates and records (in the objective apparent sense), is not the product of an individual mind, or even of a collective sociobiological human faculty. Rather, the ontological root of both thought and matter, and consequently of the three "sister disciplines"—science, art, and philosophy—is a virtual domain that functions, behind the scenes, like a vast impersonal and preindividual "brain." Both Gregory Flaxman and Arkady Plotnisky, in their contributions to this collection and elsewhere, have mapped out the specific ways in which the virtual brain, which Deleuze and Guattari generally refer to as "chaos" or the "*chaosmos*" (*What Is Philosophy?*), is actualized into individual forms of consciousness. But we find similar interpretations in the works of nonphilosophers, such as neuroscientists Gerald M. Edelman and Giulio Tononi (*A Universe of Consciousness*). As Clark Bailey argues, Edelman and Tononi have developed "a specifically *virtual* definition of consciousness, a definition depending not just on the actual activity pattern of some neuron or set of neurons, but on the simultaneous existence of objectively distinct *unactualized* patterns. . . . It is the reality of these virtual patterns that distinguishes consciousness from mere feature detection or form recognition, and that gives it the ability to find structure in a novel and uncertain world" (chapter 12).

This means that we cannot properly articulate the role of the scientific observer until we have situated it within a field of actualization, namely at the horizon of the process of becoming, or the projected point of view of a fully actualized world. As I argue in "Superposing Images: Deleuze and the Virtual after Bergson's Critique of Science," the act of observation is not "performed" by an individual, but actualized by an ideal eye that is *in things,* yet which functions by transforming intensity—or duration, to use the Bergsonian term—into an extensive "outside world." The scientific observer is simply a more "contracted" (actualized, individuated) instance of this eye and corresponds to an external reality that has been contracted along with it. The affinity of science and its object, the so-called predictive success of science, is thus analogous to the symbiosis of wasp and orchid: a block of becoming that appears, from the epistemological point of view, to follow the logic of representation (mimesis), but which confounds this logic by emerging all at once, both wasp and orchid, from a virtual plenitude by a process of extreme selectivity.[66]

Extensity, as Deleuze argues, is no more than the *"condition of the representation* of things in general."[67] Thought and matter thus simultaneously and continuously undergo a process of transformation in which sense is "seen as" form, and in which intensity is "seen as" extensity. Similarly, theoretical physicist Wolfram Schommers has argued that "a phenomenon (e.g., wave or particle) is always an *observed* phenomenon . . . without observation it is meaningless to talk about a phenomenon."[68] For Deleuze, however, the process of actualization is initiated and carried out within things themselves by virtue of a kind of "visual metaphor." This is deployed by—*but not synonymous with*—the act of empirical observation, narrowly defined as the work of the scientist vis-à-vis a testable sample. The visual metaphor, in these terms, is more general, more abstract than the scientific observer, and just one among many possible manifestations of what Deleuze and Guattari call "abstract machines"—points of singularity which, like the human eye, extract variables from the process of differentiation (sense) and actualize them as matter and intelligence (form):

> An abstract machine is neither an infrastructure that is determining in the last instance nor a transcendental Idea that is determining in the supreme instance. Rather, it plays a piloting role. The diagrammatic or abstract machine does not function to represent, even something real, but rather constructs a real that is yet to come, a new type of reality. . . . Abstract machines thus have proper names (as well as dates), which of course designate not persons or subjects but matters and functions.[69]

It is not surprising, therefore, that scientific representations match up so well to the phenomena they "imitate." In light of the Bergsonian position, which Deleuze takes as the starting point for a new "metaphysics of science," we can say that the two sides of empirical observation are always actualized together in a single block, that the number of worlds at any given moment is potentially infinite, and that this number depends only on the diversity of forces (or "lines of force") that perpetually crisscross the virtual.

Nature as Creative Process

If science lacks the metaphysics it needs, it nonetheless remains a useful antidote to a philosophy that has turned away from the creative force of thought and toward a more restrained "love of truth." The explicit aim in *A Thousand Plateaus* is to "place thought in an immediate relation with the outside, with the forces of the outside" (376–77). The result is not natural philosophy as such, but a form of thought that cannot capture nature without also being "propelled" by it: "This is Cosmos philosophy, after the manner of Nietzsche. . . . The forces to be captured are no longer those of the earth, which still constitute a great expressive Form, but the forces of an immaterial, nonformal, and energetic Cosmos" (342–43). Nietzsche, himself no great friend of science, was more critical of philosophy's turn against it, and he understood this criticism as an attempt to moderate the diversity of forces in the name of morality: "This is extraordinary. We find from the beginning of Greek philosophy onwards a struggle against science with the means of an epistemology or skepticism: and with what object? Always for the good of *morality*."[70] From this point of view, the question is not whether science is "truthful," nor whether its method transcends such philosophically determined categories as language, culture, history, etc., which appear at times (and from the skeptical point of view) to qualify its success. Questions like these are based on preoccupations of a strictly epistemological nature, and they will entangle us in the all-too-human propensity for judging manifestations of force as if they were inherently true or false (for Nietzsche, such evaluations can only emerge from the confrontation of one force by another). Rather, cosmos philosophy takes its cues from the great technological developments of the modern age; it reinvents thought as the movement of a vast "mechanosphere," transposing external causes (the mechanistic worldview of early modernity) into "invisible" internal causes (electricity as *primum movens* of the modern city). To speak of novelty, emergence, or becoming-other in the Deleuzian sense is to open thought to "forces that are not thinkable in themselves," but which comprise at the same time the virtual interiority of thought and nature. This means rejecting the epistemological or skeptical view of science, in favor of one that

(1) Emphasizes the *disunity* of science, particularly as this points to the emergence of multiple, divergent worlds, each one corresponding to a node in the intersection of creative forces (concept of the *fold*); the will to truth, on the other hand, demonstrates a desire to reduce the positive diversity of forces to "common sense," to guarantee that nothing in the logical order of things is disturbed.[71]

(2) Establishes nature or the *physis* as positive principle of this diversity, actively willing the force of the virtual in all its manifestations, and therefore never closing on itself, never complete in itself, never unified in a single world or universal law ("Nature must be thought of as the principle of the diverse and its production. But a principle of the production of the diverse makes sense only if it does *not* assemble its own elements into a whole."[72]

(3) Asserts that differentiation, the willing of difference, is everywhere co-extensive with, but ontologically primary to, differenciation, the actualization of new species and forms.[73]

For this reason, we must not understand nature as the passive contents of scientific knowledge, but instead as the willing of novelty, an abstract *natura naturans* that is the ontological root of all becomings, including scientific ones. This means that nature is the subject of science as creative process, even when it is (mis)taken for the object of science as will to truth.

The aim of the second part of this collection, "Science and Process," is to examine specific paradigms and figures of the sciences (natural, formal, and applied) with reference to the dynamic role that Deleuze attributes to nature. Do theories of biological diversification acknowledge the effects of the virtual, or something like the virtual? Conversely, does digital technology capitalize on a structural feature of the actual that escapes these effects—or do we find evidence even here of their continual re-emergence? Is there anything in the applied sciences to suggest that the virtual is an ontological necessity, so that we would find its effects already implied in the invention of new theories and applications? In sum, where and how, according to science, is the virtual manifest? We have already discussed Bergson's appeal to the common experience of movement, and the sense in which this experience (or the concept derived from it) disappears the moment we subdivide the trajectory of a moving object into a series of relative positions (Zeno's paradox). This exercise gives us the image of a virtual that binds together discrete moments of the actual, or that repeatedly intervenes in a deterministic series, the same way the body without organs intervenes in the binary-linear series of desiring-production.[74] This does not, however, effectively meet the criteria for a positive principle of diversity. What we are looking for are examples of nature as an active *willing* of difference, not just the site of its logical disappearance. When Deleuze considers the privileged position of the human species (in Bergson's conception) with respect to all other living beings, he considers how the *élan vital* actively "fills the gap" between contiguous neurophysical states of the brain:

> We recall that [the cerebral matter] "analyzed" the received exci-
> tation, selected the reaction, made possible an *interval* between
> excitation and reaction; nothing here goes beyond the physico-
> chemical properties of a particularly complicated type of matter.
> But, as we have seen, it is the whole of memory that descends into
> this interval, and that becomes actual. It is the whole of freedom
> that is actualized.[75]

It is in the interstices of a series of external causes that life animates matter, or, more generally, that the virtual reincorporates the actual into the process of actualization. Whereas Zeno's paradox helps us formulate a negative concept of the virtual, this last example brings us much farther in explaining how a series of efficient causes is interrupted by something like the will, memory, or principle of association. But we will not get far with this explanation before we are obliged to make the vitalistic assumption of a positive force that permeates matter—an assumption that even Bergson was wont to admit constitutes a new (albeit monistic) spiritualism.[76] In the interest of avoiding the imminent criticism that Deleuze's thinking relies on the same assumption, that it counters empirical truth claims with irreducibly vitalistic ones (the ontological primacy of the virtual, for example), we will be compelled to look elsewhere for an explanation.

The second section of *The Force of the Virtual* concentrates instead on points of correspondence between Deleuze's thinking and the works of such figures as Darwin, Whitehead, and Leibniz, figures who delineate the possibility—indeed, the necessity—of a creative, process-oriented conception of nature. In "Interstitial Life: Remarks on Causality and Purpose in Biology," Steven Shaviro begins with an analysis of Darwin's theory of evolution, considering how it might represent a scientific counterpart to Deleuze's principle of immanence. Indeed, it is with Darwin that science is finally able to account for diversification in nature without recourse to the position of a transcendent or "purposive" causality: "Darwin provides an immanent, nonteleological mechanism for the development of life. Given the theory of natural selection, it is no longer necessary to invoke such teleological agencies as the hand of God, or the Lamarckian acquisition of striven-after qualities, or the workings of some inner vitalistic force like Bergson's *élan vital*" (chapter 4). From a Deleuzian point of view, we are immediately confronted with the problem that evolutionary theory accounts for genetic mutation only in terms of accidental, contingent factors. In this sense, it is not consistent with Deleuze's notion of ontological or "vital" difference: "vital difference can only be thought of as internal difference; it is only in this sense that the 'tendency to change' is not accidental, and that the variations themselves find an internal cause in that tendency."[77] For Darwin, diversification arises by virtue of genetic mutations that occur *in spite of* the successful formal characteristics of a species within a given environment; it is only after subsequent changes in this environment that a particular trait is selected for, and that it is possible, in hindsight, to explain how a species has been able to adapt and survive. Darwin's theory of evolution is thus consistent with Descartes's mechanistic world-view, with the exception that it proceeds from the presupposition that genetic "mechanisms" (Darwin calls them *gemmules*) periodically misfire. On the one hand, this represents a total reduction of biological diversity to efficient causes; indeed, the main advantage of Darwinism is that it obviates all explanations of diversification along the lines of a Platonic Demiurge or "intelligent design." Why then do neo-Darwinists have such a hard time avoiding terms like *design* (of an appendage or defense mechanism), or *purpose* (of a particular trait which ensures the survival or propagation of the species)? Why must science perpetually mix the logic of *efficient* and *final* causes, when Darwin tells us that the diversity of the species arises as the consequence of a combination of chance and necessity?

In "Interstitial Life," Shaviro argues that there is more at issue in the ubiquitous language of biological purposiveness than a convenient (but misleading) metaphor. The mutually exclusive logic of *efficient* and *final* causes reminds us inevitably of Kant's distinction between (antinomy of) the empirically conditioned laws that guide nature and the "unconditioned" moral law that is grounds for the positive concept of human freedom: "The determination of the causality of beings in the sensible world can as such never be unconditioned, and yet for every series of conditions there must necessarily be something unconditioned and so too a causality that is altogether self-determining."[78] But Kant can only explain the manifest coexistence of the two kinds of causality by appealing to a transcendent principle of freedom. This principle is necessarily anthropocentric (founded on the mere form of the moral law), and it relegates all subsequent designations of purpose in nature to the status of a *paralogism* (an attribution of cause based on the forms of the understanding). This is clearly not what Deleuze has in mind when he describes evolution as the affirmation of vital difference. Yet Shaviro insists on the philosophical importance (for Deleuze) of Kant's conception of a double causality:

> A post-Darwinian Kantianism can, and should, drop the anthropocentrism altogether. It should concentrate on the structure and consequences of Kant's transcendental argument, while rejecting the idea that the argument applies only to human (or 'rational') minds . . . This amounts to saying that Kant's transcendental syntheses have a general ontological significance, instead of a merely epistemological and psychological one. (Chapter 4)

The Logic of Sense is Deleuze's attempt to formulate such a Kantianism, one that attributes the willing of difference to a particular manifestation of the virtual that emerges within the "gap" between efficient causes, providing the grounds for an aleatory point or *quasi*-cause. "Deleuze's counter-actualizing 'dancer' makes a *decision* that supplements causal efficacy and remains irreducible to it, without actually violating it" (chapter 4). The virtual is not a transcendent principle, yet it exceeds Darwin's negative conception of genetic mutation; it suggests a restlessness in things, an "appetite" for novelty, not simply the periodic accident of its emergence. This is only logical because the apparent objective *will* to survive or *will* to propagate the species can be easily contested with reference to efficient causes; yet the ontologically primary instance of mutation eludes such explanations (there is no efficient cause that would explain why nature must differentiate in the first place). Even more radically, Shaviro claims that Deleuze provides grounds for a concept of nature that "seeks transformation, not preservation," so that "an entity is alive precisely to the extent

that it envisions difference and thereby strives for something other than
the mere continuation of what it already is" (chapter 4). Evolutionary processes do not will survival; they will change.

What Darwinism implies, and what current research in the field appears
to corroborate,[79] is not simply the central role of variant genetic codes or
regulatory sequences, but an entire production of surplus code that is not
exhausted by the process of "gene expression." Deleuze and Guattari take
this as evidence of a virtual plenitude that exceeds the actualization of
traits or "phenotypes" within a given individual or population, and that
exceeds even the mechanism for actualizing such traits:

> The modern theory of mutations has clearly demonstrated that
> a code, which necessarily relates to a population, has an essential
> margin of decoding: not only does every code have supplements
> capable of free variation, but a single segment may be copied twice,
> the second copy left free for variation . . . a code is inseparable from
> a process of decoding that is inherent to it.[80]

Generally speaking, the production of code in a process of actualization
is *always* simultaneously a production of surplus code. This is what
ensures that in each case a series, structure, or organism is coupled with
an aleatory element that may initiate a process of decoding from within
the series. (It would seem, in this case, that Peter Hallward and Mark
Hansen both go too far in describing the Deleuzian principle of the virtual
as an "otherworldly" deduction, or as principle of the "marginalization of
the organism."[81] In light of the notion of surplus code outlined above, the
actual organism does not arrest the force of the virtual so much as constitute a kind of "probe head" that perpetually extends the play of difference
within a field of actualization.) Differentiation is not only imminently
possible, in these terms, but the general condition of *all* processes of
actualization. This is even true with regard to those processes like stratification, individuation, and territorialization, which proceed by minimiz-
ing variation and tend to confine matter to increasingly narrower cycles of
becoming. In other words, there is no process of production that is not at
the same time a double production—or "double articulation"—resulting,
on the one hand, in a relatively deterritorialized surplus code or matter-
flow (matter-movement, matter-energy), and, on the other, in a relative
territorialization or sedimentation that leads to more stable biological,

geological, or social structures. To return to our previous example, biological diversification involves two simultaneous movements: on the one hand, genetic mutations result in a diversity of traits, functioning as a virtual reserve for a population; on the other, natural selection reduces, shapes, and organizes a population around singular traits that derive as much from the exterior milieu (concept of *fitness*) as from the interiority of the corresponding genetic sequences. The two movements—"content" and "expression"—are relative to one another, so that, for example, a single population may function as a surplus with respect to another population in a contiguous milieu. In neither case is the virtual depleted; at most, it is left "unused."

Specializing in methods for intervening in material processes, applied science proceeds along both sides of the double articulation at once. It follows the variations that allow a matter to meet new demands on the part of technology and the forces of production: "Matter and form have never seemed more rigid than in metallurgy; yet the succession of forms tends to be replaced by the form of a continuous development, and the variability of matters tends to be replaced by the matter of a continuous variation. . . . Metallurgy is minor science in person, 'vague' science of the phenomenology of matter."[82] Metal does not individuate so much as singularize, deploying all the variations (or surplus code) intrinsic to its content. Agriculture (and applied agricultural science) strives, on the contrary, to regularize and contain these variations: "Transhumants do not follow a flow, they draw a circuit; they only follow the part of the flow that enters into the circuit, even an ever-widening one. Transhumants are therefore itinerant only consequentially, or become itinerant only when their circuit of land or pasture has been exhausted, or when the rotation has become so wide that the flows escape the circuit."[83] The circuit is a particularly sclerotic form of becoming, confronting the force of the virtual with a capacity to capture and minimize the variations of a matter, in this way obviating the need to follow it (the counterpart of agriculture is thus the movement of hunter-gatherers). Material sciences and engineering are situated between metallurgy and agriculture. The search for materials adequate to the demands of technology requires a certain itinerancy, which is further extended by the discovery and investigation of variations in the newfound or newly constructed materials (or of variations in materials already used in other fields or applications). But this process, which revolves around content, is articulated with a mechanism of expression that is developed by science proper, and that minimizes or limits the itinerant tendency. By deploying scientific models, which are like so many apparatuses of capture, the engineer intercepts the variations in new materials with fixed, idealized cycles (cycles that science, in its royal manifestation, traces and retraces as objective apparent "constants"). The applied sciences thus proceed in the manner of a Divine Craftsman or Demiurge who, confronted with the problem of primordial time, contrives to organize the cosmos into strata of interrelated orbits, thereby overcoding the play of difference with the regularity of days and seasons.[84] Applied science does not elude the force of the virtual so much as absorb it in an infinitely complex double articulation. We must not forget, however, that the strata formed by this articulation are always relative to one another, so that what is captured and expressed at one level serves as readily available content on another, and vice versa;[85] it is in this sense that deterritorialization and reterritorialization can only be defined relative to one another. The quintessential product of applied science is therefore the machine, insofar as it articulates the properties of matter (content) with ascending patterns of idealized movements and cycles (expression). Yet there is no machine, however cleverly designed, that can be entirely disengaged from the descending relations among its various strata, or that escapes the ongoing variations that take place at the level of its material substrate—an assertion explored and developed by the machine art of New York Dada and Yves Tinguely, but also by connectionist theory and neural networks.[86]

In "Digital Ontology and Example," Aden Evens tests this assertion with respect to a process that has been effectively abstracted from its material contents—indeed from the very domain of the applied sciences—tending instead toward the idealized, modal logic of the formal sciences. The digital, along with the process of abstraction it initiates in matter, can be understood in two ways. First, we see that it involves the ongoing development of silicon and other materials that would provide an ideal substrate for the storage and processing of digitized (symbolic) information: "Materials and their properties are selected precisely for their abilities to yield to abstraction, to give themselves over to the idealization that overcodes the concrete material" (chapter 5). It is this process of abstraction and idealization that has compelled robotics engineers to ask whether the digital will ever be fully liberated from matter, which tends to be the limiting factor in the ongoing endeavor to increase the speed and accuracy of data processing. Hans Moravec, an early pioneer in the development of machine vision, is optimistic: "the abacus, and the physical world in general, is just a large shared abstraction where small, independently existing abstractions like minds and numbers correlate . . . there is every reason to believe that abstractions like minds and souls, no less than numbers, exist independently of their occasional representation in a physical world."[87] The assumption here is that matter, because of its specificity, can never be more than a particular representation of an abstraction that is believed to transcend it. For Moravec, this means that the mind, in all its manifestations, exists ready-made, apart from, and prior to its actualization in material form (a notion that is rather Platonic in its formulation, and an extreme example of what Deleuze and Guattari call *decalcomania*).[88] Others have challenged the ideological assumptions underlying the ideality of the digital. Cultural critic Johanna Drucker, for example, appeals to Adorno's sociopolitical deduction of the grounds for the concept of self-identity:

> When empirical and/or positivist logic invades culture to such
> an extreme that representation appears to present a unitary truth
> in a totalizing model of thought, then that leaves little or no room
> for the critical action or agency that are essential to any political
> basis for agency. . . . My double agenda is to disclose the ideological
> assumptions in the way the ontological identity of the digital image
> is posed and to suggest that graphesis (embodied information) can
> challenge mathesis. Or, to paraphrase, I assert that the instantiation
> of the form in material can be usefully opposed to the concept of
> image/form and code storage as a single, unitary truth.[89]

Like Villani and Kelso, Drucker insists on the embodied nature of these processes, opposing her ontology of the digital as "graphesis" to an ideological demand to transcend the body ("mathesis").

Evens, too, considers various aspects of the digital that challenge its status as pure abstraction. His intention in "Digital Ontology and Example," however, is not to understand this abstraction of materials, so much as the way the digital, *as process*, confronts (and is confronted by) the force of the virtual:

> Determined in advance by a rigid and calculable logic, the digital
> displaces accident or contingency in favor of *possibility* and *necessity*.
> Every result on the computer is a necessary one, logically implicit
> in the design of the hardware and software and the values of the

particular inputs. Every input has been anticipated in advance, and every output is a necessary consequence of some input. To draw a line is only to designate a line already drawn; to choose which key next to press is to pick from a menu of key-presses. In other words, every move the computer makes is already there as a possibility. (Chapter 5)

In order to minimize contingent or unwanted outcomes, digital technology draws circuits everywhere, intercepting the *reality* of the virtual with algorithms designed to transform it into the *modality* of the possible. It is analogous in this sense to the aims and demands of royal science, which organizes each experiment around a fixed point of view that, in its ideality, is external to the variations that comprise a particular environment. This has consequences for the human user as well, who is prevented from actualizing any desire that has not already been modeled in advance by the software and interface of the computer (as in the frustratingly limited interactivity of voice response systems, for example). Without an aleatory element in the digital series, novelty and creativity are effectively precluded because, as Evens writes, "Where would creativity find its moment in a system where the output is necessarily and logically implicit in the input?" (chapter 5).

And yet, paradoxically, "The digital is creative, abundantly so, and must therefore overlap at some point with the virtual" (chapter 5). The question then is: Where do digital and virtual meet? Evens answers this question by examining the "interrupt," a feature which is not remarkable in itself (it abruptly stops the CPU from repeating one algorithm and causes it to begin the execution of another), yet which shows how "the digital escapes its own flatness" (chapter 5). Whether triggered by a human operator or by the machine itself, an interrupt articulates the digital series with a number of possible "external" factors—the lived time of the human operator, other algorithms, etc.—that, strictly speaking, are not anticipated as the *possible* outcome of any one series. What is more, each time an interrupt is signaled, the CPU is prevented from cycling all the way through a particular algorithm. This produces a "texture" of variations within the operations of the computer that Evens compares to Deleuze's concept of the "fold":

If the digital in isolation is hermetic, sealed up into its own ideality, laminated onto a plane that renders everything level, then the structure that describes its contact with the actual would be a *fold* or pleat in that plane, a wrinkle that gives the surface of the digital enough texture to engender a friction with the actual, to perturb the human world. *Anywhere the digital meets the human, anywhere these worlds touch, there must be a fold.* (Chapter 5)

The digital, in this sense, does not *un*fold on a single plane, but consists instead of numerous intersecting planes that are repeatedly and exogenously *en*folded. To work creatively with digital media thus requires a continuous effort to amplify the virtual that lurks behind the modalities of software and interface; to do this, one must maximize the positive effects of the interrupt, or else seek out folds produced by the technology itself.

This, in essence, is what we already find happening in the field of architecture, where experimentation with digital technology has resulted not only in more curvilinear, asymmetrical, and interactive forms, but in a concept of structure that emphasizes continuous variations at the level of matters-flows—what Deleuze refers to as the "objectile."[90] In "Virtual Architecture," Manola Antonioli provides context for this development, and considers what this means for architecture as a science of forms:

Urban space is generated today out of activities that are at once physical and virtual, and its physiognomy is endlessly reconfigured by material and immaterial flows, as a function of new networks and systems of communication and transportation . . . The generalization of digital technology . . . profoundly modifies the creative process. The building, in turn, is more and more often conceived as an objectile, a new type of technological object that integrates the process of variability of forms. This implies a temporal dynamic, inaugurating an evolutionary process in terms of the potential to react to environment and to climate, and to interact with changes in the environment. (Chapter 6)

Here, too, the fold has become an important concept for understanding how structural elements (matter and form) may be subjected to a process of ongoing variation, so that they "continuously fold, unfold, and refold the spaces and temporalities of experience" (chapter 6). This has allowed architects to experiment with various techniques for *virtualizing* the building, working toward what Thomas Mayne calls "Event-Architecture" (chapter 6). The city itself becomes objectile, a paradoxical territory that follows its own variations. It is evident that an architecture based on the event, rather than the means for configuring space, has profound consequences for the economic, social, and political dynamics that shape the modern city. More importantly, the three-way interactive relationship of building, inhabitants, and environment provides a point of entry for the force of the virtual in processes of subjectivation that have traditionally inhibited this diversity. It is here that the applied sciences challenge the assumptions of science proper because the temporalities of experience that constitute the fold (indeed, the infinitely many folds) correspond to a great diversity of worlds—or, in any case, to the expression of such a diversity within the process of subjectivation.

> The subject only constitutes itself by expressing a world, but the world has no existence except as the expression of a subject. This involves thinking of the relation, typical in the Baroque conception, between an interior without an exterior (the world actualized in the subject) and an exterior without an interior (as the subject is no more than the provisional site of consistency of a virtual surface that precedes it). (Chapter 6)

Whereas the aim of royal science is to orient perception around a single, static point of view, inhibiting the creative function of thought and perception, the new dynamism in applied science suggests a plurality of views and increasingly decentralized production of subjectivity. In the urban environment, but also in the digital one, this has the potential to actualize subject-worlds *as* singularities, rather than simply reproducing the claims of a de facto scientific "truth" or "common sense" that would diminish the virtual diversity which gives rise to these singularities.

The Oedipal Sciences and Impossible Real

If thought can be used to disengage the production of subject-worlds from the dominant mode of subjectivation, that is because it takes place beyond the domain of the subject. "Typically, the subject precedes chaos and is defined, a priori, in advance of experience," writes Flaxman, "but Deleuze submits thought to a becoming-anonymous (and this, after all, is the nature of all becoming) that will strip the mantle of any determinate identity or determinate experience" (chapter 7). The "becoming-anonymous" of thought is the spontaneous realignment of one kind of synthesis with respect to another: a dissolution of the cogito that takes place whenever the passive synthesis of sensations is temporarily freed from its de facto formal conditions. It is in light of this concept of becoming-anonymous that Guattari's definition of scientific breakthrough ("Machine and Structure"), together with Kuhn's work on scientific revolution, lend theoretical weight to our earlier conjecture that philosophy and science cohabit the same movement of thought, having their basis in the same zone of "ontological difference"; however, to say on this basis alone that science disengages the production of subject-worlds from the dominant mode of subjectivation is a problematic claim that we will have to work through in more detail. It is all the more problematic for the fact that several different modes of synthesis are implied in the production of scientific knowledge—a fundamentally conflicted process that cannot be resolved in a simple ideological structure.

Although Deleuze's concept of subjectivity[91] will vary, what remains consistent throughout his thinking is the anonymity and multiplicity of the thinker, corresponding to a general inversion of the subject and its contents. It is less the cogito that is at issue, in this case, than Kant's "original synthetic unity of apperception," the formal condition of all experience. As Shaviro explains,

> Kant says that my various experiences hang together by virtue of the fact that I am able, at least in principle, to claim them as *mine*. No matter what particular "ideas" or "presentations" fill my mind, and no matter what I perceive or feel, "the *I think* must be *capable* of accompanying all my presentations." . . . For Deleuze and Guattari, in contrast, the subject is the *outcome* of the conjunctive synthesis, rather than its underlying principle. It is not what drives the synthesis, but *what gets synthesized*.[92]

Subjectivity does not constitute the agency of thought, but it corresponds rather to a set of de facto relations or variables that emerge only as a result. At issue here are the ongoing and reciprocal interactions between actual, quasi-material or "biopsychical" excitations, on the one hand, and the virtual, symbolic "events" on the other that bind them to a particular set of possibilities, including that of orthodox subjectivity. We will have to understand these interactions better if we are to understand how scientific thought "works."

In his earlier thinking, from the monographs through *Anti-Oedipus*, Deleuze situates these interactions predominantly in the social field, with a special emphasis on the role played by language. In *The Logic of Sense*, for example, he names three categories of synthesis within the basic machinery of the proposition: "the connective synthesis (if . . . , then), which bear upon the construction of a single series; the conjunctive series (and), as a method of constructing convergent series; and the disjunctive series (or), which distributes the divergent series: *conexa, conjuncta, disjuncta*."[93] Elsewhere in his works (*Empiricism and Subjectivity, Difference and Repetition, Anti-Oedipus*), the first (reproductive or passive) synthesis, or *conexa*, refers to a contraction of sensations within bodies and things. In *Difference and Repetition*, for instance, Deleuze arrives at the concept of the first synthesis by working through the Humean theory of habit.[94] According to Hume, something in the mind, which corresponds neither to memory nor to understanding, allows particular sensations to be contracted into a general case. The experience of repetition is thus founded on the capacity for this presubjective faculty (Hume calls it the imagination) to generate connections among sensations, these sensations having no inherent connections and thus lasting no longer than the shortest moment in time. Indeed, excitation and time are simultaneously produced by the first synthesis. ("Time is constituted only in the originary synthesis which operates on the repetition of instants.")[95] Deleuze discovers a similar principle in the basic biopsychical events that, for Freud, consist of the prolongation of excitations in sensory organs and erogenous zones: "Excitation as a difference was *already* the contraction of an elementary repetition. To the extent that the excitation becomes in turn the element of a repetition, the contracting synthesis is raised to a second power, one precisely represented by this binding or investment."[96] Libidinal cathexes or investments, the drives as such, represent the same capacity to generate connections as the Humean imagination; this time, however, the excitations themselves are connected in a series, so that we can say that the libido is no more than a contraction of a contraction (or contractions).

In *Anti-Oedipus*, Deleuze and Guattari will considerably expand this concept of the first synthesis to include both natural and technological processes of production.[97] It is here that we see social production, which includes the production of scientific knowledge, emerge from the first synthesis as principle of its formalization. The second (active) synthesis, or *disjuncta*, refers to a "recording process" that continually writes and rewrites a set of rules for the distribution of energy consolidated in the first synthesis. For the most part it operates on the basis of an exclusion, or law of noncontradiction, which opens one channel or possibility while closing or denying another. Truth claims work precisely in this way, anticipating a world or monad that comes to exist in such and such a way by virtue of the claim itself. They are not descriptive but *proleptic*. It is only in the third synthesis, *conjuncta*, that subjectivity as such emerges, or "*gets synthesized*," as Shaviro writes. Taken as a mutually determining network of individual process, the three syntheses, from affect to thinking subject, account for all the changes that take place within a particular field of actualization. It is also evident that when we speak of "a world" it is always in reference to an assemblage that shifts and changes according to the ongoing interactions not simply among the three syntheses, but among all the various processes of actualization that intersect with that assemblage. To think *about* the world is therefore to take part in its actualization. But does this mean that thought is also generated by one or more of the syntheses? Or does it represent a special case—an event that occurs everywhere at once in the field of actualization? What does thought consist in, and how does it influence the production of "a world"?

At this moment, we will be tempted, no doubt, to refer once more to the force of the virtual. But not everything that takes place in the virtual domain constitutes an act of thought or a becoming-anonymous, and Deleuze will repeatedly characterize the latter as a special case. In contrast to the simple connection of one sensation to another, thought proceeds like a dream or a dice throw. It is a game "without rules, with neither winner nor loser, without responsibility," and aims to "*affirm all chance and to make chance into an object of affirmation*."[98] It is an act of "pure creation" on the part of an "involuntary intelligence,"[99] or a "groping experimentation" that belongs to the order of "dreams, of pathological processes, esoteric experiences, drunkenness, and excess."[100] This is what separates thought from subjectivity: the involuntary, indeterminate, and creative nature of the one in contrast to the determinate and created order of the other. But thought is also distinct from habit, and from the recording processes that make up the social field. It is not a harmonious movement, an instrument of reason, but an event that disturbs the common order of connections, destabilizing common sense together with the world it actualizes. Opposed to this is the set of axioms that we traditionally refer to as thought: problem-*solving* (especially in the case of science). Solving a problem consists merely of containing the irruption of thought and normalizing the impossible connections that emerge between disparate sensations.

This is to say that thought cannot be fully assimilated to any of the three syntheses outlined by Deleuze. We must consider, rather, how thought designates an event that reconfigures the order among the three syntheses, disrupting their collective production of a logically noncontradictory world. The result is a world characterized by infinitely multiple disjunctions—a multiplication of worlds sustained all at the same time by the force of thought. We might even say that thought is the milieu in which this multiplicity of actual worlds already existed before being homogenized, formalized, and made to appear as property of a transcendent subject. This is what Deleuze and Guattari make out of Artaud's conception of a form of thought that "operates on the basis of a *central breakdown*," and that develops "in a pure milieu of exteriority, as a function of singularities impossible to universalize, of circumstances impossible to interiorize."[101] Thought has the double character of that which is always outside, peripheral to the thinker, but also of an event that takes place locally, in the sense that it necessarily emerges from the intimate and immediate experience of difference.[102] "It might be worth recalling Jacques Lacan's wonderful neologism, '*extimité*,' in order to convey the intimate exteriority that characterizes the outside," writes Flaxman, "because the impulse to overreach ourselves that the outside demands invariably returns us to that which is folded into the unfathomable depths of the soul" (chapter 7). But thought also has the character of that which reveals the extimacy of *everything* we think, experience, and perceive. Even those experiences that convey a pure and immediate sense of interiority now appear to emerge directly from a cosmic unconscious or "magnetic field." Whether we are speaking of the syntheses proper or thought as a special case of the syntheses, we are referring always to singularities that operate on the milieu of the emerging subject, which is to say, on an entire population of subjects as they emerge together under a single truth-claim or "social experiment," with the result of a sedentary assemblage or nomadic one, of an individual or a pack (as in the case of the Wolf-Man).[103]

The critique of psychoanalysis is therefore prepared, in advance of any explicitly political argument, by a basic inversion of the customary distinction between inside and outside, between subject and world. Oedipus first appears on the scene with the classical image that places all possibility of thought on the side of the subject—when thought is conceived of only as an interiority. Attendant on this is the assumption that thought operates an internal reconstruction of events that have already taken place in the outside world. By making this assumption, we create a criterion for evaluating thought as a good or bad copy, and, more generally, for retroactively matching percepts and affects to a set of rules concerning their legitimate or illegitimate use. This alone is enough to bind the entire psychic apparatus to the structure of representation. What is more, if the subject is the locus not only of conscious thought but, more profoundly, of unconscious drives and fantasies, if all these forces that act on the subject are now considered to comprise its contents, the whole activity of the organic individual is now implicated by the regulatory principle of legitimate use. At such a distance from the outside world, the unconscious appears to want more than reality can offer, as well as some things that do not even exist. It is on this basis, for example, that Freud justifies the transition from pleasure principle to reality principle, from the unrestricted expression of unconscious drives to the "educated" and "reasonable" mediation of the ego:

> The sexual instincts, from beginning to end of their development, work towards obtaining pleasure; they retain their original function unaltered. The other instincts, the ego-instincts, have the same aim to start with. But under the influence of the instructress Necessity, they soon learn to replace the pleasure principle by a modification of it . . . The ego discovers that it is inevitable for it to renounce immediate satisfaction, to postpone the obtaining of pleasure, to put up with a little unpleasure and to abandon certain sources of pleasure altogether.[104]

For Deleuze and Guattari, Oedipus assumes its place by a similar transition but not in the sense that Freud intends here. It does not emerge from an interpersonal or intrapersonal conflict, much less from the impossibility of giving the unconscious what it "wants." We can say, rather, that Oedipus involves the hidden moral function of Necessity when it is placed in the position rightfully occupied by the creative force of thought. To this extent, at least, Oedipus is just one more example of the *ressentiment* that animates the history of Western metaphysics from behind the scenes.[105]

What is at issue here, and not just in *Anti-Oedipus*, is the whole notion of the syntheses and their subordination to social production in its function of codifying the norms of judgment. Like Freud's conception of the two "psychic instances," which account for the distorted form of wish-fulfillment in dreams,[106] synthesis and judgment each designate something between a moment and an agency in the production and expression of desire. In Deleuze's conception, synthesis and judgment are not specifically psychological in nature; they represent an immaterial, preanthropomorphic form of desire that is closer to Whitehead's concept of appetition[107] than to the Freudian libido. Nor do they emerge, in Deleuze's earlier thinking, from the Freudian schema of the ego and the id. Instead, we find the first use of the terms in a working through of the Humean conception of relations of identity in *Empiricism and Subjectivity:*

> The relation always presupposes a synthesis, and neither the idea
> nor the mind can account for it. The relation, in a way, designates
> "that particular circumstance in which . . . we may think proper
> to compare [two ideas]." "To think proper" is the best expression
> for it; it is, in fact, a normative expression. The problem is to find
> the norms of this judgment, of this decision, and the norms of
> subjectivity.[108]

Deleuze aims, first, to establish the grounds for a synthesis that takes place outside the faculties of memory or understanding, beyond the structure of rational subjectivity. He finds these grounds in Hume's notion of the principles of association. ("We can see now the special ground of empiricism: nothing in the mind transcends human nature, because it is human nature that, in its principles, transcends the mind.")[109] The rational subject occupies a secondary position, emerging as the consequence of a judgment. It is no use to say that the moral basis for this judgment is redundant, that the real criterion for thinking properly *(Eudoxus)* lies in the recognition of necessary relations among objects. For Hume, these relations are not on the side of objects but on that of the subject, corresponding to the way the imagination has been affected by the principles of association. More radically, in Deleuze's reading of Hume, there *are no objects,* except those which emerge in the synthesis presupposed by judgment: "Not only are perceptions the only substances, they are also the only objects."[110]

Deleuze resumes his discussion of synthesis and judgment in *Difference and Repetition*, when he works through Freud's distinction between the pleasure and reality principles—or what he calls the "passive" (or "reproductive") and "active" syntheses:

> On the one hand, an active synthesis is established upon the foundation of the passive syntheses: this consists in relating the bound excitation to an object supposed to be both real and the end of our actions. . . . Active synthesis is defined by the test of reality in an "objectal" relation, and it is precisely according to the reality principle that the Ego tends to "be activated," to be actively unified, to unite all its small composing and contemplative passive egos, and to be topologically distinguished from the Id.[111]

Deleuze's notion of judgment or active synthesis thus corresponds to what amounts to a Freudian theory of objectal relations: a new orientation of the libido toward objects determined by "reality testing." But there are two aspects of Deleuze's formula that pose some difficulty with regard to the psychoanalytic determinations of this expression. For one, he will maintain a claim we saw earlier in *Empiricism and Subjectivity*: that the subject is continuously organized by principles that transcend it. The active synthesis cannot be a function of the subject in the process of unifying itself because the subject only comes about as a result of that same process. Simply put, subjectivity has no agency (in *Anti-Oedipus*, Deleuze calls it a *residue*).[112] The active synthesis must therefore be based on criteria that exist prior to and outside the subject. This is all Freud wishes to convey in *Beyond the Pleasure Principle*, when he describes the ego as a topological limit that separates and negotiates between the impossible demands of the id and the necessary relations among objects.

From the Deleuzian point of view, however, the topological definition of the ego will introduce a more fundamental problem. The limit that is drawn during the formation of the ego does not distinguish between an interiority ruled by desire and an exteriority ruled by necessity. It merely imposes a set of rules concerning the way we will be obliged to designate bodies and things "in reality," joining our individual productive energy to that of the social field. In other words, the norms of judgment are what produce necessary relations among objects, what produce reality as such, not the other way around:

The fact that there is massive social repression that has an enormous effect on desiring-production in no way vitiates our principle: desire produces reality, or stated another way, desiring-production is one and the same thing as social production. It is not possible to attribute a special form of existence to desire, a mental or psychic reality that is presumably different from the material reality of social production.[113]

As Deleuze and Guattari will proceed to argue, the organization of a perceptual field according to the norms of judgment is an ongoing process, susceptible to frequent disturbances of all kinds. Under very special circumstances, it may even be reversed, as we see in Deleuze's analysis of Tournier's revisionist Robinsonade, *Friday*: "In the Other's absence, consciousness and its object are one. There is no longer any possibility of error, not only because the Other is no longer there to be the tribunal of all reality—to debate, falsify, or verify that which I think I see; but also because, lacking in its structure, it allows consciousness to cling to, and to coincide with, the object in an eternal present."[114] The norms of judgment cannot therefore be based on necessary relations among objects supposed to be real; it is judgment itself that establishes these relations.

What then does it mean to submit a collection of impressions to the test of reality—to any kind of standard at all? What role does science have, if it can no longer be said to derive its principles from (already existing) reality? Before judgment, there is only "bound excitation," the contraction of sensations in an eternal present. We must therefore say that the subject is structured correlatively with objects, and that our impressions are transformed directly into necessary relations by "the test of reality in an 'objectal' relation," according to determinations within a historically and culturally specific social field.[115] We must therefore de-anthropomorphize the common (psychoanalytic, linguistic) notion of social production: we must think beyond the latent humanism of the human sciences. The example Deleuze borrows from biology—"An animal forms an eye for itself by causing scattered and diffuse luminous excitations to be reproduced on a privileged surface of its body"[116]—is not metaphorical. The passive synthesis produces an eye-light assemblage ("The eye binds light, it is itself a bound light"); the second and third syntheses create a cleavage that puts the eye on the side of the subject and light on the side of a determinate object. A subject that sees emerges out of a particular "use" of a bound excitation. It is by force of a cleavage in the field of actualization that a series of excitations, first contracted in an eye or antenna, develops a metaphysical, interior surface, a site for the construction of a subject to whom these excitations can now be said to "belong." The need for such a cleavage can be explained by referring to its specifically regulatory function. Indeed, the apparent gap between subject and object, as one possible outcome of the process of actualization, has no other function than to shape desire into the contours of a reality that matches the libidinal investment of the social field. For it is only with this gap that the norms of judgment can acquire the sanction of a presumably external reality (Necessity), axiomatizing the production of desire on the basis of what is presumed to be real.

If thought is a special case of the syntheses, it is because it allows us to be diversely affected by any number of singularities at a given moment, and not simply those sustained by the law of noncontradiction. Or, to put it another way, the exterior milieu of subjectivity, the domain of thought itself, simultaneously includes all singularities, all monads or incompossible worlds. The scientist catches a glimpse of these other singularities every time an experiment turns up data that is at the same time marginal and inconsistent with the dominant scientific paradigm of statistical variance. The exterior milieu of subjectivity thus comprises what Deleuze and Guattari call "unthought" or "the impossible real"[117]: impossible because it sustains multiple disjunctive worlds; real because it refers to a domain that, like the data dismissed as statistical variance, is always present in its latent or unactualized state.[118] Conversely, processes of subjectification (*subjectivation*) establish a network of axioms for validating and articulating our perceptions according to what is conventionally viewed to be "proper," "truthful," or "real." A fundamental tension thus emerges between thought and subjectivation, according to which transversal relations among desiring-machines in the impossible real must either pass muster with the norms of judgment ("synthesis of recording"),[119] or dissolve them altogether. This is what Kuhn discovers in his theory of paradigm shift. We cannot say then that truth is a structure that transcends subjectivity or scientific observer; rather, it is the product of an ongoing machinic process, by which the impossible real is transformed into a possible reality—but "possible" precisely in the sense that it exists as an axiom that is constructed alongside actual desiring-machines, and that regulates the way they connect to other machines.

This is all that Deleuze and Guattari mean when they say that desiring-production is actual but Oedipus is virtual: *"it is the Oedipus complex that is virtual, either inasmuch as it must be actualized in a neurotic formation as a derived effect of the actual factor, or inasmuch as it is dismembered and dissolved in a psychotic formation as the direct effect of this same factor.... It is a serious mistake to consider this formation in isolation, abstractly, independently of the actual factor that coexists with it and to which it reacts."*[120] The virtual is not the same as the possible; but the possible can only be constructed within the virtual, as an axiom for the synthesis of connections among *actual* desiring-machines. Here again we see that science occupies a gap, applying its methods and instruments not to existing objects, but within a production of subjects in anticipation of objects that it *also* produces. To the extent that this same process is applied (often through popular conceptions of science) to sanction and validate the collective means of judgment, it does not function according to a modal logic ("it is necessary that," "it is possible that"), but on the basis of a moral imperative, or what Nietzsche refers to ironically as "the obligation to lie according to a fixed convention."[121] It is not informative or proleptic, in this sense, but imperative, formulating its truth claims in a discourse that is meant to be obeyed, not believed (*mots-d'ordre*, "Let people say that . . ."").[122] The discourse of science is all the more powerful for the fact that it guides the connective synthesis of desiring-machines, through the production of technology and technological production, on a massive and global scale. Axioms based on science are not qualitatively different from other truth claims; they are only more powerful than other forms of orthodoxy because they carry the presumed or declared validity of whatever universal truths they are in the process of creating.

But axioms are hard to define. They are not *ideologies*. As Deleuze and Guattari repeatedly warn us, they are the product of competing political and social forces, of ascendant regimes and coup d'états, but also of group fantasies and other manifestations of the collective libido—music, television, cinema, advertising, not to mention science and technology. In each of these cases, what we find is not so much the repression but *instrumentation* of desire, in the sense that an axiom does not inhibit desiring-production, but channels and extends it toward a particular use.[123] Oedipus is the general case for all axioms that cause the productive principle to pass as a case of representation, with the secondary effect that Oedipus itself is made to look productive (*wage labor*, for instance, as principle of control, results in *capital:* the apparent objective creative force of the market).[124] Sometimes an axiom is simply not strong or flexible enough to bind the spontaneous synthesis of connections among machines. At this point, desiring-production becomes revolutionary. It is no use to say that revolutions, too, have their axioms, that they produce slogans and doxologies concerning the truth of this rather than that revolutionary cause. This may be the case, but desiring-machines—according to Deleuze and Guattari, the only real revolutionary force—do not desire truth, only free variation of the connective synthesis.[125] Even when this free variation is effectively prevented, the passive synthesis retains its original function unaltered: "The desiring-machines are always there, but they no longer function except behind the consulting-room walls."[126] Because desiring-production never ceases, the social field, for its part, must vigilantly and endlessly build and repair these virtual walls.[127] In the case of both the neurotic and the psychotic, the free variation of desiring-production has found a breach or a fissure in the wall and threatens the integrity of the entire recording-process. But the schizophrenic formation is unique, inasmuch as it causes a segment of desiring-production to pass completely through and to set up its own unauthorized recording-process on the other side (the "disjunctive syllogism"). [128] What prevails in this case is not the *truth* of desire over a symbolic order presumed false, but of the impossible real over the exclusive true-or-false disjunction introduced by judgment.

By realigning our view of science and nature in such a way that these appear as creative processes rather than a collection of laws and constants, by changing our whole image of thought, we have been left with a new set of problems with respect to *ethics*. It would be too simple to turn now to this or that use of scientific knowledge—to follow, for example, the development of moral opposition in the field of nuclear physics, from the Manhattan Project to the Russell-Einstein Manifesto of 1955.[129] The problem runs much deeper than this and involves a tracing and retracing of de facto relations at the expense of the diversity of creative forces that comprise the impossible real. The relevant questions are: What are the practical consequences of the assertion that we inhabit the impossible real? What happens once we feel ourselves implicated by the ongoing work of the syntheses—how do we *use* them? And what does it mean to *use* something that categorically exceeds the power and sovereignty—the "empire"—of the subject? This is the primary nature of the problems addressed in *Anti-Oedipus:* "Given the syntheses of the unconscious, the practical problem is that of their use, legitimate or not, and of the conditions that define a use as legitimate or not."[130] The third section of *The Force of the Virtual*, "Science and Subjectivity," seeks to answer these questions with respect not only to science proper, but also to the position occupied by science and scientific truth claims within popular culture. How does popular culture voluntarily call on science to regulate its own syntheses? For it is evident now that the question posed by Mullarkey concerning judgment ("according to which values should we prefer the thoughts of Deleuze over those of, say, Gottlöb Frege? *How* did *he* get it right?")[131] invokes no other problem than this. As a machinic system of values, Deleuze's thinking, his Bergsonism, is not simply the ongoing critique of science based on an ontology of the virtual. It is a more comprehensive effort to bring the basic conditions of a monistic ontology (actual *and* virtual) as far as possible toward a Nietzschean ethics. It is with Nietzsche that the notion of values supplants the question of truth, and where, as a consequence, Necessity no longer occupies the place of thought. (This is also the implication of Michael Hardt's argument, in *Gilles Deleuze: An Apprenticeship in Philosophy*, concerning the monographs on Bergson and Nietzsche. According to Hardt, these represent the essential foundation on which Deleuze's entire philosophical doctrine will be constructed;[132] but they also represent two very different tendencies within that doctrine: an ontology and an ethics. It is the book on Spinoza, *Expressionism in Philosophy*, that joins the two tendencies together, moving both centripetally, from thing to God, and centrifugally, from God to thing, an approach that joins actual and virtual together in a single circuit.)

In "The Subject of Chaos," Flaxman considers how judgment, and the whole problem of ethics, is implicated in Deleuze's attempt to establish the grounds for a relationship halfway between subjectivity and the "chaos" it represses:

> On the one hand, when we refuse to recognize our need for sense, we supply too little order—as if we could simply except ourselves and thereby encounter chaos in a pure and untrammeled state . . . But on the other hand, and even more troublingly, when we go in the other direction, mustering the indelicate urge for order, we run the risk not of chaos but of the reaction (reactive) formation it induces. In the midst of the infinite variation of mind and matter, the development of the subject amounts to the static appearance

of things, to the encrustation of identities, and to the assignment of values. Thus, for Deleuze, as for Nietzsche before him, the molarization of chaos is redoubled by its moralization whereby the formation of semistable extensionality and semistable intentionality reciprocally determine each other, the former projecting the latter, the latter introjecting the former. (Chapter 7)

Deleuze (and Guattari) will increasingly identify the impossible real with chaos, or in any case with the passive support on which chaos first emerges: "Chaos does not exist; it is an abstraction since it is inseparable from a screen that makes something—something rather than nothing—emerge from it."[133] The passive support or "screen" has two effects: (1) it allows chaos to subsist as such by providing the necessary conditions for the inclusion of any and all disjunctions (Deleuze calls it a "membrane," interposed between the *Many* and the *One*), and (2) it introduces an aleatory element into any subsequent order that might emerge from the chaos. It is this element that, like Raymond Roussel's method of composition, generates novelty by turning exclusive disjunctions in the symbolic order back into inclusive ones. Any new order is thus susceptible to a reversal that would bring its constituting elements back into proximity with chaos.

As Flaxman points out, it is because of the imminent nature of chaos, the fact that it is always near us, always so "extimate," that we submit voluntarily to processes of subjectivation that guarantee our protection (as well as our confinement). In the first instance, science would be no more to blame than philosophy; it is chaos—or, in any case, our own reasonable fear of madness in the face of chaos—that shapes us into subjects. Under such circumstances, "we quickly divest ourselves of chaos altogether in favor of so many tacit presuppositions that provide the model of what it means to think (rightly, for instance, according to transcendent values) and how thought proceeds (properly, for instance, according to logical methods)" (chapter 7). To think rightly and to think properly: again it is a question of self-preservation, a fear of madness, but it proceeds by way of judgment, values, interest—even a certain submission to the slogans of official culture—all with the presumed "goodwill" of the thinker ("a matter of discovering and organizing ideas according to an order of thought, as so many explicit significations or formulated truths").[134] If philosophy and science have no initial role in compelling us to think this way, they will nonetheless expand and systematize this image of thought, establishing a hierarchy of values based on a consensus regarding (already) formulated truths. This is what Deleuze and Guattari mean by "common sense": a form of thinking that proceeds on the presupposition of transcendent values, and from the point of view of a "generic subject experiencing a common affection."[135] As Flaxman demonstrates, this point of view does not only presume an agreement among individuals, but also the "agreement of the legislated between faculties" (chapter 7). Common sense would be the general form under which, as in philosophy proper, the subject attains internal consistency by identifying with a universal subjectivity.

But this form of thought is still only an empty form, a set of rules that apply to the thinking subject. Philosophy needs science to give formal consistency to its contents—that is, to the objects of thought. Deleuze and Guattari call this "good sense." Classical "normative" physics responds to this need by condensing all the instances and variations of a process into a single, fully determined image or model. Such is the case, for example, with Schrödinger's argument against the Copenhagen interpretation of quantum physics ("The Present Situation in Quantum Mechanics"), in which he recapitulates the formal and practical necessity of a scientific standard based on the production of fully determined models: "The representation in its absolute determinacy resembles a mathematical concept or a geometric figure . . . if a state becomes known in the necessary number of determining parts, then not only are all other parts also given for this moment . . . but likewise all parts, the complete state, for any given later time."[136] But even Schrödinger is not able to say whether the belief that a given process is fully determined in itself precedes or follows the scientific standard: "Perhaps the method is based on the belief that *somehow* the initial state *really* determines uniquely the subsequent events, or that a *complete* model, agreeing with reality in *complete exactness* would permit predictive calculation of outcomes of all experiments with complete exactness. Perhaps on the other hand this belief is based on the method."[137] Deleuze's concept of good sense corresponds to the latter interpretation: belief is based on the method—indeed, the method is designed to generate belief. "The modulation of objects in time, or movement, is thereby determined in accordance with the habit of the subject itself," writes Flaxman, "a mobile milieu, which anticipates itself in the prolepsis of its own common sense" (chapter 7).

From the form of knowledge, we thus arrive at a method for applying it universally, for insuring—rather than presuming—that the universe of thinking subjects will have a basis for agreement. Science provides the social field with a virtual site in which to construct its Habitus, the spatio-temporal dimension of its becoming. It is only then that the subject, triumphing over the body and over a field of excitations that borders on hallucinatory, establishes its own internal consistency. It is only then that the subject of common sense develops a legislated and legislating faculty. But there is another possible outcome that should not be overlooked in a discussion of science and the force of the virtual. We have already seen, in Deleuze's working through of the pleasure principle, that the formation of the subject is based on other processes ontologically primary to it: the desiring-production and synthesis of recording. Before the form of the subject comes on the scene, these syntheses do not belong to any subject, insofar as this refers to the subject of common sense. Even more radically, we can say that an excitation, before it is consolidated and prolonged, does not belong to any body, insofar as this refers to an object of good sense. To whom then do these excitations belong? In an eye that binds light or an antenna that binds vibration and odor. But who sees? What senses? To what interiority can we ascribe thoughts, imagination, or dreams?

Flaxman compares good sense to the passive synthesis, a brute repetition of associations on the part of the individuated consciousness (chapter 7). This opens the way to a very different notion of science and what it is capable of building on the plane of reference. For if the task of science is to establish the terms of good sense, and if the passive synthesis binds excitation in sensory organs and erogenous zones, we can say—not at all metaphorically—that each experiment, each act of empirical observation, operates a consolidation of collective excitations on the surface of the social body. Science forms the eyes and antennae of a social field, not only in the sense that it helps negotiate a proleptic external reality (recording synthesis) but in the ongoing passive synthesis of desiring-machines on the surface of the "full body of the socius." In other words, the productions of science do not need common sense. They do not rely, for their operation, on the form and sanction of the rational subject. Content simply to transform differences in intensity into objects, science is capable of doing so at a rate that far exceeds the demand for good sense (and that ultimately threatens the stability of its own objective parameters). This unauthorized synthesis takes place at the very heart of science, not at its fringes, but the image it engenders in the popular imagination is one that must be pushed to the margins of scientific production, beyond the presumed goodwill of the thinker. In the figures of Dr. Frankenstein and Dr. Jekyll, for example, or even Nikola Tesla and Wilhelm Reich, we see a capacity for the scientist to accelerate the succession of ruptures in a particular field to the point where it begins to produce monstrous new organs. "Lacan explains well how, in terms of the crises and ruptures (*coupures*) within science, there is a drama for the scientist that at times goes as far as madness,"[138] write Deleuze and Guattari. In its orthodoxy, science presents the image of an intellectual curiosity that is detached to the point of indifference. In its excesses of production, however, science betrays a creative synthesis that borders on delirium. Detached curiosity and schizophrenic excess constitute the reverse side of the royal or *Oedipal* sciences, and indicates a resurgence, under the auspices of scientific discovery, of the impossible real.

In the final chapter of this section, "Numbers and Fractals: Neuroaesthetics and the Scientific Subject," Patricia Pisters explores the excessive nature of scientific production, looking specifically at the film *Pi* (1998) by Darren Aronofsky. In form and content, Aronofsky's film exhibits both the "limitless beauty and power of numbers and geometric figures" and the "neuroaesthetic" image of madness that this power invokes. At the same time, Aronofsky explores the unsettling proliferation of various mathematical figures, namely π, spiral logarithms, and the Fibonacci sequence (chapter 9), and the transversal connections among natural and social systems that such a proliferation implies. Based on the foregoing analysis, it will be meaningless to ask whether these numbers really inhere to DNA sequences, seashells, whirlpools, etc., or whether they are projected there by science—that is, by the protagonist of *Pi*, who represents our own (hysterical? paranoid?) desire to make sense out of them. As we have already seen, science does not stand apart from the world it "observes." The problem is elsewhere, in the moral dilemma these figures provoke when they lead rational thought into a domain of unreasonable, or even "impossible" connections, connections that are nonetheless *empirically* real. In the case of *Pi*, the method of science, which was supposed to ground the eye/I of a transcendent subject, has led the protagonist to its obverse virtual side, where the rules of nature do not simply exist, but are continuously generated anew. Aronofsky explains how he himself caught a glimpse of this groundless and subjectless production during a trip to Mexico, at the site of a Mayan temple colonized by ants: "he saw the groundlessness of the 'I,' which is also a groundlessness Max discovers the closer he comes to the mysteries of the universe"; as Pisters writes, it is a discovery that "resonates with Deleuze's conclusions of *Difference and Repetition*, where he develops a nonrepresentative, preindividual way of thinking about difference" (chapter 9).

In what does this groundlessness consist? The proliferation of π, like that of all singularities, is unsettling to the extent that it does not point to a higher meaning or higher order, as some of the characters in Aronofsky's film believe (or as Luca Pacioli and Leonardo da Vinci, in their search for the Golden Measure, proposed). It refers instead to the groundlessness of this proliferation, and consequently to a generative principle that has no relation to the transcendent and universal subject. It thus presents us with a new image of thought, characterized by the externality of ideas, and by the virtual, immanent singularities that may effect new connections in the actual. This is what allows Deleuze to define transcendental empiricism as the action of thought outside the limits of a certain illusion: "The ultimate, external illusion of representation is this illusion that results from its internal illusions—namely, that groundlessness should lack differences, when in fact it swarms with them. What, after all, are Ideas, with their constitutive multiplicity, if not these ants which enter and leave through the fractured I."[139] Under the sign of the virtual, processes of subjectivation do not always result in the well-behaved, self-identical subjects that law and custom demand; this is precisely because they have no ground other than the eternal insertion of difference. This is what leads Deleuze to ask: "[does] losing one's memory or being mad belong to thought as such, or are they only contingent features of the brain that should be considered as simple facts?"[140] The Oedipal sciences, in contrast, draw strength from the illusion that thought relies on a ground—that thought, too, like the Demiurge, must be sanctioned by the eternal forms or essences of some transcendent domain, the universal form of the subject, a self-same and eternal truth, in the absence of which no semblance of order in the sensible world would be possible.

The Strata of the Brain

In the foregoing analysis, the distinction between the actual and the virtual acquires new significance with regard to processes of subjectivation. The virtual refers not only to the immanent and inexhaustible creative energy of becoming, but also, and more specifically, to numerous competing recording processes that aim to channel this energy. Taken as a whole, these processes comprise what Deleuze and Guattari call social production.[141] We must think of the social field as a real but virtual multiplicity that, by bringing so many singularities or events to bear on a targeted field of desiring-production (sometimes merely anticipated or imaginary), results in the emergence of a population of subjects. It is not a homogenous field of forces, but it constitutes its own "impossible real," with Oedipal alongside *anoedipal* libidinal investments, the political slogans of one party or corporation alongside the slogans of many others. A neurotic group fantasy may resolve in schizophrenic lines of flight or in a microfascist investment—or even, as Guattari claims in the case of punk music, in an unconscious reproduction of dominant systems of expression.[142] In all these cases, social production is suffused with conflict and difference, animated by the convulsive force of the virtual, which is the event and domain of thought itself. The relation of one social force to another constitutes a kind of presubjective agency, the agency of the virtual, which is animated by the dynamic differences among all forces in the social field: "The relation of force to force is called 'will' . . . Power, as a will to power, is not that which the will wants, but *that which wants* in the will (Dionysus himself). The will to power is the differential element from which derive the forces at work, as well as their respective quality in a complex whole."[143] The differential element among presubjective agencies shapes the actual into a heterogeneous totality.[144] But this totality is fundamentally open because it does not exhaust the forces that shape it. What is more, it is a contingent and provisional totality because it comes about, in each case, at the intersection of dynamic forces that retain their differential element, their will to power.

The actual itself seems less dynamic by comparison, referring only to the shifting outcome of these forces. It proceeds by means of passive rather than active syntheses. It does not act but is acted upon. But these are superficial and misleading distinctions. In *Anti-Oedipus*, the actual does not refer to this or that psychic or social formation but to the *entire domain of desiring-production*. Taken by itself, the actual is like a massive assembly line, but with no line, and therefore only the most rudimentary means of effecting an assemblage (the contraction of habit or the passive syntheses). Rather than a factory, we must think of a virtual articulation of flows, a will to power that does no more than plug one flow into another. A sawmill, for example, is no more than a particular organization of flows, a forest "plugged into" a river. Even after they are joined together in the production of dimensional lumber, both river and forest maintain their status as partial objects and flows; *as process,* the virtual factor gives them a new organization without changing this status, and this is true for all the successive organizations in which the lumber will be used (at some point, a house, *as process,* must also be connected to both a forest and a river). Similarly, the *"Sainte famille"* is no more than a connective synthesis that gives a new organization to existing processes of desiring-production. It only differs from other syntheses in that it has been selected out and reinforced by the dominant mode of subjectivation, with the aim of organizing a readymade field of desiring-production (children's bodies and their flows) into a ready supply of obedient workers-consumers. Which is to say that family relations do not produce anything; they only (1) actively diminish the dynamic differences in the social field that make up the will to power, and (2) join together the microproduction of desire in an emerging child-subject with macroproductive forces in the economic field.

What is at issue here is the genesis of desire, where it comes from, and what forces are necessary to make it productive. Deleuze and Guattari will insist that desire is already productive in itself: it is nothing but the principle of this production. The family, the education system, advertising, and cultural production in general are all contingent and provisional processes that submit actual desiring-production to this or that organization without altering the original function of desire (indeed, they depend entirely on this function). It is only in the most superficial sense that advertising, to take a particularly salient example, "creates demand." Strictly speaking, advertising does not create demand. It only creates a site on which to assemble desiring-production in the form of a "consumer-machine." Each machine assembled in this manner maintains only a virtual status with relation to desiring-production. This is why Deleuze and Guattari will be compelled to establish a hierarchy among the syntheses, based on the ontologically grounded claim that desire comes first, every other manifestation of productive energy being derived from it: "Conforming to the meaning of the word 'process,' recording falls back on (*se rabat sur*) production, but the production of recording itself is produced by the production of production. Similarly, recording is followed by consumption, but the production of consumption is produced in and through the production of recording."[145] The increasingly complex organizations that make up society all depend, or "fall back," on the primary and ongoing production of desire—that is, on the production of production as such.

What is interesting in the case of consumption is that there is always some part of the recording process left over, a residual segment of code that rises to the surface to provide the basis for a unique instance, a *"for whom."* This residue, the subject, is both a surplus and a constraint: *surplus,* because it is born each moment from a segment of code that is not taken up by the process of consumption; *constraint,* because it comprises exclusive disjunctions originally forged by social production (recording process). Advertising does not create consumption any more than it creates demand. It aims rather to intervene in desiring-production by augmenting the surplus and reinforcing the constraint: which is to say, it invents a whole virtual population for whom a product is necessary or desired. But an advertisement only works if we feel ourselves implicated by our own virtual image. Desire must be caught in the snare of an image or slogan, so that the determinate order actualized by virtue of an image seems to have always belonged to us, to the ontologically primary process that is desire itself.

Something similar happens when we turn to the sociological or psychological determination of the subject, and to the internal sense in which the experience of enjoyment is reconstructed in pseudo-humanistic terms. In these cases, the subject, or something internal to the subject, is said to *want* something: it may be imminent or remain repressed, it may be orthodox or perverse—for Deleuze and Guattari, such determinations are of little consequence. What is decisive in each case, rather, is that the principle of production and the whole order of its assembly have been reversed. The concept of "want" essentializes the "for whom" of subjectivity. The transcendent principle of structure displaces and obscures the productive one of desiring-machines. All further analysis will now be bound to the de facto social or psychological structures of dominant mode of subjectivation, i.e., the production of neurotic subjects. It is here that Deleuze and Guattari take psychoanalysis to task, criticizing its essentialist determination of family relations and the Oedipus complex. From their point of view, Oedipus is *"actualized in a neurotic formation as a derived effect of the actual factor"* or *"dissolved in a psychotic formation as the direct effect of this same factor."*[146] By inverting the ontological relation between the actual (desire) and the virtual (Oedipus), psychoanalysis arrives at the claim that desire is only possible once the various drives of the organic individual have been organized in a neurotic formation. For Deleuze and Guattari, Oedipus can only be a derivative, virtual organization of a system of production that begins with desire.

We must also see, however, that Oedipus, at least in the broad political and social terms of *Anti-Oedipus,* is an *active* misrepresentation of social production as the "actual factor." We are Oedipal precisely to the extent that we give credit to social production—to the principle of regulation itself—for syntheses that belong on the side of desiring-production. This is what makes the concept of the actual so conceptually demanding. It is as if Deleuze and Guattari, in their first collaborative work, wanted to rework all the social sciences—psychoanalysis, anthropology, linguistics—one at a time, restoring the actual to its proper complex of relations. Above all, *Anti-Oedipus* is a work that makes way for, and insists on, the practical and ethical problems posed by desire, i.e., "given the syntheses of the unconscious, the practical problem is that of their use."[147] Taken by itself, the concept of use is only an analytical problem; *given the syntheses of the unconscious,* however, which is the actual factor, use becomes a practical problem. Or, put another way, Oedipus as a *slogan* is only virtual and poses no ethical problem as such; Oedipus as a *complex,* on the other hand, represents the entire problem of the actual-virtual system: a social force propelled by the energy of desiring-machines it organizes and conceals.

This leads our discussion of the force of the virtual to something of an impasse. Our line of inquiry began (in the last section) with the practical relationship between synthesis and judgment: According to what criteria and toward what end do we define the legitimate or illegitimate use of the passive syntheses? What choices do we make with regard to scientific truth (as a function of good sense, for example) that influence our own actuality? Now we will have to be more critical of the "for whom" implicit in each of these questions. We will need to take a step back from the anthropomorphic determinations of subjectivity, so that we may ask instead: In what ways do we choose this or that use? Who—indeed *what*—chooses? Where do we situate the agency that directs the actual-virtual system toward one outcome or another? For Deleuze and Guattari, each side of this system represents a kind of agency-in-multiplicity: the virtual, as affirmation of difference, is the will to power; the actual, as desiring-production, constitutes a creative force and a field of immanent connections (machines to be plugged into machines, to be plugged into still more machines, in any configuration whatsoever). Given that "our own" agency is drawn equally from both sides of the cosmic actual-virtual system, given that we are implicated at all times by both an ontology and an ethics, what will constitute the pivotal moment in the organization of "our own" desire?

A working through of the respective roles of subject and object seems pointless in this case because, as Antonioli writes, "Between that which one customarily calls the 'subject' and that which is defined as the 'object,' there exists no causal link or direct action because subject and object are no more than the result, always provisional, of a process of individuation and actualization" (chapter 6). Subject and object correspond to the de facto status of desiring-production as it has been organized, at any given moment, by social production, as well as other competing forces intersecting in the virtual. Indeed, the closer we look at these terms in Deleuze's works, the more it seems that they have lost their power to invoke the problems that are most important to him. The problem of desire, for example, its ethics and ontology, cannot be engaged simply by destabilizing the sovereignty or identity of the subject, or by retroactively introducing the "agency" of the unconscious drives (though, admittedly, this is a step in the right direction). For Deleuze, it is the whole psychological determination of the actual-virtual system that has come to falsify desiring-production. It is the subordination of desire to meaning—to the presence-as-absence of the unconscious—that gets it all wrong:

> *Wo es war, soll Ich werden.* In vain has this been translated as: "There where it was, there as subject must I come"—it's even worse (including the *soll*, that strange "duty in an ethical sense") . . . The Freudian formula must be reversed. You have to produce the unconscious . . . and it is not easy, it is not just anywhere, not with a slip of the tongue, a pun or even a dream. The unconscious is a substance to be manufactured, to get flowing—a social and political space to be conquered. There is no subject of desire, any more than there is an object. There is no subject of enunciation. Fluxes are the only objectivity of desire itself.[148]

But how do you write a sociology of fluxes? How do you psychologize a process that has no "for whom"?

Deleuze made the above statement in 1977, between the two *Capitalism and Schizophrenia* books. It thus marks a transitional moment, where the dissolution of the essentialist subject-object dyad in *Anti-Oedipus* anticipates and prepares the way for the more explicit scientific formulations of *A Thousand Plateaus*. Attendant on this transition is an attempt to describe particular connections and assemblages more and more as virtual processes, insofar as they would organize partial objects and flows—the "objectivity of desire itself"—without changing the fundamental status of the actual *as* partial. Ten years later Deleuze would be even more emphatic:

> It's not to psychoanalysis or linguistics but to the biology of the brain that we should look for principles, because it doesn't have the drawback, like the other two disciplines, of drawing on ready-made concepts. We can consider the brain as a relatively undifferentiated mass and ask what circuits, what kinds of circuit, the movement-image or the time-image trace out, or invent, because the circuits aren't there to begin with.[149]

The actual brain begins as an "undifferentiated mass" with respect to the circuits that will be traced there by the virtual. These circuits do not emerge from the subject-object designations of the social sciences, but from events in the virtual that are projected on the actual—quite literally, "the brain is the screen."[150] It is not that the social sciences fail to describe human behavior in detail, but that the details themselves are given an ontologically primary position with respect to the processes that shape them. This has the result that machinic processes are anthropomorphized. Molecular behavior is thus misleadingly described as aberrant with respect to de facto molar assemblages. Deleuze and Guattari already indicate this predicament in *Anti-Oedipus* when they write: "The Oedipal uses of synthesis, oedipalization, triangulation, castration, all refer to forces a bit more powerful, a bit more subterranean than psychoanalysis, than the family, than ideology, even joined together."[151] In order to disengage molecular forces from their misleading anthropomorphic and representational determinations, Deleuze and Guattari organize their next work, *A Thousand Plateaus*, around a new "Geology of Morals," which advances a concept of the actual as process of stratification.[152] It is worth working through this difficult concept here because it comprises an intermediary stage in Deleuze's transition from the notion of social and desiring-production in *Anti-Oedipus* to an image of thought in *What Is Philosophy?* that is irrevocably "beyond" the impasse of the social sciences—a kind of strata of the brain.

As we saw in preceding chapters, natural processes are fundamentally open-ended, incomplete, or partial, even in those cases when we would expect them to terminate in, or at least alternate with, a fully actualized state of affairs. Logically speaking, the actual can never be completely determined, and no object perfectly whole, so long as we define these terms according to ascending and descending strata of partial objects and flows. Strata and stratification do not take place on the plane of consistency (the virtual) but in an already actualized state of affairs. Except that, according to Deleuze and Guattari, an actual object is simultaneously more partial or *molecular* in the direction of the descending substrata and more unified or *molar* in the ascending direction. To return to the (rather simplistic) example we used before, a house, or the event of a house *as process*, is situated on a stratum above the sawmill; the sawmill, in turn, is on a stratum over the forest and river, and so on. A higher stratum or molarity is not more actual than the stratum below it, nor does it annul the molecular flows that comprise its ontogenetic substrate; rather, it tends to bring these flows together in a system of interconnected rings or surfaces: "there is a change of organization, not an augmentation . . . The materials furnished by the substratum constitute an *exterior milieu* for the elements and compounds of the stratum . . . [which] constitute an interior of the stratum."[153] The higher stratum draws from all the lower ones, recomposing them in a network of outward-facing surfaces, like an inverted Russian doll (faces painted on the inside). We see this, to take another example, in the various organic cycles that are brought together to form a single agricultural machine. A farm does not merely cut across the organic strata it captures, but it tends to duplicate those strata within a composite system, reproducing in its own unity of composition a cross section of the Russian doll: "A stratum necessarily goes from layer to layer . . . from a center to a periphery . . . Flows constantly radiate outward, then turn back."[154] The ascending series of strata are not more organized than the descending series; they simply change the principle of organization, by degrees, from singularization to individuation.

Why then do we speak of *a* process, of this or that individual object? Are these terms not completely relative, to the point where the concept of identity, as applied to an organic or inorganic process, to a person or thing, leads us to yet another paralogism—the attribution to things of a quality that refers only to an anthropomorphic mode of thought? Or is there some mechanism within the infinitely open series of strata that would allow for a self-same object to emerge, a thing-in-itself? Rather than considering the problem as a matter of passive and active syntheses, Deleuze and Guattari will now base their argument on a relationality between interior and exterior milieus, and on a dynamic process that simultaneously involves both singularization and individuation. The argument in *A Thousand Plateaus* begins with a basic distinction: on the outside, molecular materials; on the inside, molar elements and compounds.[155] The process of DNA transcription, for example, is exterior to the process that organizes it in chromosomes. This, in turn, is exterior to processes of duplication, which are exterior to cell division, and so on. One singularity is taken up by another, in the same way that the orbital movements of planets around a star are taken up by the compound movements of a galaxy. What is difficult to grasp in Deleuze and Guattari's concept of stratification is that the molecule is always exterior to the molarity. DNA forms a milieu, within which the individuated organism emerges. A planetary or star system forms part of the galaxy's *exterior* milieu; the galaxy, conversely, is *interior* to the solar system. The subject, too, as principle of enjoyment, is only an interior limit within an endless flux of perceptions and affections. These fluxes— the subject's specific contents—move centripetally away from the center and have no outer limit. "The Oedipal wad does not absorb these flows, any more than it could seal off a jar of jam or plug a dike. Against the walls of the triangle, toward the outside, flows exert the irresistible pressure of lava or the invincible oozing of water."[156] As we see in Pisters's chapter on cinema and neuroaesthetics, the freedom of the schizo is based on a propensity to follow these flows over to the other side, to set up an unauthorized synthesis of recording that is based on the nonexclusive disjunctions of the impossible real (the cinematic apparatus itself, in this sense, must be regarded as a schizophrenic machine).

Evidently, outside and inside do not refer to spatial relations, but to interrelated processes, cycles, and patterns. What is more, the complexity of a system does not increase as we go from outside to inside of a process. What changes, rather, is a general dynamics based on relative proximity to the innermost layer within the unity of composition. So at some critical point within the process of stratification, partial objects and flows are submitted to a new principle: a topological limit that turns outgoing flows inward and repels others away. This kind of organization can only emerge by virtue of a feature—a membrane or "central ring"—that stabilizes variation within the process of becoming.[157] A higher stratum tends to organize its molecular contents according to a network of membranes, each one designating an interior surface that faces outward toward the variations in its milieu. By the same token, a molar system must enter into progressively more dynamic relations with these variations along all of its surfaces because the molecular systems it appropriates from other strata (on which it subsists) are themselves subject to constant variation. Thus, agriculture exists only by virtue of its ability to limit the variations in an exterior milieu, reduplicating only specific segments from every organic cycle it captures. Organisms themselves emerge only by acquiring the means to regulate the exchange of waste and energy, thus expressing a qualitative difference between interior and exterior milieus. Inorganic "bioids" or "chemotons" also exhibit a feature of this kind: chemical and crystal oscillators, tornadoes, and tsunamis or "solitary waves"—or, indeed, star systems and galaxies—these all must have some basic mechanism for designating an interiority.[158] A provisional and contingent form of identity gradually emerges with the ability to select for certain flows, in a manner not unlike Maxwell's demon, drawing energy inward and deflecting nonenergy sources away (principle of "associated" or "annexed" milieus).[159] More importantly, we can say that identity is mutually determined with the threshold of variation that draws a provisional line around an interiority, so as to stabilize and prolong a single principle of selection.

How does a system acquire this feature? What causes partial objects and flows to submit to the formation of membranes and thresholds? This is not a question of ontology *or* ethics, but of the interdependency of actual and virtual on each stratum. Nor will it simply be a matter of dividing up a preexisting world into smaller and smaller objectivities. Deleuze approaches the whole problem in the opposite direction, tracing the formation of objects from flow to flow, from outside to inside, and always with a view to the ongoing variations on a cascading series of lower strata. This is best illustrated by his interest in the function of the brain, which will be fully developed in the last section of *What Is Philosophy?* but is already present in *A Thousand Plateaus*:

> What we are trying to say is that there is one exterior milieu for the entire stratum, permeating the entire stratum: the cerebral-nervous milieu. It comes from the organic substratum, but of course that substratum does not merely play the role of a substratum or passive support. It is no less complex in organization. Rather, it constitutes the prehuman soup immersing us. . . . The brain is a population, a set of tribes tending towards opposite poles.[160]

The brain comprises an ascending network of flows that behave in a particularly stratified manner, without closing itself off to variations—indeed, with the result that these variations play a decisive and ongoing role in its organization. Thus, it demonstrates how a molarity may, at the same time, be extremely adaptable in its relation to changes in the exterior milieu. Indeed, it is no more than the construction of such a milieu, a continuous connective synthesis, in which the most disparate partial objects and flows are joined together and split apart in rapid succession. This is true not only in terms of *organic* processes, in the strict sense of the word (dopamines, neurotransmitters, synapses, etc.), but in terms of every kind of event on the plane of immanence. The light contracted by an eye is imminently connected to a linguistic sign, a word to an erogenous zone, and so on. As a consequence, the brain operates like a galaxy unto itself, continuously distributing and redistributing the strata of actual and virtual cycles around it, drawing and redrawing the limits between *Umwelt* and *Gegenwelt*, organizing and reorganizing the play of identity, difference, and world.

Rather than speaking of social formations as such, Deleuze and Guattari will increasingly turn to the figure of the brain to work through problems in philosophy, science, and art, with the explicit aim of situating these problems "beyond the human."[161] *Can* losing one's memory or being mad belong to thought? We can now pose the same question in perfectly nonanthropomorphic (yet far from materialistic) terms: How does the brain, from its own undifferentiated resources, provide a screen for events in the virtual? As subject of knowledge, how does it "plunge into chaos"? And how does it respond productively to this chaos—using what strategies, what tools or machinic processes? This line of inquiry will bring the ethical problems posed by *Anti-Oedipus* all the way to a new system of reference, a domain of quasi-scientific variables that intersect with the metaphysics that science lacks, and with aesthetic principles that traverse both disciplines by virtue of the sensations they compose ("Every territory, every habitat, joins up not only its spatiotemporal but its qualitative planes or sections: a posture and a song for example, a song and a color, percepts and affects").[162] The fourth and final section of *The Force of the Virtual* develops the scientific and philosophical (and artistic) importance of the brain, with chapters by Arkady Plotnitsky ("The Image of Thought and the Sciences of the Brain after *What Is Philosophy?*"), Andrew Murphie ("Deleuze, Guattari, and Neuroscience") and Clark Bailey ("Mammalian Mathematicians").

There will be two issues of importance here. First, how does the brain's engagement with chaos necessarily point to an intersection among philosophy, science, and art? Quite distinct from anything human—as both Plotnitsky and Bailey argue—the brain creates a plane of composition that helps to organize an environment, prolong perceptions and affections, and provide a universe in which other forms of thought and activity can begin to develop. Such "interferences" among the various planes, as Plotnitsky writes, "make it no longer possible for a given field to maintain its identity as defined by its particular mode of confronting chaos, and make thought enter more complex forms of this confrontation, for example, by shifting it to other modes of thought: philosophy to science, art to philosophy, science to art, and so forth" (chapter 10). Second, this section responds to Deleuze's claim that the unconscious is a "substance to be manufactured" and that "fluxes are the only objectivity of desire." How does the brain "manufacture" an unconscious with respect to its environment? Murphie looks closely at the probabilistic nature of synaptic activity, of the capacity of a sign to jump from one series to another: "The material domain and action of signs is not that of the symbolic processing of representations. It is an ongoing synthesis of syntheses, of complex durations and mixtures. It is, again, 'probabilistic' and uncertain. More than this, it involves both active and passive synthesis, a work on habit. It is therefore often 'discontinuous' and calls for 'jumping in a probabilistic regime'" (chapter 11). It is the discontinuous nature of the brain that accounts for its particularly dynamic behavior, and, conversely, for mirror-like processes in which "a movement in an abstract field" causes two or more fully individuated organisms to engage the same synaptic process (chapter 11). There is no need to speak of identification with a universal subject; such jumps and fissures within the brain, and the same processes as they occur in a population of organisms, points to a "global transconsistency" that is only relatively universal, i.e., contingently and provisionally based on an event that affects an entire population at once. Cognitivist schemes for recognizing objects are less relevant, in this case, than Deleuze's insistence on an impossible real, or Guattari's formulation of a "machinic unconscious." The brain does not "contain" such a reality or unconscious, but it actively produces it.

It is this continuous production of an unconscious that we must keep in mind whenever we attempt to follow a scientist investigation of Deleuze's thinking, especially in such groundbreaking works as DeLanda's *Intensive*

Science and Virtual Philosophy. Like any strictly material process, we cannot limit ourselves to a neomaterialist ontology of symbolic knowledge because we are always dealing with a set of topological and abstract mechanisms that actively create new connections, habits, jumps and ruptures. Which is to say simply that we are always working, even with the most material system, within the parameters of an immaterial and yet immanent cosmic unconscious. As Bailey suggests, any neomaterialist investigation of the functionality of the brain is bound to cover much more, and at the same time much less, than consciousness in the strict sense (chapter 12). Inevitably, it traces out an entire "voyage in intensity," a field that is always more abstract than the results of any particular process of individuation, but which gives us a language (and a plane of consistency) that is capable of covering the same disjunctions, the same impossible connections, and effectively introducing the productive principle of the unconscious within the theoretical apparatus itself.

By insisting on the unconscious as force of the virtual, Deleuze effectively precludes any sense in which his philosophy could be interpreted as either materialist *or* idealist, at least insofar as each of these terms refers only to the exclusion of the categories comprised by the other. The force of the virtual is always the result of a relationality between the two systems, articulated by the positive—perhaps even *willed*—principle of chance. As Plotnitsky writes, "taking chances and betting on outcomes, and hence interacting with chaos as chance is something that our brain does all the time but that our culture is reluctant to accept at least at the ultimate level" (chapter 10). To think in the way that Deleuze proposes involves above all the engagement of an immanent creativity that effectively reverses the Freudian formula, and an affirmation of the productive difference that animates all natural and social processes. It means producing and never ceasing to produce a fundamentally subjectless subjectivity, as in the revised formula: *Wo Ich war, soll es werden!* (There where I was, there as subject must it come!)

· I ·

The Virtual in Time and Space

CHAPTER 1

The Insistence of the Virtual in Science and the History of Philosophy

Arnaud Villani
Translated by Thomas Kelso and Peter Gaffney

> ... *it is a matter of evaluating every being, every action and passion,*
> *even every value, in relation to the life which they involve.*
> *This is the opposite of a cult of death.*

—Gilles Deleuze, *Cinéma 2: The Time-Image*

THAT THE VIRTUAL PLAYS A STRATEGIC ROLE in Deleuze's conceptual apparatus is not at first clear, and it is difficult to see how it would alter his interpretations of philosophy, the history of philosophy, and science. In this article, I would like to broaden the scope of the investigation, beginning with the nonmodal character of the virtual, continuing with the notion of the "broken totality," and finishing with an examination of the virtual as irrational cut (*coupure irrationnelle*). I hope in this way to show that the impact of the concept of the virtual on philosophy's relationship with itself and with science is in both cases considerable.

A Nonmodal Virtual: The "Second Hypothesis"

Let's begin with the basics. What is the difference between the virtual and the possible? They are not of the same order. The notion of possibility arises as soon as we mistake representation for presence. The possible, the real as modal reality, and the necessary are not *real*; rather, they constitute a logical point of view on being. We are speaking here of Aristotelian modality. The human real does not exist except for a consciousness capable of asserting it. Yet this does not mean that the real depends on representation. One may well ask, however, where are we to situate the real if not according to a series of representations to which we add an operation that consists in subtracting the representational itself? Do we not find here the supreme naïveté of thinking that one should be able to represent, then to *depresent,* and believing that the *remainder* would be the real?

If the real exists only through representation, this displaces philosophies of the will and returns the logical approach to primacy (philosophies of language, analytic logic, semiotics, pragmatics, and modality). If, according to this hypothesis, one says "real" with the possibility of meaning something, one accepts the degrees of modality that go with it: the possible on one side and the necessary on the other. But when I say, "I am," is it because I am conscious of a representation or rather that I let the body speak (cenesthetics and kinesthetics, Heidegger's *Grundstimmung,* sensation of the reality of the real, feeling of existing)? Surely, "I am"—and Nietzsche knows this well—is not a *first truth.* It allows for interpretation on at least two levels: the first is "I say that I am"; the second, "I feel that I am" (sensualist cogito). The order that relates these phrases is unalterable. To make any such utterance in a real-life situation, I must first have a body. *So long as I am neither ill nor in danger,* I can always advance the intellectualist hypothesis. Conversely, I can always bring to my aid what, without my consent, does not cease to feel.

In the first hypothesis, the real is a logical construction, and the factual depends on modalities. Man exists because he has a *mind* that "governs" him. Since Plato, this has been the object of the *great detour* that aims at ceasing to "blind the eye of the soul." In philosophy, one likes to think that everything comes back to the sovereign rule of intelligence, that culture is the end of nature. The West has built itself on this assumption, so much so that, in the space opened by a subjectivity that neutralizes its grounds, ethical or technical *praxis* submits to an "infinite task" that requires the category of the "possible."

But the first hypothesis hinges on the second. The latter, as unnoticed as grass, serves as the foundation: Antigone is the basis for Creon, for she inscribes an insurmountable fault in her effort, no matter the lengths to which he may go in order to annul this foundation. There is a kind of *Nemesis* in that which, nearly trampled to death, returns. Schopenhauer and Nietzsche represent this attempt to overthrow (the Will as ground, the will to power as the being of becoming). To bet on the second hypothesis means no longer believing in *a great unity of being and knowledge*—which is to say, in the image of thought dismissed by Deleuze. But philosophers show little tolerance for this reversal. Like Plato vis-à-vis Heraclitus and the Sophists, they are compelled to choose between two attitudes, one as perverse as the other: to distort the opposing discourse, or to claim that it is unintelligible. Yet we will see that it is apt (*juste*), when philosophy is divided along the lines of these two hypotheses, to let the second have its word again. This is what Deleuze does with the virtual option.

And if Nietzsche, as emblem of the second hypothesis and of its mar-·ginalization, is dispossessed of that which made him so explosive, there is no longer anything to stop the field of philosophy from being occupied exclusively by neuroscience, cognitivism, and language. The harsher the Nietzschean move, the more important it became to hide it and to explain "why we are not Nietzschean," why "the century" will not be Deleuzian. To say that Deleuze is Nietzschean does not mean that he wrote about Nietzsche, but that he is on the other side, that he is "outside," that he marks the moment when predicative logic must give way to one that is verbal—an *antimodal* logic.

For a reason that is simple, but goes unnoticed. Frege showed that, before him, logic had yet to draw attention to *the act of judgment* in the proposition. A proposition so forgetful of the act that founds itself could then dissociate itself from logic and graft itself onto life, giving the impression that being and knowing are identical. Didn't Plato, the most important example of this tendency, reject "the implication of the speaker in his proposition," something the Sophists had not taken care to do? To graft the philosopher directly to the voice of being has given rise to the notion of a "godlike" voice. To understand that *the virtual is not the possible in the modal sense* makes obvious the *ventriloquy* of philosophy speaking before the "good will of the thinker." These are all pitfalls from which we cannot completely extricate ourselves. Properly understood, however, the theory of the virtual can allow us to focus (*détourer*) before we turn away from it (*détourner*).

Consciousness and Virtuality

The notion of science as a "closed field" pertains to space, but even more so to time. *Scientific* pertains to that which is sheltered from the temporal. One might object that physics is exceptional in this regard; however, we are not speaking of an abstract and measurable time, but of real time, wearing away at bodies, acting on bodies—what Plato wanted to exempt himself from by making a safe place for the *episteme*. If movement as the reality of time is beyond the grasp of science, it is because science occupies itself with the eternal present. That is its strength. Everything is actual in science; time never begins. There is at most a fourth dimension, never a "worldly" fabric, the flesh of things, the essence of being. If the virtual does not belong to science, that is because it escapes it entirely. We can go a step further. That which is present, measurable and measured time, is given to a consciousness whose operation implies recognition, accompanied by succession in the form of causality. Here, Kant has indicated the essentials with his schematism and the three syntheses. A consciousness is *clear* if it grasps the links of logical causality and initiates movement (*amorce des gestes*) "as a consequence." "This is not that," and "if this, then that." It is the same move when Descartes makes a method out of mathematics and attempts to grasp the essence of a consciousness that neither dreams nor meanders. The *mathesis* of the clear and distinct derives from the awakened consciousness and gives it the support of an intelligible world.

The actual thus depends on a vigilant consciousness, serving as the object of science, and voluntarily awakened. The virtual carries with it a floating consciousness, dreaming if not unconscious. It is thus absolutely logical to associate *virtual becoming* with an attitude that allows itself to be traversed by the movement of time and by the time of movement. The virtual emerges from a consciousness inseparable from movement, or else it is movement that results in consciousness (what Ehrenzweig aptly calls "unconscious scanning" in *The Hidden Order of Art*) and not the consciousness that divides movement. The virtual appears as the sign of an affect of consciousness that becomes time.

A Virtual Capable of a "Good Movement"

Something remains unexplained. For it is not that we need to make a place for the virtual in a world that would define the actuality of presence (*ousia*), but rather, if we want to think at the level of things, that we need to give a place to the actual in a virtual that surpasses it. The virtual is everything that surrounds the actual *in all directions*. Whether it designates the "pure past," the "subjectivity of duration," the Idea *perpliquée* or the "intensive multiplicity," it is this determined real that is missing nothing, that is differentiated but does not differentiate, and that does not actualize itself. This is not because *a pure movement has no forms* (noun), but because it "*forms*" (verb).

Philosophy, and philosophy alone, allows us to rigorously think about the disparity between science and reality, which arises in the first case from the difference between the *discontinuous* (the concept, number, divisible) and the *continuity* of life. Change, as a whole or as a block, does not affect the triangle or curve of a mathematical equation. The "line that goes to infinity" in the Baroque characterizes that which is susceptible only to change. Movement, from beginning to end, is of a single nature. If it is arrested, one passes to a discrete set of elements, leaving the real behind. Kant said it with his opposition between the "logical" and the real (*Essay on the Introduction of the Concept of Negative Grandeur in Philosophy*), by once more taking up Gaunilon's objection to Saint Anselme.

There would thus be, on one side, an actual, abstract insofar as it separates itself from movement, susceptible to plusses and minuses; on the other side would be the virtual, which escapes all measure because its intensity involves a *global process that affects*. The virtual, in the way opened by the "geometry of sensible qualities," is recovered by a rigor that does not apply to discontinuous quantities, but works through that which *affects as a block*, and, by way of this affect, *produces totalities* that are "indivisible" (without changing nature). To feel the precise point, every second, where time changes me is of the order of signs, affects, percepts, blocks of sensations, the virtual.

Insofar as it is the seat of this continuous and infinite development, the inorganic must therefore become, as Whitehead understood, the object of metaphysics. Whether the soul is *ruah, pneuma, anima,* or *breath,* it is a *continuous and unstoppable movement* that can only acquire *form* through a slowing down. That which traverses and constitutes a continuous movement—such is the soul. For Hegel, what never stops is dialectical movement, the Spirit. And the Greeks' conception of *equilibrium in movement* prepares the way for this idea. In short, there would only be one object of metaphysics, and Deleuze makes this clear with the power of the continuous fold: the impossibility of stopping, for even an instant, signaling in reverse fashion the Aristotelian notion of *anankē stēnai.*[1]

When Deleuze introduces his idea that the Baroque or the fold is that which goes to the infinite, he could say that it is the *inorganic* life that extricates itself from its actual status. *The virtual is that which creates the link, rhizome, line of flight, everywhere in every direction that permits the continuity of a body that is "full like an egg."* A body that does not cease moving—this is yet another way to define the *body without organs.* The "full like an egg," filled with virtualities that stop up the fissures where the separation of the nonliving could be introduced, is also that which prepares the way for the *"faille de fuite"* that must follow, veritable wandering Jew (*Laquedem*), its path infinite and nonlinear, always pointing toward other systems of folds.

To say that the virtual is a life means that this life is singular, nontranscendent, quotidian—a life that does not exist except as a block and that owes to the virtual *its ability to never stop.* The real is not made of static moments but of intervals, of passages. Where does this blindness before the evidence come from, if not from dangerous situations that threaten to eliminate a *fixed* view of the real? How do we think movement? If every form, in a deafening ballet, is *in passage,* must not thought be a thought *in* passage and not merely *about it?* A thought about passage need not change. It is the *"suave mari magno"* of he who looks at the storm, then takes up once more divisions, the order and rites of propriety, and exclusive valorizations. Every "closed field" makes "real in storm" (Nietzsche) "negligible." But a thought *in* passage enters into movement and does not come out unscathed.

If a thought about movement has sense, it does not stop at points but describes continuous curves—not only successively, but *in the instant.* It is here and there. One could say that this thought is characterized by *absolute speed* (to say "the absolute speed of deterritorialization" seems redundant to me). And, because this thought is a single curve without points, the closest is also the farthest away. Riemann says as much with his *sheets* and *varieties* (*Mannigfaltigkeit*). Several of Deleuze's propositions now become clear:

(1) The *faithful traitor,* or how to "make children behind the back" of someone: to be faithful to the effect of displacement of a philosopher, to stay by his side, where he improvises out of tune.

(2) The *plane of immanence* or *"body without organs":* to restore continuity, the Leibnizian "here just as elsewhere." A thought *in* passage can only be *univocal.* If the body is taken in passage, the organs represent divisions which *arrest* the wave of sensation. All points, organic or inorganic, unleash the block of the infinite virtual.

(3) The *line of flight,* the *absolute survey,* the *lightning bolt:* they give us a sense of the inseparable movement called the fold, in the extreme vicinity of all that can be distinguished there. We go

from hawthorn to clock towers *without raising a hand,* in the blink of an eye.

(4) *"Becoming"* is the verbal noun that refers to this operation. A block of becoming means that all becoming forms a block. "Between" does not imply any point of departure or arrival. Even more than the Reals, the virtual stops up the breaches of repetition and produces the *Aïon,* of all time taken as a whole, which leaps over the rupture of the repetitive present.

The Virtual as Variety

But what are we saying? For classical thought, isn't the fold *indistinction* itself? And how does one think "at high speed" (*"en vitesse"*)? These hesitations arise from believing that continuity is *homogenous.* When three or four lines of intensity draw their intensity from one another, they form a fold. For this reason, we must leave homogeneity to the discontinuous, and accord to the continuous a heterogeneity that, by becoming, produces a "block."

Let us continue to interrogate the history of philosophy. Plato, beginning with the *Philebus,* tasks himself with conceiving of living Ideas. The Idea will be "the relation that remains identical." But the relation, isn't it already the identical *par excellence*? In his ball of wax, Descartes seems to have found the royal road to science: *l'inspectio mentis.* The wax is neither form, nor color, nor flavor, but extension, figure, and movement. One is compelled to ask what fruit the mountain bears. Reading Gaston Bachelard's analysis, in *The New Scientific Spirit,* of a drop of pure wax heated at slightly varying temperatures, producing "gas structures"— that distinguish it from iron or ash—is enough to be convinced that this "extension" is no better than "the night where all cows are black."

In seeking the idea as relation, we come once more to Plotinus and his sphere that is dematerialized but that *conserves the same relations.* Now, one may very well translate a cat into a *pretty picture;* the inverse does not result once more in a cat. Thus, the relation of identity is not the differential relation. Variations or variables are not variety. The relation is universal because it is immaterial and incapable of individuality (*eidos,* as the Stoics say). The relation is the unique signature of individuation (the Stoics' *hexis*). This is why Deleuze, in his Ideas, does not look for relations of identity but rather the interstitial varieties within pairs of movements. In this way, what makes a tick's body precisely *this* form, *this* consistency, *this* color, *this* disproportion between the head and abdomen, *these* silly legs, is the ensemble of the relation "butyric acid + hairless spots + mammal blood," which are the *stimuli* which guide it and the totality of its reactive milieu. The typical series of active characters and of perceptive characters, embedded in behaviors of the same order according to a functional circle (Uexküll), is called a "tick." This *ensemble,* this is the virtual (sensory-motor reactivity being the actual).

The Sense of Detachment, or the Enigma of "Letting Go"

One suspicion comes into focus. Would not phenomenology, rather than reforming science, merely constitute another field of human studies *(Geisteswissenschaften)*? How could Husserl ever have done with the paths of Reason (his objectivist naturalism) if he does not cease, in the conference of 1935, to extol *sans examen* the merit of "infinite tasks," Western thought's volontarist point of departure? The barbarism evoked by Husserl, which will not hesitate to crash over Europe: is it reasonable to think that it could be avoided by an *ameliorated* rationality, while sustaining uncritically an idea of the Progress of Spirit that does not include the virtual?

The problem is not limited to a denunciation of objectivist naturalism. We must understand what Deleuze means with the idea *of the breach of the Outside and the suffering of thought.* This comes back to our earlier discussion of Plato's principle that the *"noûs* governs," and of the notion, maintained by the Stoics, Plotinus, and Leibniz, that the spiritual principle is superior; with Husserl, the transcendental ego is the condition *sine qua non* of a world, once more valorizing the acting consciousness. We can go farther. When Bergson—and we will see this at times with Deleuze—distinguishes between *qualitative* and *quantitative* multiplicities, it is still a spiritual consciousness as *will* of "more" and of "better," an inadmissible *Bildung* that prevails.

But shall we deny that Deleuze as well gave proof of a persistent *will* in the appropriation of knowing? Shall we deny that his critical *lucidity* is total? And that a *Deleuze-subject* appears in his work, does this not present us with its unmistakable mark? Why is it with Deleuze that qualitative multiplicities seem to take precedence over other kinds? If, on the other hand, I have believed this (notably in *La guêpe et l'orchidée*), today I see only illusions. It is because we continue to carry along this voluntary Spirit, bent on hegemonic perfection, preferring the mathematical form that allows autocracy, that we, in the name of a history of philosophy that Deleuze is in the process of modifying from top to bottom, and that we reject modifying in a like manner, project on him that which is not him at all.

Bergson maintained the possibility of a contemplative or spiritual life. But what Deleuze borrows from Plotinus is not the interior "statue" to be sculpted, but rather the *spontaneous, unconscious, and tireless nature of superabundance*—the "you have become that"—of a metamorphic *contemplation*. In this sense, we must speak of a Deleuzian absence of the will. To let go, to regress through the various steps of psychic evolution described by Hughlings Jackson, and then to keep going, until one reaches the animal, the vegetal, the mineral, this means that there is within reach a degree of intensity to connect. To become detached thus consists in *letting the virtual swarm*. To refuse form means to enter a process which, going to the infinite, is defined by "that which grows from the middle."

The late François Zourabichvili was right to insist on a Deleuzian *involuntarism*. That is why, in order to distinguish himself from a spiritualism whose main issue is a struggle against mathematicalism and the theory of observable measurements, Deleuze, as a good Humean empiricist, will speak of *intensive* multiplicities. The Soul, spirit, consciousness, intellect, and the *episteme* are only phantoms (there's a bit of Stirner in him!). If he takes up the Riemannian multiplicity, he adds to it that which, before Deleuze, was rarely encountered in Western philosophy: an *appropriating depropriation*. The alternative is no longer to seize (the grasping of the concept, or *Begriff*) or to let be (*Gelassenheit*). This has to do with letting oneself act, to let the virtual infuse, without forcing it. The block is the consequence of the acephalous.

The virtual imposes an attitude (the term "posture" appears to me to be too voluntary here) whereby the more becomes the less. Nothing of the aristocrat or the elitist with him: what an error, what a projection of our orthodox beliefs—our *toxic* beliefs! This would be something of a ruse, a Stoic who lets the world invade him, a Spinozist who *floats* on the infinite attributes, a Taoist or Tantric, Powys's lizard lying under the sun, in other words, a jellyfish! *To become* has this at stake: it is a "transitive" verb that "lets itself be traversed by." "To become" indicates *all at once* the virtuals that we did not see so long as our attention left us blind. There is, in flight or in deterritorialization, a *disappropriation that gives*. One would be inspired to search for Deleuze's inspiration among mystics of the Quietist type. And why not? We might even think of Cordelia (Kierkegaard) before Johannes (Christ), or the insane gesture of Job, the *chevalier de la foi* who wins everything because he has conceded to lose everything.

In order to resituate his aim in the most ancient (the newest), we must add an anthropological remark. According to Hamayon and Hénaff's analyses, hunter-gatherers do not "master" nature, but take it *as it is*. It does not matter that the hunter finds game nor, in the case of the gatherer, that the year is prosperous. *They let go to time*. All-powerful time puts into play the nonwill. On the contrary, the nomadic herdsman moves the flocks at will, and the farmer brings about the harvest. They are *masters* of time, so that we pass here to the rule of the will. Hunter-gatherers allow nature its own rhythm: patience of the gatherer, speed of the hunter. Speed does not interrupt the order but is wedded to it. One must become animal before becoming hunter, must develop a certain cunning to adapt to various speeds, including that of language, such as the play on words that turns Ulysses into *outis* (it is "nobody" who tricks the Cyclops) or *mêtis* (in Greek, meaning both "for fear that"—that is "for fear that some Cyclops would cast huge rocks upon us"—and "cunning"). Everything changes with the pastoralist who divides and distributes the pastures at random. Everything changes with the farmer who "initiates" and pursues his successes. This is the first "rational," technico-scientific attitude in history.

In the first case, nature is "still moving," in the second, it is man who moves—nature remains inert. Thus, the affect of movement's nature, its seething reserve of the virtual and its positive infinity, is extinguished. This brings about the forgetting of the *physis*, reducing it to *natura naturata*, fixed, and ready at hand. This has considerable consequences for the history of philosophy. As soon as the speed that indicates the infinite disappears, it is human understanding, with its slow seduction, its discontinuities, its measurable stases, which take its place. This signals the end of cunning, the burial of the infinite, the oblivion of the symbolic, and torture imposed on the Many (*torture imposée au Trop*). Plato, always in a hurry to expulse, "to erase, exclude, suppress" (*The Republic*) the fisher and the hunter, as well as the politician, the sophist, the doctor, the poet, the strategist, and those who use *mêtis*—Plato knew what he had to do: banish *metamorphic* becoming, the infinity of the virtual.

Deleuze is an attempt *to bring an end to the conjunction of the intellect and the will against nature.* This extends to politics as well: the Deleuzian attitude consists in "letting the virtual grow on him," in the same way that breasts grow on Schreber's body. *The people that are missing* appear as soon as we have become acephalous enough. Science, *by striving to see everything and let nothing pass it by,* strangles the real. True knowledge (*savoir*) consists in letting oneself be carried away by the virtual. And there is enough of the virtual to make this possible: this is what accounts for Deleuze's optimism. It is time that we took this task up once more, to relearn our detachment.

Inevitable Recourse to Chaos: "Philosophy of Emulsion"

The Deleuzian virtual presents a limit to science and philosophy. Philosophy is neither a reactionary spiritualism nor the purported materialist stance of the avant-garde, satisfied by a hackneyed sensualism. The object of philosophy is the "*Internel*"[2] infinite of the virtual, *nonbiological life.* The real is that which never ceases to transform itself. The finite *floats* like a piece of straw on the sea of the infinite. To draw a connection between the *Philebus,* section 26d ("to impose a measure on the Many"), and Maldiney's debate with Hegel regarding sensible certitude, one perceives the gesture of philosophy since Plato as an "exclusion of the infinite."

And yet the world is but a glass palace with sons of the virgin floating on a strong current (Nietzsche, *Truth and Lies in an Extra-Moral Sense*). The infinite is a fragile bridge. The senses are not mistaken when they see a cylinder in the place of a square tower, but they are when they compel me to believe in an in-itself outside of me. Outside, there is only the invention of the new. It is not that one goes from one actual to another, but from one virtual to another by means of a virtual momentarily grasped as actual (Nietzsche would say *Herrschaftsgebilde*). The illusory imperative to see everywhere the actual, which plays the same role in "constitution" as it does in the "position of universality," the belief in a static actuality of things (Whitehead's "presentational immediacy") is only one of the more pernicious and comfortable inventions with which one may mask the real. The virtual is not something we add to philosophy so as to least disturb its order. (Philosophers believe that they are obliged to wear "skates" when they venture into philosophy's salon.) On the contrary, it forces us to reconsider the question that has agitated it since the beginning: what to do with the excess, the *apeiron*, the *arrythmiston*, and what to do with the "sixty thousand fathoms of water," with chaos?

But what should we see in this "great malady of the universe"? Gorgias remains lucid on this question: nothing presents itself as being; if there is being, it depends neither on knowing nor on speaking. Just as a consciousness does not exist except as "consciousness of," it must also be that what lies before us is *determinable*. Here we see a double error appear:

(1) To deprive ourselves of the "nothing" by calling it the "absolute contrary" of the object, which means progressing by means of "dead and not opposed extremes." *Thanatocracy by stasis.*

(2) In order to escape this objecticity that seeks refuge in stasis, there is a tendency to "wildly destratify," to flee toward the hole of abolition (the hole in the sink for Bacon), to opt for an unspeakable fluidity. *Thanatocracy by flux.*

The solution of the myth had the advantage of posing *at the same time* the Open (that which exists before the "world") and the cutting edge (that which gives a world). In order to bridge the gap between Spinoza (the infinitely infinite), Nietzsche (Dionysian chaos), and Bergson (qualitative interval), Deleuze seems to benefit from a return to myth. But he translates these metaphors in rigorous terms. It is this configuration, trembling with power, which he describes as *terra incognita*.

It is surely to get at what is most at issue, on the part of Badiou, when he claims that Deleuze can only pose the notion of the infinite (which Badiou disqualifies) by way of misreading Cantor. Either by ignorance or by bad faith, he has missed his one and only chance to convince us. But who cannot see that Badiou's reference to sets produces an *infinite death before ever being born*? The actual infinite of sets is *the death certificate of the virtual infinite*. And because the virtual is for Badiou a sanctuary for the ignorant, the discussion is over before it has even begun.

Deleuze's resistance vis-à-vis this attempt to make him say what is claimed he meant to say "without knowing it" is exemplary: a perfect illustration in Badiou of the "tyrant" described by Hegel in the *Phenomenology of Spirit*. It is this resistance that animates a large part of the philosophical domain: put briefly, the "symbolic" versus a rational asymbolic. There are moments that call for polemic, for we will not let ourselves be deluded by irenism. One may very well consider oneself to be "at peace" and prepare in one's thoughts terrifying wars. War is in man, in his head, and not in the real. Anti-real, it has the strange characteristic, not being of the real, to destroy it.

Let's consider whether it be the root *pel- in *pallô*; to shake, *palmos*; or that which balances, *polemos*. *Polemos* is the "jolt." A fragment of Heraclitus evokes the mixed liquid that one may never stop shaking (never stop the movement). Oil and vinegar may vainly try to separate; what interests us is the vinaigrette. One could find no better translation of *polemos* than the word *emulsion*. Is there really any single entity in the world that would confirm the tautology A = A? How could A be separated from what makes it what it is? The tick is *nothing* without its stimuli; man is *nothing* without his society, language, and history; consciousness is *nothing* without its world. Emulsions. Pure secondness. The Heraclitean duality of opposites maintained as one, and each one absolutely dependent on the other, is a summary of the world. It launches philosophy in the direction of the world, and not toward science alone. Taking account of the virtual that is intermediary to the "two in the one," it promises the infinite intersecting that animates contraries, or what Hölderlin called "the all-living in its thousand articulations." From Heraclitus to Deleuze, *polemos* in a *fold*.

To be fair, let's also try to understand the inverse of this. A is not A except by way of the *stasis* of movement that links it to *n*. An analysis of A requires that we concentrate on a partial view. Thus, for a hand, the best point of view is that which does not separate it from the organism (the hand is only valuable to the organism; Aristotle's "wooden" or "dead" hand is only a hand as a homonym). Yet if A is only such by virtue of its surroundings and the relations between the two, it is also A because we *isolate* it as A. The two perspectives are interchangeable. Philosophy is double, and the symbolic depends on the asymbolic, and vice versa. Philosophy is the endless debate between these two points of view, symbolic and asymbolic, virtual and actual, infinite and finite, global and local.

To bring this antilogical philosophy to life, to enter into the violence that sets fire to thought: that's what it means to think. Classical philosophy entails an aspect that needs "awakening," its feel for the inseparable as a whole *and* for its necessary division, for the *differentiated* infinite, and for the *broken* Totality, which reinscribe Deleuze in the secret history of the struggle between the two parts, and which alone merits the name of philosophy. Yet it is important to better understand this "two in one," concealed by mainstream philosophy, whose two sides, continental and analytical, far from battling one another as it may have seemed, share a common interest in keeping it secret.

The actual is that which provides a view on the virtual. If it is true, however, that the actual is traversed by virtuals, it remains the case that actuality—"to be a thing"—belongs to the transcendental form of every perception of becoming. Or better, actuality is the virtual as it is manifest to the understanding. Let me not attempt to represent a world to myself by means of common sense, primary qualities, or an eye-hand connection; rather, I'll be a jellyfish, a melody, the virtuality of a little phrase from the "Emperor" concerto as played by Edwin Fischer. As soon as it has been heard, it may at any point return. The refrain appears to carry me aloft like a swing, or a barcarole. What constitutes the object of Deleuze's thought is *the unthinkable ensemble of the actual and the virtual that we nevertheless must think,* with neither preference nor priority. It is in this sense that he reconnects us to the pre-Socratics much better than Heidegger.

The Nature of the Whole

The Deleuzian "reduction" thus permits us to discover an inevitable *addition.* If, for Frege, it is the inseparable dyad of function and argument, or for Husserl, that of intentionality and world, for Deleuze, it is the actual and the virtual. Yet the outcome is completely different. Husserl deploys sketches, silhouettes, beginnings, chiasmi, and infinite tasks, in other words, a *world to be described,* while Deleuze gives us a will-less world, floating on the infinite, *indescribable.* A serious account of reason, a "look around the property" reveals, according to Kant's humorous metaphor, the presence of moles, of *subterraneans.* Deleuze calls them the Outside, the involuntary, flight, a series of "nonplaces" that escape language and create space.

To foreclose the Many, to have done *with that which is never done,* to stop the gesture, that's what really stands out. Beyond that, a science and a philosophy removed from their grounds could deploy only the most impoverished form of thought, a Pyrrhic victory indeed, a flashy way to bash in doors already opened. Having first dissected the world as a simple given, the philosopher and the scientist were destined to leave it behind, walking arm in arm toward greater and greater technical victories. Did they promise to compensate for these incisions (*schizes*) by reconsidering the life of the Whole? It is clearly on the status of this Whole that every-thing now depends.

The Virtual as Irrational Cut

Starting from the Deleuzian virtual, it seems that we have now reached the question of the distribution of knowledge. But have we really understood the difference between Deleuze and Husserl? No. We have only avoided the false path; making our way through the history of philosophy, we have arrived at a clearer definition. Yet there is still something missing. Deleuze has drawn attention to the attempt, on the smallest (Leibniz) and largest (Hegel) scales, to "atone for difference." And he has argued for the enlarge-ment of Leibniz by the concept of *divergent* series. We must try now to understand the difference between a unified and a "*broken*" whole.

It is in *Cinema 2* that we find the best illustration of the relationship between the virtual and disconnection. Here, Deleuze is seeking to reach the point where transmission is suspended, where it disengages, in other words, the "dead point." What he finds is this: if it is true that organic conti-nuity implies causality and resemblance, if it allows for adaptation through the sensori-motor reflexes, then the same kind of continuity marks its limitations as soon as it has to do with thinking. Thinking begins with the "solution of continuity" that invents hypotheses and jumps directly into the problem. Thinking reveals the switch (*aiguillage*) that leads to the out-side. The connections give another power to the continuous: that of the labyrinth, of intersections that give rise to the *heterogeneous*. The connec-tions manifest the sum of "differences of potential."

One does not create force with the consensus of what follows from "assuming an orderliness among [thoughts] which did not at all naturally seem to follow one from the other" (Descartes, *Discourse on the Method*). One becomes powerful by virtue of the "and," by way of asyndeton and parataxis. *Cinema 2* seeks out situations that involve, because of a turn-ing point in the conception of the image and of cinema, what I would call thought's "cartilage of the disjunction." Let's slightly rectify the earlier discussion of modality. "To tell the truth," is not necessary to question modal logic. More to the point, it is logic as an encounter with truth and judgment (i.e., Plato's critique, the groundwork for Aristotle's, of Sophistic artificialism) that Deleuze attacks, "in order to replace it by the forces of life."[3] Against the power of the true, Deleuze introduces the "powers of the

false," the "jumble of vanishing centers" (*Cinema 2*, 142), the "false movements" of the divergent series and multiplication of vectors (143). He
turns "shape" against "form," as a way of revealing forces and privileging
the creation of the new (147). This great turning point is founded on the virtual. In effect, what was
blocking our attempt to think the difference between Husserl and Deleuze
is that we were still endorsing, by habit, the omnipresence of the sensori-
motor schema of the movement-image, one that belongs to the "grand
narrative" of the West, where resemblance, causality, and consequence
prevail. With Deleuze, we find an "if . . . then" in which there is no "then," a
sensori-motor without the motor or a "antirhythmic suspension." The key
words in the second volume of the *Cinema* books will be "to disconnect,"
"to unlink" (183), "to disjoin" (172), "to bifurcate" (49), which gives the
Whole its seductive "fusion of the tear" (268). Let's concentrate on the
Deleuzian "atom," when, in neorealism for example, there is no longer a
narrative that would compel us to react "organically," and when nothing
remains but an optical and audio sign. An image appears before us; if it is
logically connected to the following image, the result will be the immediate formation of a conditional action reflex. But let's consider the same
image as it is first unlinked, then *relinked* (in the sense given by Raymond
Ruyer) to the following image. Between these two images, there is both
continuity *and* a crack (*fêlure*) (86), fissure (180), rupture (181), and irrational cut. "The cuts or breaks in cinema have always formed the power of
the continuous" (181).

But how shall we think a *disjunctive whole?* Let's consider it according
to information theory and its concept of sequential dependencies. A large
quantity of information is reduced in a sequence of letters because the
addition of each new letter makes it more probable that a word (sense)
will appear among all possible words (sense). It is this that ensures that
sense (signification) is also the loss of sense (affect of surprise). At any
moment, an unforeseeable bifurcation can release a powerful thrust of
inorganic (nonfunctional) life, the emergence of the new. At the juncture
that Derrida called the "break" (*brisure*), or at the "limit,"[4] life can be seen
as disorder, surprise, flight toward chaos, and death can be seen as order,
redundancy, or *déjà vu*—or rather, all of these at one and the same time.[5]
In other words, in as much as cinema presents us with an art of the virtual, it is the *virtual atom* that decides: a swarm of differences in degrees
of power. "When Spinoza directed his eyes toward things he saw neither

forms nor organs, neither genera nor species. He saw differences of degrees of power."[6] Deleuze will call the bifacial pair of the actual and virtual the "crystal," where the "interstice" becomes primary and "proliferates."[7] Insofar as this interstice gives us a glimpse of the way the actual (the already-there of the present) and the virtual (the past and future that evade the present) battle one another and then recover, everything depends on force, on art's own capacity for "resistance."[8] In this combat, it is important to delay, to suspend, and to block easy connections, to prohibit the chord's resolution and agreement, "to care otherwise for discord" (Hölderlin).[9]

Perhaps art is this supreme domain of the virtual where one gets a glimpse of what lies beyond representation, without will, in the infinite. Art as attainment of the virtual is the "limit" between the rigor and necessity of science and the indivisible movement that belongs to philosophy. Not only does Deleuze put his finger on the turning point between the symbol and the asymbolic, where philosophy and its history are redefined, but he also brings the debate farther, to the point where the secondness of the "time-crystal" authorizes us to think a movement that does not end. Ananké mê stênai . . . and yet, on such a good path, one mustn't stop.

CHAPTER 2

Superposing Images: Deleuze and the Virtual after Bergson's Critique of Science

Peter Gaffney

On this new ground philosophy ought then to follow science, in order to superpose on scientific truth a knowledge of another kind, which may be called metaphysical. Thus combined, all our knowledge, both scientific and metaphysical, is heightened.

—Henri Bergson, *Creative Evolution*

I N THE EXACT SCIENCES, "repeatability" and "reproducibility" refer to the validity of experimental findings with regard to successive attempts to create the same results under identical (or at least similar) circumstances.[1] This gives scientists a standard for the production of knowledge based on patterns that are conventionally believed to inhere to the object itself and that are broadly conceived as natural laws or universal constants. It also shows how efforts on the part of scientists to understand the world result in a virtual image or representation that develops over time, approximating the actual world with increasing detail and accuracy. (We will see later how Erwin Schrödinger appeals to this common sense view of science in his riposte to the Copenhagen interpretation of quantum physics.) There are two problems that immediately confront this notion of reproducibility. First, the historical development of scientific knowledge appears to be susceptible to sudden schisms and ruptures; it does not progress toward a more and more complete image of the world so much as fragment and subdivide into infinitely new configurations of its own virtual image. This is what leads Kuhn, for example, to describe the production of scientific knowledge as paradigmatic (and, more problematically, *gestaltic*),[2] a position that also accounts for more recent claims of a "disunity" of science.[3] Second, modern physics introduces a new figure into the image itself: that of the observer, whose position in time and space, and whose methods, instruments, and objectives all have some bearing on the outcome of the experiment. The state of a physical system is thus inextricably linked to the status of this figure. What are the observer's aims and intentions? In connection to what matrix of relations is the observer measuring results? Indeed, with Einstein's theory of relativity and Heisenberg's uncertainty principle (which lays the groundwork for contemporary quantum mechanics), we see that experimental data is as much a portrait of the observer as it is of the world, not in the sense of an object that remains inseparable from a *subjective* point of view (as in Locke's "secondary qualities"), but in that of an observer whose virtual image remains part of the *objective* field of vision. In this case, the scientist cannot consider natural laws alone to account for the reproducibility of an experiment. Indeed, repeatability and reproducibility acquire a subtler meaning here, suggesting a point of view from which the world, including the observer, *can be made* to reproduce its own behavior.

It is this subtler meaning that underlies Deleuze and Guattari's distinction between two different attitudes in science, corresponding to two distinct roles for the scientific observer. On the one hand, "reproducing" organizes thought and matter around laws or constants (the "legal" model); on the other, "following" implies a continuous realignment of the observer's point of view around changes internal to both the observer and the visual field, or "field of individuation." This difference is elaborated in this passage on royal science and "nomadology" in *A Thousand Plateaus*:

> Reproducing implies the permanence of a fixed point of *view* that is external to what is reproduced: watching the flow from the bank. But following is something different from the ideal of reproduction. Not better, just different. One is obliged to follow when one is in search of the "singularities" of a matter, or rather of a material, and not out to discover a form; when one escapes the force of gravity to enter a field of celerity; when one ceases to contemplate the course of a laminar flow in a determinate direction, to be carried away by a vortical flow; when one engages in a continuous variation of variables instead of extracting constants from them, etc. And the meaning of Earth completely changes: with the legal model, one is constantly reterritorializing around a point of view, on a domain, according to a set of constant relations; but with the

ambulant model, the process of deterritorialization constitutes and extends the territory itself.[4]

It is important to note that Deleuze and Guattari do not suggest anywhere that the truth claims of positive science are illusory. Indeed, their conception of science as either reproducing or following offers a robust framework in which to understand the predictive success of normative physical science with respect to the so-called laws of nature. At most, this conception might seem to invoke an instrumentalist point of view, one that opens notions of "celerity" and "vortical flow" to philosophical considerations that exceed (and outweigh) the unproblematic scientific acceptations of these terms. Yet the instrumentalist interpretation is also misleading. Deleuze and Guattari do not claim to do science of *any* kind, and they will insist that "it is always unfortunate when scientists do philosophy without really philosophical means or when philosophers do science without real scientific means (we do not claim to have been doing this)."[5]

What is at issue for Deleuze and Guattari, and what I would like to explore in this chapter, is not a way of doing science, but a manner of describing the ontological dimension in which a scientific observer takes part in the actualization of a world or of multiple divergent worlds: that is to say, in the creation of an actual object out of its virtual image. This notion far exceeds our common understanding of science, narrowly defined as the work of rational subjects collecting data and building representations of the "outside world." Rather, it tends toward the broader notion of a spontaneous converging of heterogeneous elements in the material, social, and infrapersonal fields, or what Deleuze and Guattari call an "assemblage." Much more than a general expansion of abstract scientific knowledge into a succession of uncharted territories, the new notion suggests the creation of territories that were not there before, and which exist only by virtue of an ongoing process of actualization. Rather than thinking of the observed as the imminent basis for a representation in the mind of the observer, we can follow the formulation of the actual observer/observed dyad back to its root in a virtual unity that is ontologically primary to the act of observation.

This is an important consequence of Deleuze's philosophical project as a whole. As Eric Alliez has written, Deleuze aims throughout his works to establish a rigorous basis for "integrating the physico-mathematical phenomonology of scientific thought into a superior materialism founded on a general dynamics."[6] If the integration of these diverse planes is possible, if we can describe thought as a process of actualization that begins with the virtual unity of observer and observed, it is because "the act of knowing tends to coincide with the act that generates the real."[7] Deleuze and Guattari are profoundly influenced here by Bergson's notion of intuition as a philosophical method that effectively bridges the gap between the social and exact sciences. Consequently, they will deploy it everywhere as a strategy for *superposing* metaphysics on scientific truth, that is, for formulating a rigorous correspondence between the production of scientific knowledge and the virtual as central figure of a philosophy of immanence. To understand this strategy is also to understand reproducibility and the predictive success of physics in terms of an ontologically determinable realism, rather than taking them once more as the grounds of a reactionary epistemology (the function of truth as such).

In light of the new notion of science, we will have to look elsewhere than the regularity of essences or forms (hylomorphism) to account for the fact that scientific models *do work*. In this chapter, I will be focusing primarily on the predictive success of physics with regard to the manner in which the conventional scientific model or "image" distributes a succession of instants that all belong, relatively speaking, to the same synthetic arrangement in space—that is, to the same present moment—which is to say that this conventional image works by suppressing the condition of change (*difference*), and by reterritorializing the process of actualization around a fixed point of view that is external both to what is reproduced *and* to the lived experience of the observer. In *What Is Philosophy?* Deleuze and Guattari describe the work of science as the "slowing down" of a virtual chaos:

> Philosophy proceeds with a plane of immanence or consistency;
> science with a plane of reference. In the case of science it is like
> a freeze-frame. It is a fantastic *slowing down,* and it is by slowing
> down that matter, as well as the scientific thought able to penetrate
> it with propositions, is actualized.[8]

This is not an act of vision, in the proper sense of the word, nor does it belong primarily to the figure of the observer. Rather, it is what Deleuze and Guattari mean when they speak of a "distribution of thresholds and percepts" on a plane that "renders perceptible without itself being perceived."[9] At most we can say that something emerges within the process (or field) of actualization—an eye, or something analogous to an eye—that makes it possible for both knowledge and its object to assume a more or less determinate spatial arrangement, all the while remaining inseparable from (indeed *coexisting with*) the condition of qualitative change that it arrests and distributes.

In these terms, the difference between reproducing (royal science) and following (nomadology) does not involve a differential of speeds, but different directions of development within the field of actualization. In the first case, all the resources of intelligence, comprising an assemblage or bloc with the corresponding material forces, are committed to the contraction of a single image. There is movement and also growth, but only with respect to the complexity and detail of the same "snapshot" in time. It is not that matter freezes, or that the subjective, internal sense of time is arrested; it is a case, rather, of the absorption and nullification of difference with respect to a single set of extensive relations, brought about by the protracted emergence of a single, unified world. Deleuze and Guattari give historical examples for this phenomenon, citing various attempts in the nineteenth century to reduce the laws governing matter and energy to a single principle of gravitation or attraction;[10] but we have seen this more recently in the search for the Higgs boson or "God particle" (the theoretical origin of mass).[11]

The nomadic or "ambulant" sciences, in contrast, are not capable of producing images of such exacting complexity and detail. Here, "Everything is situated in an objective zone of fluctuation that is coextensive with reality itself. However refined or rigorous, 'approximate knowledge' is still dependent upon sensitive and sensible evaluations that pose more problems than they solve: problematics is still its only mode."[12] This assertion leads us to pose several questions about the complicated manner in which nomadism engages scientific realism: What is the reality of the problems or problematics generated by "following"? How do physical processes themselves pose problems, and in what way are they coextensive with the zone of fluctuation? How are these processes actualized, or otherwise engaged by scientific observation? If nomadic sciences represent a different kind of development—a constantly shifting point of view—what does this mean with respect to the processes that animate matter? These questions will necessarily lead to more general ones about the way Deleuze's (and Guattari's) thinking intersects with science, and what this means for the familiar dyads that populate the later works: the actual and the virtual, science and metaphysics, the precision of the scientific image and the "philosophical openness" (as James Williams writes)[13] in which thought attains its full expression. As I will argue here, the Deleuzian project can be considered as an attempt to reconcile each of these (only) apparently opposing pairs by "inhabiting" the process of actualization, and by taking account of one's own position vis-à-vis the forces that represent its laws and limits, as well as its ruptures and breakthroughs. We see a formulation of this kind, for example, in Deleuze's books on cinema, particularly in his conception of the crystal-image as "the indivisible unity of an actual image and 'its' virtual image. . . . The past does not follow the present, it coexists with the present it was. The present is the actual image, and *its* contemporaneous past is the virtual image."[14] Here, the actual and the virtual do not refer to different modalities of Being, much less to the epistemological binary of thought and matter, but to two directions of the same symmetrical movement or coimbrication that defines the process of actualization. To put it another way, by inhabiting a particular point of view, this process is revealed as the properly ontological dimension of both science and philosophy, in the sense that it constitutes the dimension in which force, as the positing of a point of view, engages the virtual image in the production of an actual object.

This emphasis on the ontology of the image takes us rather far from the implications of both modern physics and Kuhn's paradigm-induced gestalt shift, while preserving elements of both. Indeed, as I will argue here, what Deleuze and Guattari offer us in their distinction between reproducing and following (and in their interpretation of scientific thought generally speaking) is a notion of visuality that elaborates and extends what Bergson called the "superposition" of metaphysics and science. This connection is important not only because Deleuze borrows much of his conceptual system from Bergson, but more specifically (and more importantly) because he resumes the philosopher's attempt to "superpose" metaphysics on the scientific image. It is in this sense that Deleuze's (and Guattari's) thinking on science necessarily exceeds and elaborates the standard acceptations of scientific terms. In the next part of this essay, I will try simply to understand what Bergson means by this superposition, and how it relates to two very different ways in which thought confronts its own indeterminacy: on one side, as epistemological impasse, a consequence of what Deleuze calls a "*retrograde movement* of the true";[15] on the other, as virtual image. Finally,

I will consider how the virtual, as framed by Bergson's critique of science and by early attempts to reconcile quantum with classical physics, provides the theoretical grounds on which Deleuze will establish a new set of practices with respect to thought. Understood properly, to think, to see, does not mean opposing the virtual to the actual, but proceeding always from within the "field" of actualization that joins the two movements in a continuous circuit. Every act of thought, every act of observation, consists first and foremost in this "leap into ontology," even if the leap remains more or less veiled from the point of view of the thinker.

The Scientific Image

I have chosen to use the term "scientific image" rather than the more current "scientific model" or "modeling" because it serves as a reminder that science itself must first formulate a way of seeing—a kind of positivist visual culture—before it may constitute a body of knowledge. This is literally the case for Bergson in *Time and Free Will* (1889), where measurements, numbers, sets, etc., comprise the superposition of observable phenomena in a single image: "we must retain the successive images and set them alongside each of the new units which we picture to ourselves."[16] Deleuze and Guattari will use a similar metaphor in their theory of science and the partial observer (*What Is Philosophy?*): "perspective fixes a partial observer at the summit of a cone and so grasps contours without grasping reliefs or the quality of the surface that refer to another observer position."[17] We can reconstruct this notion of the partial observer in a diagram (Figure 2.1) where an eye (a) at the summit of a cone (b), which is intersected by three surfaces (c, d, e), sees only a single contour (Figure 2.2, f) without grasping how it would appear from "another observer position" (as c, d, and e appear from our own point of view in Figure 2.1). The cone, in this diagram, represents the perceptual metaphor or projection by which thought is able to organize the heterogeneity of lived experience (affect) into the contours of a single world, at the same time organizing a body of intelligence with regard to that world (what Deleuze and Guattari call the "plane of reference"). In this way, the planes c, d, and e are superposed in the way described by Bergson. Conversely, this superposition of planes allows thought to project a point of view outside of lived experience, one that corresponds to the ideal actuality that we see, for example, in the Lockian category of "primary" qualities. The selection and convergence of rays of

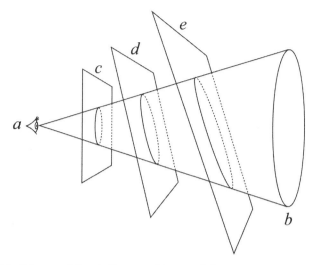

Figure 2.1 Deleuze and Guattari's notion of the partial observer.
Illustration by author.

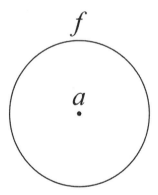

Figure 2.2 Sections of the cone, as seen by the partial observer.
Illustration by author.

light that is necessary to establish a clear and stable field of vision is analogous, in this sense, to a world that has been totally actualized by a rigorous process of selection.

In thinking of each of the above scenarios (Bergson's quantitative multiplicity and Deleuze and Guattari's partial observer), it would be wrong to characterize the scientific image as a simple matter of perception; rather, it is a process by which consciousness acquires form *alongside* matter (*"Plus la conscience s'intellectualise, plus la matière se spatialise"* [The more consciousness is intellectualized, the more matter is spatialized]).[18] Yet this process necessarily follows the visual metaphor insofar as thought must "picture to itself" all changes—even its own—as if they were extensive properties of the same schema; it must regard every event as if it took place in the same present moment and on the same plane of reference. Without recourse to the visual metaphor, such a synthesis would be impossible, as would any attempt to take measurements or gather data; thought and world would remain a virtual chaos. This holds true even in regard to phenomena that have no visual component as such (Bergson speaks extensively of sound and auditory perception in *Matter and Memory*).[19] We could say then that science is no more than an extension of the infrapersonal faculty for schematizing consciousness in terms of relations in space.

Starting from this assertion, we arrive at all the familiar terms of Bergson's critique: by adopting the scientific point of view, philosophy (mis)construes all differences in terms of the schema that science has placed before it.[20] There is even a tendency to regard changes between qualitatively different states of being or "intensities"—from passion to sorrow, from sorrow to joy, from one lived state A to another B—as so many superposable images (i.e., as simple elements of one homogenous image) whose properties are completely determined by their numerical relations.[21] "The question, then," writes Bergson, "is how we succeed in forming a series of this kind with intensities, which cannot be superposed on each other";[22] "For there is nothing in common, we repeat, between superposable magnitudes such as, for example, vibration-amplitudes, and sensations which do not occupy space";[23] "all measurement implies superposition . . . there is no occasion to seek for a numerical relation between intensities, which are not superposable objects."[24] Failing to take into account a categorical difference between quantitative and qualitative "multiplicities," various new disciplines interpose themselves between science and metaphysics. Bergson cites the example of psychophysics (Alexander Bain and Gustav Fechner), but we could also mention August Comte's "social physics," or even modern-day demographics, pharmacology, or neuroscience. Each one strives for a more complex picture of life, but only by virtue of a scientific image that reduces all phenomena to a common denominator, to a single "plane of reference."

Bergson's critique is not of science as such, but of this tendency in thought to assume a point of view from which *everything* appears as an extended thing, and in which intensities, which constitute the experience of difference as such, have been superposed in a single image. Conversely, the mechanistic view of nature is not simply a philosophical attitude toward the world, but a convention that establishes, by means of the constantly growing social, political, and technological means at its disposal, real mechanisms for transforming the movement of life into pure material extensity. We see then that the practical application of a collective body of knowledge (legal institutions, communication, transportation, agricultural and industrial production, etc.) plays a central role in fixing life within a purely symbolic matrix of relations, and in veiling from thought the visual metaphor through which it was able to assume this form in the first place. Deleuze calls this the retrograde movement of the true: "In short, there is a point of view, or rather a state of things, in which differences of kind can no longer appear."[25] Starting from the illusion of *primary* qualities, thought is no longer able to grasp its ontological being in affect (this has already begun to happen in Locke's concept of "secondary" qualities). A dangerous illusion to be sure, but one that, in Deleuze's interpretation, "belongs *to* the true itself" and results not only from individual or collective human nature "but from the world in which we live, from the side of being which manifests to us in the first place."[26] We will later examine how this culminates in the two founding errors of physical realism and epistemology (and a crisis in the social sciences). For now we should note that Bergson understood the mechanistic view of nature neither as a mere philosophical misstep, nor as the overreaching effects of science. For him, it was the symptom of a growing imbalance in the organization of human life considered as a whole, and one that could only be properly diagnosed by grasping thought and matter as features of a single system.

In his 1914 address to the Académie des Sciences Morales, Bergson expressed his concerns about the developing war in Europe by invoking the apocalyptic image of a machine that has become completely detached from the practical aims of its maker:

> What would happen if the mechanical forces, which science had brought to a state of readiness for the service of man, should themselves take possession of man in order to make his nature

material as their own? . . . What kind of a society would that be which should mechanically obey a word of command mechanically transmitted; which should rule its science and its conscience in accordance therewith; and which should lose, along with the sense of justice, the power to discern between truth and falsehood? What would mankind be when brute force should hold the place of moral force? What new barbarism, this time final, would arise from these conditions to stifle feeling, ideas, and the whole civilization of which the old barbarism contained the germ?[27]

The old "scientific barbarism" by which thought first finds itself alienated in an external point of view thus culminates in a prosthetic technology—an "artificial organ"—that behaves as if it had no more use for the living body. In other words, a *méconnaissance* occurs simultaneously in the social-psychological constitution of a point of view *and* in the material conditions that give thought its shape and content. On both sides, internal and external, we find "mechanical forces" at work, transforming dynamic qualitative multiplicities (life, thought, intensities, etc.) into a static "numerical" image.

Given the nature and urgency of Bergson's critique (and of the present-day expansion of science and technology as an end in itself),[28] we might expect the only remedy to lie in a careful disentanglement of the two multiplicities. On one side, science would then be free to explore the determinate relations that constitute inert matter; on the other, philosophy could pursue a less determinate but more holistic approach to questions regarding life. To restore thought to its starting point within lived experience would require no more than a turning away from science based on a philosophically grounded rejection of symbolic, quantitative knowledge. But this is not what Bergson has in mind (nor the path that Deleuze would later follow). Instead, he proposes a deliberate *entanglement* of the two disciplines through a different kind of superposition:

If science is to extend our action on things, and if we can act only with inert matter for instrument, science can and must continue to extend our action on things, and if we can act only with inert matter for instrument, science can and must continue to treat the living as it has treated the inert. But, in doing so, it must be understood that the further it penetrates the depths of *life*, the more symbolic, the more relative to the contingencies of action, the knowledge it supplies to us becomes. On this new ground philosophy ought then to follow science, in order to superpose on scientific truth a knowledge of another kind, which may be called metaphysical. Thus combined, all our knowledge, both scientific and metaphysical, is heightened.[29]

We may well ask, if the function of the scientific image is to reduce differences of kind to differences of degree—to pare down the contents of experience so that they may be deployed as a material instrument of thought—what could it possibly mean to undertake a superposition of the kind Bergson prescribes here? What kind of attitude toward science and scientific knowledge does this superposition imply?

The very term *superposition* reminds us of the process we saw earlier, by which thought first translates the heterogeneity of its milieu into terms commensurable with the exigencies of a particular course of action. In the case of quantitative multiplicities, for example, the point of view of the partial observer flattens out disparate images on a single plane so as to establish among them a network of numerical, spatial relations (analogous to a constellation of stars that occupy, relative to one another, widely varying positions in time and space). The ability to think in this way, to deploy the visual metaphor in the synthesis of instrumental knowledge, is a necessary and logical corollary of the various physical interactions that comprise a large part of human experience. It is a process of selection, retaining only a variable that may be articulated within the schema of another image. We must remember, however, that it is just one of two forms of thought—or, in Bergsonian terms, of two forms of *memory*: "the one, fixed in the organism, is nothing else but the complete set of intelligently constructed mechanisms which ensure the appropriate reply to the various possible demands. This memory enables us to adapt ourselves to the present situation; through it the actions to which we are subject prolong themselves into reactions that are sometimes accomplished, sometimes merely nascent, but always more or less appropriate."[30] True memory, on the other hand, remains "suspended" above the contingencies of the present moment, "truly moving in the past and not, like the first, in an ever renewed present."[31] Bergson illustrates this relationship with his illustration of a cone (Figure 2.3), in which the base AB represents true memory and the point S an image of the (perceiving) body; the

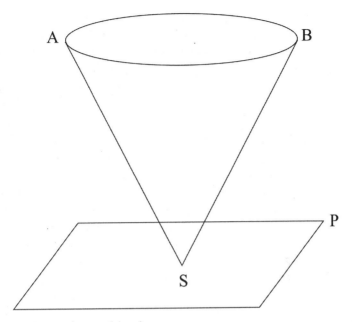

Figure 2.3 Bergson's inverted cone diagram.
Illustration by author.

latter functions in turn to "receive and restore actions" emanating from the images on plane P.

Space and duration, extensity and intensity, thus remain linked by virtue of a continuous effort on the part of consciousness to transform true memory AB into a form of instrumental knowledge appropriate to new challenges, obstacles, and interactions confronted by the material body S in its present environment P. Conversely, we see that Bergsonian intuition is that which carries consciousness immediately away from the present and back into a duration AB, such that the articulations effected in matter by instrumental knowledge are joined once more in lived experience: "That which is commonly called *fact* is not reality as it appears to immediate intuition, but an adaptation of the real to the interests of practice and to the exigencies of social life. Pure intuition, external or internal, is that of an undivided continuity."[32] Here again, we see how science performs the same function within social life that sensory-motor perception performs for the material organism; yet the coexistence of memory and perception, the mutually determining role played by one with respect to the other, implies a superposition of a different kind. In Bergson's diagram above, pure memory (what Deleuze describes as the "ontological past") cannot be combined in a simple image with the perception that "contracts" it. Nor is this contracted image, at the summit of the cone (S), to be confused with the ontological grounds from which it has arisen as a particular distribution in space. The two instances are only the ideal limits of a continuous movement that can be described along the same lines as the visual metaphor discussed earlier; yet they remain necessarily distinct.

If, on the one hand, Bergson's strategy of superposing scientific truth with metaphysical knowledge merely duplicates this metaphor, we see, on the other hand, that it also extends and elaborates a relationship that already exists between pure memory and sensory-motor perception. The effect of this superposition, in the second case, would not result in a homogenous quantitative multiplicity, but instead would restore a heterogeneity to thought that was formerly eliminated in the synthesis of a scientific image. The heterogeneity at issue here is not only, or not primarily, lived experience as duration, but the body of knowledge produced by science in response to specific social, political, economic, and technological imperatives. The movement of philosophical thought is not the product of a historical development, but it takes place within a dimension that Deleuze and Guattari refer to as "stratigraphic time," an image of the world that "expresses before and after in an order of superimpositions."[33] But does this kind of superposition or superimposition also mean visualizing simultaneously a succession of images that do not "belong" together, and for which, by the same token, we can determine no properly *quantitative* relations? Deleuze and Guattari answer this question indirectly when they write that "Philosophy can speak of science only by allusion, and science can speak of philosophy only as of a cloud."[34] In the following section, I will try to explain what makes philosophical and scientific images irreconcilable with respect to their varying attitudes toward temporality and indeterminacy. We may then look more closely at what happens when they are nevertheless brought together in the way Bergson prescribes. But we find another answer in Deleuze's conception of the crystal-image as "the indivisible unity of an actual image and 'its' virtual image."[35] Superposition in this case makes something else out of the superposition of actual and virtual, an image that can be situated at neither extreme but must be considered as the entire field in which sense and form are joined together in a single circuit. It is a virtual chaos that has already begun to coalesce in facets, like multiple competing planes of reference. It is, as Bergson described his own project, a "positive metaphysics."[36]

Historical Adaptation or a "Blurred" Image of Reality?

When Deleuze and Guattari describe science as predominantly about seeing,[37] it is not in the sense of *seeing what is* but in that of *seeing what happens*. A scientific image is only complete to the extent that it anticipates the successive states that make up an event. It is therefore confronted, even in its simplest manifestations, with the problem of mixed composites (space with duration). We see this clearly in the definition of the scientific image outlined by Schrödinger in his paper on the "cat paradox" (1935):

> [This] representation in its absolute determinacy resembles a
> mathematical concept or a geometric figure which can be com-
> pletely calculated from a number of *determining parts;* as, e.g., a
> triangle's one side and two adjoining angles, as determining parts,
> also determine the third angle. . . . Yet the representation differs
> intrinsically from a geometric figure in this important respect, that
> also in *time* as fourth dimension it is just as sharply determined
> [so that] if a state becomes known in the necessary number of
> determining parts, then not only are all other parts also given for
> this moment (as illustrated for the triangle above), but likewise all
> parts, the complete state, for any given later time.[38]

The superposition of quality and quantity in the scientific image is a function of the need to go beyond a mere snapshot of matter into this "four-dimensional" geometry—an example par excellence of spatialized time. As we saw with psychophysics and other disciplines, the experience of qualitative change, initially eliminated to make way for quantifiable relations in matter, is now reconstructed out of those same relations, "surreptitiously bringing in the idea of space."[39] In Schrödinger's conception of the scientific image, time has lost all resemblance to Bergson's duration; it has been replaced with an extensive property of the image.

But it is his *meta*-scientific conception of time that is most relevant to our present discussion. In constructing a particular representation,

Schrödinger warns us that "one must not think so literally, that in this way one learns how things go in the real world";[40] the scientific image is no more than a thinking aid that must be submitted to "the historical process of adaptation to continuing experience."[41] Even when this process is infinite, as he concedes it is likely to be, scientists must proceed with the same representation, on the assumption that any new knowledge acquired by experience might simply be added to it. From this point of view, it is as if both the world and its image unfolded within the same eternal moment, the first being infinitely complex and the latter necessarily lagging behind. In other words, the world *represented* by science may only approach the complexity of the real world asymptotically, as shown in the following graph (Figure 2.4). Here we see that time—or something like time—really does enter the picture, but obliquely and external to it, as if

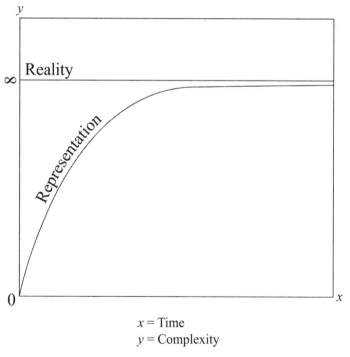

x = Time
y = Complexity

Figure 2.4 Scientific representation approaches the world asymptotically.
Illustration by author.

(in true Kantian fashion) its purpose was merely to provide human understanding with a dimension in which to reconstruct all the contours of a self-identical world.[42]

Starting with the assumption that the world follows a determinate order, physical realism (classical Newtonian science) proceeds inevitably to the belief that its representation should be determinate as well. The failure of such a representation to predict the outcome of a particular experiment is then attributed to that degree of indeterminacy that is to be slowly (but never fully) diminished through the process of historical adaptation. "If in many various experiments the natural object behaves like the model, one is happy and thinks that the image fits the reality in essential features. If it fails to agree, under novel experiments or with refined measuring techniques, it is not said that one should *not* be happy. For basically this is the means of gradually bringing our picture, i.e., our thinking, closer to the realities."[43] This view will remain for the most part unchallenged until an attempt to picture events at the subatomic level, starting with Max Planck at the turn of the century, leads scientists to a meta-scientific position that is fundamentally inconsistent with the realist assumption. With quantum mechanics, as Schrödinger writes, "The classical concept of *state* becomes lost, in that at most a well-chosen *half* of a complete set of variables can be assigned definite numerical values. . . . The other half then remains completely indeterminate."[44]

We reach an adequate understanding of the polemic raised by Schrödinger (the cat paradox) simply by considering the shifting meaning of indeterminacy in the quantum model. Because we are still dealing with numerical relations (the relative positions and velocities of point masses) and with consecutive changes from one set of relations to another, we can say that the quantum interpretation of time is still fundamentally spatialized in the Bergsonian sense. In trying to produce a complete image of wave particles, however, one finds that only half of these numerical relations may be determined at any given moment: the position of a particle may be deduced only by forgoing precise knowledge of its velocity, and vice versa (Heisenberg uncertainty principle). Measurement of one set of variables thus necessarily distributes the others across a range of states, so that we can no longer accept as self-evident the function attributed to historical adaptation. With the Heisenberg uncertainty principle (and, more importantly, with Niels Bohr and the split in wave-particle physics that followed the 1927 Copenhagen interpretation),[45] a physical system could be seen to pass through a moment in which the numerical relations of many states were sustained at the same time. This put scientists into a quandary of a distinctly *philosophical* nature. Should the synthesis of a scientific image extend only as far as measurement permits, or may it introduce an element of indeterminacy? Schrödinger's famous thought experiment is devised to resolve this quandary by recourse to direct observation at the macroscopic level:

A cat is penned up in a steel chamber, along with the following device (which must be secured against direct interference by the cat): in a Geiger counter there is a tiny bit of radioactive substance, *so* small, that *perhaps* in the course of the hour one of the atoms decays, but also, with equal probability, perhaps none; if it happens, the counter tube discharges and through a relay releases a hammer which shatters a small flask of hydrocyanic acid. If one has left this entire system to itself for an hour, one would say that the cat still lives *if* meanwhile no atom has decayed. The psi-function of the entire system would express this by having in it the living and dead cat (pardon the expression) mixed or smeared out in equal parts.[46]

Because the behavior of subatomic particles has no visual reference as such, the scientist can only indirectly "see" what happens. The quantum image of reality is the product of measurements that, strictly speaking, do not correspond to phenomena in the visual domain.[47] But what happens when these measurements lead to a *determinate* image of what has not been completely actualized? Schrödinger sees this question and its implications as the result of a philosophical misstep:

It is typical of these cases that an indeterminacy originally restricted to the atomic domain becomes transformed into macroscopic indeterminacy, which can then be *resolved* by direct observation. That prevents us from so naively accepting as valid a "blurred model" for representing reality. In itself it would not embody anything unclear or contradictory. There is a difference between a shaky or out-of-focus photograph and a snapshot of clouds and fog banks.[48]

Has the quantum photo been prematurely snapped, before further experience can bring it into focus—before one can discover "hidden variables"[49] that account for the paradox? The Copenhagen interpretation leads to a confrontation within the visual culture of science that has prompted its realist contingency to limit indeterminacy to the epistemological status of the observer. The theoretical necessity of the cat paradox lies in the need for an instance of direct sensory-motor perception to bring science back into the fold—i.e., to shore up the position of an eye that is "not in things." Bergson, on the other hand, claims that the objective world is itself "a work of adjustment, something like the focusing of a camera. . . . Little by little it comes into view like a condensing cloud; from the virtual it passes into the actual."[50]

A number of scientists since the 1970s (notably Stuart Freedman, John Clauser, William Wooters, and Wojiciech Zurek) have responded to this polemic by constructing experiments capable of imaging the successive states of wave particles. These experiments have resulted in evidence that bears out the Heisenberg principle, and that demonstrates how one may indeed observe a physical system that sustains multiple states at the same time, a phenomenon referred to as the "superposition principle" (one physical state superposed on another). More recently, the move from classical to quantum mechanics has been aided by the research of Bernard d'Espagnat, Philippe Eberhard, Franco Selleri, and Wolfram Schommers, scientists who are especially interested in the mutually determining relationship between observation and reality—that is, in the possibility of a science that does not simply *reproduce*, but *follows* the play of difference. In his note on objectivity in *Quantum Theory and Pictures of Reality*, Schommers writes: "It is meaningless to talk about the physical properties (e.g. wave or particle) of quantum objects without precisely specifying the experimental arrangement which determines them. . . . In other words, a phenomenon (e.g. wave or particle) is always an *observed* phenomenon (Copenhagen interpretation); without observation it is meaningless to talk about a phenomenon."[51]

As a result, one may not speak of a reality that precedes the scientific image, but may only say, with Schommers, that "Reality is projected on space and time and we obtain a picture of reality," and that "physically real processes do not take place in, but are projected on, space-time."[52] Where then do these processes take place? The question brings us back to Bergson's superposition of metaphysics and science, and not only in the sense that quantum theory tends to engage experimental data from a meta-scientific or philosophical point of view. Where Schommers emphasizes the role played by "experimental arrangement" in determining quantum objects, Bergson describes the synthesis of instrumental knowledge alongside the spatialization of matter. Where quantum theory considers a superposition of states, Bergson, too, speaks of a process of superposition, first in the sense of a fully actualized quantitative multiplicity, and then in the sense of a new distribution or combination of images that are not commensurable, and therefore cannot be fully actualized. The two instances of Bergsonian superposition are not divergent, as initially seemed to be the case, but refer simply to diverging "stages" in the field of actualization, as illustrated by the elaborated cone diagram from *Matter and Memory* (Figure 2.5).

We can now make certain claims with respect to historical adaptation and the "blurred" representation of reality. If a scientific image undergoes a process of development, it is not in historical time but in the nontemporal relay of thought and matter that spans the interval AB↔S. As Bergson specifies, thought (or the *idea*) sustains the process of actualization by

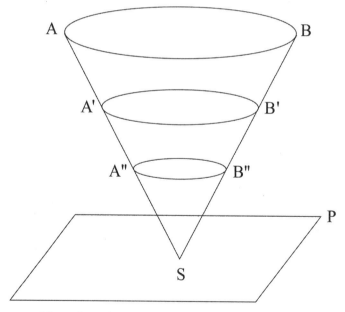

Figure 2.5 Elaborated cone diagram, representing diverging "stages" in the field of actualization.
Illustration by author.

oscillating continuously between "the clearly defined form of a bodily attitude or of an uttered word [and] the aspect, no less well defined, of the thousand individual images into which its fragile unity would break up."[53] This takes place instantaneously in a double movement or "current" that gives actuality to memory at the same time that it gives continuity to a series of discrete perceptions. As a consequence, we find that more or less "contracted" durations may coexist within the interval AB↔S, as indicated above by A'B' and A"B". The scientific image is no more than a variation of this process. Like the idea, it moves between duration and perception (it, too, is a way of seeing); but also, with respect to the object it determines, it moves between two forms of matter. Rather than speaking of memory and perception, we may say that the image begins its trajectory as infinitely dilated experiential data (an indeterminate chaos) and ends in a more or less determinate diagram of numerical relations. Like the idea, the synthesis of a scientific image inscribes a double movement or "current" into the field of actualization, transforming general experience into the appropriate form of instrumental knowledge for a particular challenge or obstacle confronted by the (collective) body S as it negotiates its shifting environment P.

It is at this point that scientific thought diverges from the general movement of the idea. As we have seen (with Bergson's concept of independent "mechanical forces," for example), it is the specific nature of scientific knowledge to fix an environment with respect to a single social, political or technological regime (as when the global economy relies predominantly on fossil fuels as a source of energy). The process of actualization, in this case, tends to coincide with the same present moment, even though it may result, historically speaking, in the synthesis of a more and more complex image (e.g., increasingly sophisticated methods for extracting and combusting fossil fuels). This is how the superposition of quantitative multiplicities arrives at a diverse numerical expression for a particular physical process, giving greater and greater "resolution" to a single image. Yet this is not where real physical processes take place. Indeed, as Bergson argues, the more quantitative, the more symbolic our knowledge of these processes, the less anything *can* happen. So we must say that scientific knowledge does not refer to a world in constant change, but it fixes the world within an infinite present of its own making. The place of reproducibility in the exact sciences is assured not by the need to establish the validity of a particular theory, but so as to unite a particular scientific community with respect to a single present as plane of reference.

Schrödinger's concession to the infinite process of historical adaptation is entirely consistent with this view of things, so long as we consider it to unfold in a single duration that is infinitely contracted (Figure 2.6). The scientific image moves in a circuit within its own ontological dimension, between total indeterminacy and a more or less focused image of reality. Yet if the world seems infinitely complex from the point of view of physical realism, it appears, from this point of view, as an image that can always be brought into greater focus. At no point is a duration fully actualized. As we move from left to right on the *x* axis (field of actualization), the physical system obtains more and more variables, spatial relations within it prove to be endlessly divisible, and the photo is snapped with ever greater resolution and detail.

We seem to be speaking metaphorically here. Doesn't the very definition of instrumental knowledge imply that science takes place in actual, linear time, that the scientific image is continuously disrupted from the

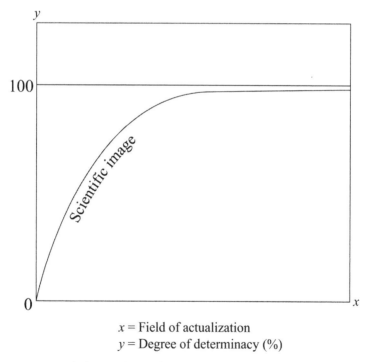

x = Field of actualization
y = Degree of determinacy (%)

Figure 2.6   A single duration that is infinitely contracted.
Illustration by author.

outside as changing material circumstances on the plane P impinge on the appropriateness of this or that course of action (e.g., shifting trends in energy technologies or the depletion of fossil fuels)? And do we not also find that classical physics, with its "four-dimensional geometry," can successfully predict those changes, if not in terms of society and politics then certainly with regard to isolated physical processes? Finally, is it not the case that advances in scientific thought, as well as the technology of observation, indicate that we will be able to make predictions with more and more accuracy as time passes? Questions like these will almost certainly lead us to a reconciliation of classical and quantum physics in favor of physical realism. Time *does* pass in a series of moments, the scientific image *does* work, and this alone—the "success" of physics—points to a mind-independent world that follows a predictable and determinate order. We can see it with our own eyes!

In this case, however, it is the visual culture of science that has changed. As Bergson anticipated, and as Schommers claims outright, science reaches a point where it can no longer pretend to see the interval between one state of affairs and another, and must therefore forgo a particular way of looking if it wishes to see at all. In effect, the image can no longer refer directly to *what happens* or to "real" physical processes; these are always elsewhere, in a movement or duration that escapes the eye of the scientific observer. The event can only be brought into focus (actualized) to the same extent that "mechanical forces" within experiential reality succeed in extracting it from the dimension of time, so as to synthesize, by virtue of the visual metaphor, a more or less determinate (external) objectivity. In this way, where everyday experience suggests a single approach—*I see what happens*—here, there are necessarily two: *What does science see?* and *What happens?* Each of these questions is inseparable from the other. Yet their validity is inversely proportionate, just as a duration A"B" is only determinate to the extent that it contracts another duration A'B', and so on (Figure 2.5). We can speak neither of the accuracy of a scientific image with respect to reality, nor of changes in reality with respect to the scientific image except—and this is where Deleuze gains considerable insight from Bergson—by referring to varying degrees of duration that are superposed within the field of actualization. Here, the two questions necessarily coincide. Science is superposed with knowledge of a specifically philosophical constructivist or intuitionist nature.

The Force of Science: Reproducing and Following

If Bergson's critique of science gives Deleuze an ontology (duration) and a method (intuition), it also gives him considerable room to maneuver with respect to the diversity of thought and univocity of being. By referring to the field of actualization rather than any de facto actuality, Deleuze is able to pose questions that deal with the ontological conditions of becoming as such, but by the same token, with infinitely multiple, divergent actualities. This is the sense in which Bergsonism, for Deleuze, represents a particularly effective strategy for overcoming epistemological realism as a function of normative physical science. But Deleuze also sees Bergson as an ally of science, or in any case, as a philosopher who spoke for the nascent intuitionist tendency within science itself:

> [He] did not confine himself to opposing a philosophical vision of duration to a scientific conception of space but took the problem into the sphere of the two kinds of multiplicity [of degree and of kind]. He thought that the multiplicity proper to duration had, for its part, a 'precision' as great as that of science; moreover that it should react upon science and open a path for it that was not necessarily the same as that of Riemann and Einstein.[54]

Similarly, in *The Movement-Image*, Deleuze speaks of Bergson's "profound desire to produce a philosophy which would be that of modern science (not in the sense of a reflection on that science, but on the contrary in the sense of an invention of autonomous concepts capable of corresponding with the new symbols of science)."[55] Deleuze and Guattari will draw a sharp distinction between philosophical concepts and scientific functions,[56] and they will maintain Bergson's claim that while science may produce multiplicities of space, number, and time, it is incapable of expressing the inseparability of variation proper to duration.[57] But they will also go to considerable length to show that "philosophy has a fundamental need for the science that is contemporary with it," a need that cannot be satisfied simply by reflecting on the nature of science, or by making reference to particular examples from the history of science.[58] Rather science and philosophy need one another because they are each one coextensive with a field of actualization in which they continuously intersect and interact:

It is true that this very opposition, between scientific and philosophical, discursive and intuitive, and extensional and intensive
multiplicities, is also appropriate for judging the correspondence
between science and philosophy, their possible collaborations, and
the inspiration of one by the other.[59]

The most important theoretical step, in this case, will be in showing that
the illusion of physical realism with respect to the unity of the scientific image does not derive from the incommensurability of thought and
world, or from any other merely epistemological problem. The retrograde movement of the true belongs to the true itself—that is, to the
side of being that relates to itself as to a mathematico-physical diagram
of relations. This is why the Heisenberg uncertainty principle is not to
be confused with a purely psychological or phenomenological entanglement of subject and object (the so-called "observer effect"). Like the
quantum principle of superposition that will be its logical consequence,
uncertainty involves a very special point of view or "demon" that, as
Deleuze and Guattari write, "does not express the impossibility of measuring both the speed and the position of a particle on the grounds of a
subjective interference of the measure with the measured, but . . . measures exactly an objective state of affairs that leaves the respective position of two of its particles outside of the field of its actualization."[60] This is
what Schrödinger fails to grasp in his appeal to macroscopic verification
(an appeal that d'Espagnat has called the "macroscopic diversion.").[61] In
quantum theory, it is never a question of whether the scientific observer,
by looking, may deliberately orient a physical process toward this or that
outcome (or toward both outcomes at once). It is rather that the scientific image in this case is the product of an act of observation that has
exceeded the precision of matter itself, but only by sacrificing its own
determinate status in the process. It belongs to a position within the
process of actualization that has begun to exhibit the nomadic potential of science—something already exhibited by its sister disciplines of
art and philosophy. In their analysis of the artistic plane of composition,
Deleuze and Guattari write: "We are not in the world, we become the
world; we become by contemplating it. Everything is vision, becoming.
We become universes."[62] The visual metaphor, such as it is deployed by
quantum physics, intersects with the rigorous plurality of universes that
are specific to the becoming of percepts and affects.

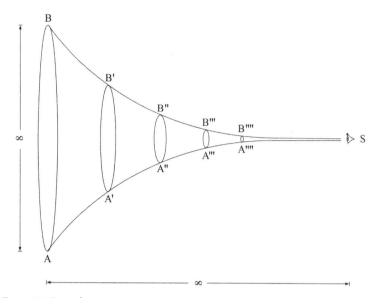

Figure 2.7 Reproducing.
Illustration by author.

We may now return to the primary distinction that Deleuze and Guattari outline between royal science (reproducing) and nomadic science
(following) in order to better understand how the two cases differ with
respect to the contraction of an image. The diagrams I have presented
throughout this article can now be made to coincide in order to show how
a partial observer is established in each case at an endlessly receding point
of view (Figures 2.7 and 2.8). At the beginning of this article, we saw the
manner in which conic sections may be flattened out, from a certain point
of view, and joined together in a single contour (Figure 2.2). This provides
a simple analogy for the way science as a visual culture establishes a point
of view that eliminates heterogeneous difference—indeed, that opposes
its own internal condition (difference) in order to construct a world made
up entirely of external relations. The eye at S is what Deleuze and Guattari describe as *"a reference capable of actualizing the virtual."*[63] This does
not represent a historical process of adaptation (Figure 2.3) but a more
and more narrowly defined present (Figure 2.4). Yet it is a present, as in
Bergson's elaborated cone diagram (Figure 2.5), that necessarily coexists

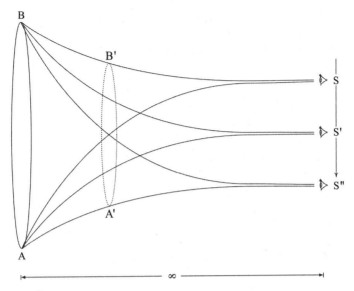

Figure 2.8 Following.
Illustration by author.

with relatively more or less contracted states of the same duration; a relation existing in the duration AB and actualized in a present perception S
is simultaneously contracted at other levels: A'B', A"B", etc. This then gives
us the basic outline for the first composite diagram (Figure 2.7), representing a single line of force drawn into the field of actualization by the partial
observer, that is, reproducing.

The second diagram (Figure 2.8) shows how this point of view may
describe its own movement (S→S'→S") on the plane of reference by contracting the same infinite duration AB with respect to difference itself, i.e.,
to the process of differenciation or individuation that joins together the
disjunctive syntheses of discrete, actual worlds. The positions S→S'→S"
are not simply different ways of looking at the same thing—a kind of subjective (or even scientific) relativism. They are different lines of force, or
as Deleuze and Guattari write, they are "axes" that actively break up a virtual whole into a block of becoming that subsequently passes through a
series of qualitatively different worlds. The actual, in this case, is traversed
by intensities that are evoked by the scientific image but that elude its

explanatory power. This sense of objective indeterminacy exceeds the nomadic sciences as such, and it can be produced merely by extricating our point of view from any historically specific scientific paradigm, so as to grasp the history of sciences as a series of irrational cuts (*coupures irrationnelles*):

> The history of the sciences is inseparable from the construction, nature, dimensions, and proliferations of axes. Science does not carry out any unification of the Referent but produces all kinds of bifurcations on the plane of reference that does not preexist its detours, its layout. It is as if the bifurcation were searching the infinite chaos of the virtual for new forms to actualize. . . . This is what happens when Newton is derived from Einstein, or real numbers from the break, or Euclidean geometry from an abstract metrical geometry—which amounts to saying with Kuhn that science is *paradigmatic*, whereas philosophy is *syntagmatic*.[64]

When we compare this notion of paradigms to the field of actualization as the becoming of material bodies, things, or states of affairs, it too seems overly simplified because there are relative differences between axes that can still be contracted into a single image, while others only intersect in an image that is so indeterminate it remains invisible. In the diagram for following (Figure 2.8), the dotted line of duration A'B' indicates a unity of individual images that has already begun to break up under the pressure of new intensities.[65]

The genesis and function of even the simplest scientific image, of the most rudimentary intelligence, is complicated by these sudden ruptures but also by the logical necessity of moving along the two axes at once:

> [Intelligence] is acquaintance with matter, it marks our adaptation to matter, it molds itself on matter; but it only does so by means of mind or duration, by placing itself in matter in a point of tension that allows it to master matter. . . . It might therefore be said that its form separates intelligence from its meaning, but that meaning always remains present in it, and must be rediscovered by intuition.[66]

This means that even the most determinate image is susceptible at all times to the force of the virtual (as sense), at all levels within the field of actualization. We can even imagine quite complicated processes (Figure 2.9) in

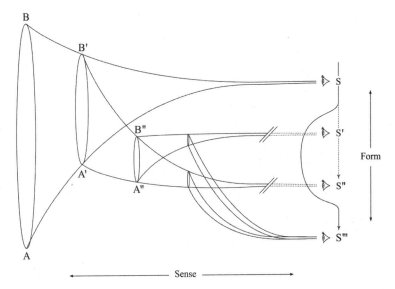

Figure 2.9 Overcoding images.
Illustration by author.

which an image actualized by a particular paradigm at S diverges at A'B' into two or more systems of reference (S' and S"); these in turn may enter into relation with a new paradigm or block of becoming at S'", which captures or overcodes them, without, for that matter, ever eliminating them. Or they may remain entirely separate from yet another independently actual "world" (or monad) at S'", and so on. The pages in *A Thousand Plateaus* that deal with the concept of double articulation and stratification provide a complete outline for this process in terms of interrelated strata, epistrata, and parastrata.[67] In particular, they help us understand how each paradigm in a series S'→S"→S'" remains fully actualized even while it functions as the interior substratum or "content" with respect to an emerging stratum: "The materials furnished by a substratum are no doubt simpler than the compounds of a stratum, but their level of organization in the substratum is no lower than that of the stratum itself. The difference between materials and substantial elements is one of organization; there is a change in organization, not an augmentation."[68] It is in this sense that science, no less than the physical world with which it is coactualized, constitutes a "geology of morals."

At this point, the diagram itself loses its explanatory power, without, for that matter, failing to evoke the complicated sense in which the act of observation necessarily engages a play of forces at different levels within the field of actualization, as well as numerous collisions and collusions between blocks of becoming. In other words, there is no simple image, nor is there any limit to the number of image-worlds that can be actualized from the same virtual image, even if we concede, with Deleuze, that the functioning of this complicated system represents a "virtual Whole."[69] But my intention with these diagrams is not to synthesize a set of relations in which all the various points of view represented in this article can be made to coincide. It is rather to show how Deleuze and Guattari's thinking obliges us to see physical process from a point of view that restores heterogeneity to the scientific image. This superposition of science with metaphysics, while coinciding with Bergson's philosophical program generally speaking, obliges us to broaden the scope his categories, or even frame them in different terms. Indeed, Deleuze is at pains to reconcile a dualism at the heart of Bergson's system between life and matter, to recast them as symmetrical movements within a single Time.[70] Time, or "real" time, has no correlate in linear movements as we understand them, but only in difference as the condition for movement (Deleuze's "third hypothesis" regarding the monism of Time): "There is only a single time, a single duration, in which everything would participate, including our consciousness, including living beings, including the whole material world."[71] Similarly, in *Negotiations,* he claims that "Bergson's always saying that Time is the Open, is what changes—is constantly changing in nature—each moment. It's the whole, which isn't any set of things but the ceaseless passage from one set to another, the transformation of one set of things into another."[72] This interpretation of Bergson enlarges the concept of duration considerably, making way for a more holistic understanding of life *and* matter at all levels of contraction/dilation. Which is to say that the duration AB, in the diagrams I have presented here, stands for something more than an indivisible, internal variation; it is a virtual domain that incorporates all images, all new and imminent worlds, as well as the play of forces that emerges through and between them. It is Spinoza's notion of God as a substance with infinite attributes.

What this superposition restores to science is the properly political and social matrix in which it acquires its true meaning and form (exceeding the instrumentalist interpretation of these contexts). By the same token that observation no longer refers primarily to a perceiving subject, science no longer involves a passive accumulation of data, much less historical progress in the rational humanist sense. Rather, it represents in every case (both reproducing and following) an instance of force within the field of actualization:

> A well-defined observer extracts everything it can, everything that can be extracted in the corresponding system. In short, the role of a partial observer is *to perceive and to experience,* although these perceptions and affections are not those of a man, in the currently accepted sense, but belong to the things studied. . . . Partial observers are forces. Force, however, is not what acts but, as Leibniz and Nietzsche knew, what perceives and experiences.[73]

We may establish a direct lineage, in these terms, that begins with Bergson's *mechanical forces,* those that perpetuate and elaborate the "real" and necessary illusion of space, to Deleuze and Guattari's *abstract machines* as pure "matter-functions."[74] The field of actualization is swarming with machines of this kind, each one situated like an eye at the summit of a cone to contract heterogeneous multiplicities into aggregates of spatial relations, even when there is no scientist to behold or embody them. If, on the one hand, this force is proportionate to the ability of the observer to initiate a process of contraction (and selection) within the field of actualization, it also gains in scope and power by combining this or that block of becoming into a single assemblage. The unity of science can be accounted for in this way: by noting how it reproduces the conditions under which matter and thought form a single assemblage on an increasingly global scale. In this case, there is a massive accumulation of force—though never a total one—that allows science to designate *what is* or *what happens* from a de facto position of authority. The sanction of this authority is not in the hands of science, but in those of the political, economic, and social assemblages whose interests science is made to serve.

CHAPTER 3

The Intense Space(s) of Gilles Deleuze

Thomas Kelso

DETERRITORIALIZATION CAN BE IDENTIFIED as the most famous spatial concept invented by Gilles Deleuze, but his work is full of many more ideas about space: plateaus, the fold, smooth and striated space, the cartography/tracing opposition, *l'éspace quelconque* [any kind of space whatsoever], nomadology, and many others.[1] It is my contention that Deleuze's reconceptualizations of space—in philosophy, in war, in cinema, in art, and in science—all derive from *Difference and Repetition,* which claims that intensity should be seen as more fundamental for thinking about space than the Cartesian notion of extension. Deleuze's intense spaces are much more interesting than centimeters or any other units of measurement. In his terms, depth is not the same as length. For instance, according to Deleuze's analysis, it is intensity that accounts for the fact that we can correlate shades of color with depth in a two-dimensional painting, even without perspectival geometries, for instance in the work of Paul Klee (see Figure 3.1). Deleuze illustrates this by noting that if we lay down the line of a framework for single-point perspective (see Figure 3.2), we have no way of determining whether the lines point toward or away from us until we add to the picture familiar objects, or at the very least, shadings of color. M. C. Escher's prints (see Figure 3.3) exhaustively exploit this zone of indeterminacy. This paper will examine Deleuze's original treatment of space as intensity, so I will start rather obliquely with an intense quote:

> "Whither is God?" [Nietzsche's madman] cried; "I will tell you. We have killed him—you and I. All of us are his murderers. But how did we do this? How could we drink up the sea? Who gave us the sponge to wipe away the entire horizon? What were we doing when we unchained this earth from its sun? Whither is it moving now? Whither are we moving? Away from all suns? Are we not

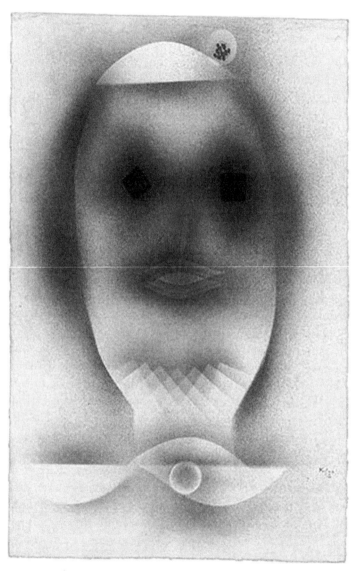

Figure 3.1 Two-dimensional painting without perspectival geometries in Paul Klee's Wintry Mask (1925).
Reproduction by author.

plunging continually? Backward, sideward, forward, in all directions? Is there still any up or down? Are we not straying, as through an infinite nothing? Do we not feel the breath of empty space?"[2] (See Figure 3.4)

In response to the madman's query, I will offer a Deleuzian formula: Absent God = empty space = the virtual. The virtual, in turn, equals what Deleuze calls the "spatium," and paradoxically, but necessarily, it is *not* empty at all, but "full of intensive ordinates" like the univocal being of Spinoza.[3] Chances are when you think of the word "virtual" you think of some sort of interface

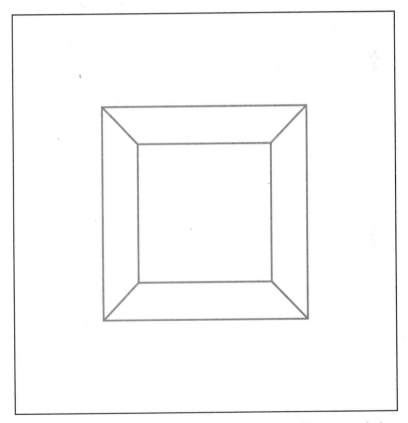

Figure 3.2 In a single-point perspective drawing, we have no way of determining whether the lines point toward or away from us.
Illustration by author.

with cyberspace, a representational double of the experienced world. The absent god problem requires something quite different. To account for existence as we know it, ever changing, ever shifting, and certainly not empty, do we not still need some sort of differentiating mechanism, an "abstract machine,"[4] to take the place of the absconded God? And, if it is not to be God, what can it be? Deleuze finds that the answer is not a structuring lack, nor a primordial division, but rather what he calls "the virtual," something emphatically nonrepresentational that for the moment we will define as the

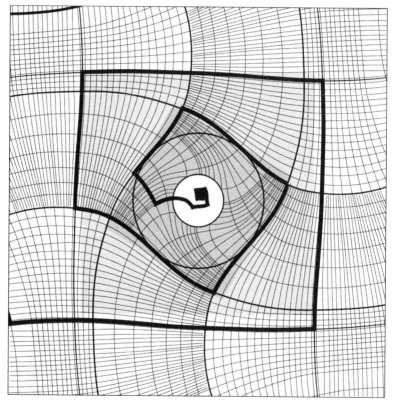

Figure 3.3 A grid similar to the one used by M. C. Escher to create "Print Gallery" (1956). By infinitely "twisting" the grid at its center, a single frame can be mapped onto the shape of a spiral, with the effect that objects inside the frame are always simultaneously outside the frame (called the "Droste effect"). At the center of the grid, one finds what Deleuze calls a "zone of indeterminacy," a region where the resolution of inside and outside regions is infinitely deferred. Diagram by Peter Gaffney.

Figure 3.4 *"Do we not feel the breath of empty space?"*
Photograph by author.

transcendental conditions, not of possible, but of real experience. Deleuze's questions are not the questions of postmodernism. They are questions of ontology, questions of metaphysics: questions such as "What is?" Or "What is a thing?" Or, finally, "What is space?"

Space is a vast topic. I don't know all about it, nor can I cover it all in a few pages. Actually, I had quite a hard time merely figuring out where to start. Consider this: of the last three sentences, precisely none avoid the eminently common conceptual metaphors that confuse space with time, or with thought. Conceptual metaphors such as these, described by George Lakoff and Mark Johnson in their work *Metaphors We Live By,* attempt to understand one conceptual domain by restating it in terms of another. We do so because space (the concept) is not easy. Deleuze doesn't make it any easier, but his ideas do make it more interesting than a mere "primary quality." Since my favorite quote from Deleuze implies a certain deeply paradoxical concept of space, I will begin to characterize Deleuzian space with it. In *The Logic of Sense,* Deleuze says of the Zen (or Stoic) archer: "the bowman must reach the point where the aim is also not the aim, that is to say, the bowman himself; where the arrow flies over its straight line while creating its own target; where the surface of the target is also the line and the point, the bowman, the shooting of the arrow, and what is shot at."[5] Where could such an event conceivably occur? What kind of space could host all of these entities on "the surface of the target"? The easiest answer is this: such a thing could never occur in actual space, the quantifiable domain of extension, but it *always* occurs for the archer (or the dancer or the mime, for that matter) who successfully inhabits the *virtual* space of the event. That's why Deleuze says: "incompatibility is born only with individuals and worlds in which events are *actualized* but not between [virtual] events themselves."[6] In other words, the actualized coordinates of an object in motion are linked together in virtual space.

Of course, this easy answer requires further explanation. After all, what is virtual space? Readers of Deleuze will know that his predilection for Bergson will lead him to invent lots of concepts that extend his predecessor's idea of duration. Since duration itself implies the real but virtual coexistence of the present and the entirety of the past, linking duration to the idea and logic of the event is intuitively an obvious move. Transposing this logic from time to space might seem less so, but that is precisely what Deleuze does in chapter V of *Difference and Repetition:* "The Asymmetrical Syntheses of Perception." I will approach his actual/virtual opposition by indicating the two separate, but interlaced, kinds of space that it implies: actual and virtual.

Actualized space can be, and generally is, characterized as extension. It is quantifiable, divisible. It has limits and borders. As Jean-Clet Martin says, this space is subject to "the transcendence of number,"[7] so we can fit it onto the gridwork of Descartes's coordinates, and if we are feeling Kantian, sometimes—but not always—we can synthesize it according to units of measure that we generate by comparison to the proportions of our own bodies. In this latter case, transcendence might be said to reappear by way of the subject (cf. the emphasis on perspective in phenomenology by Martin).[8] Like the subject, like essences, too, our concept of actual space is *abstract.* So what? one might ask. Well, the problems with abstract space, with extension, are actually quite numerous. Look at Wittgenstein's rabbit/duck (see Figure 3.5), or myriad other optical illusions that confound ground and field; they all imply that positionality does not adequately describe space. Why, or how, do we "surf" in flat media? Why does a certain identical distance cause some people vertigo, but not others? Why, in daily life, do we routinely navigate through space without visualizing it? Rather, simply by relying on proprioceptive perception, we retrace routes that exist virtually in our bodies, but not abstractly in a map or a diagram in our minds. That's how we do things without looking. Constantin Boundas points to yet another problem. Referring to Descartes's "proof of God from preservation" in the *Meditations,* he observes: "a world of extended magnitudes cannot, without further ado, account for the synthetic operations that make it a world in the first place. Descartes's fate is

Figure 3.5 The duck-rabbit illusion.
Reproduced from J. Jastrow's "The Mind's Eye" (1899).

instructive: if the 'now' of the cogito is an extended magnitude, it cannot have the ratio of its retentions and protentions in itself; only a God [can] guarantee the temporality of duration."[9] Another problem with extended space "is that it presupposes dimensionless points as its primary material, but then cannot account for their interconnection."[10] This point leads us naturally to yet another archer, the archer in Zeno's paradox. This archer, it seems, cannot shoot at all. In his *Physics*, Aristotle formulates Zeno's paradox as follows: "if it is always true that a thing is at rest when it is opposite to something equal to itself, and if a moving object is always in the now, then a moving arrow is motionless."[11] We might also say, following Brian Massumi: given the infinity of points between any two given points, positionality swallows up movement and immobilizes the arrow. But of course arrows really do fly, and Bergson and Deleuze account for this by insisting, not on the punctual *representation* of space, but on the primacy of *process*, on the dynamic unity of the arrow and its real movement in space. From this perspective, only the arrow embedded in its target actually occupies a position, and as Massumi puts it: "the points or positions [of its trajectory] appear *retrospectively*," as "logical targets, or possible endpoints."[12] This point has huge ramifications: "Bergson's idea is that [extensive] *space is itself* a retrospective construct of this kind. When we think of space as 'extensive,' as being measurable, divisible, and composed of points plotting possible positions that objects may occupy, we are stopping the world in thought. We are thinking away its dynamic unity, the continuity of its movements. We are looking at only *one* dimension of reality."[13] In Deleuze's own words: "When you invoke something transcendent you arrest movement, introducing interpretations instead of experimenting."[14] If this sort of thinking were restricted to relatively obscure philosophical paradoxes, I suppose it wouldn't matter much, but Deleuze goes to great lengths to show how retrospective formations also impose similar gridworks over problems ranging from the scientific and political to the linguistic and the social. Think, for instance, of the primacy of the idea of subject position in the notoriously problematic discourse of identity politics. Deleuze's project ("experimenting") moves in the opposite direction by focusing on the processes that define individuals according to structured relations between the actual and the virtual, and these in turn are meant to work with, rather than against, the creative potential of such individuations.[15]

So, if extensive or actual space is in fact constituted by the retrospective projection of a grid onto process, what, or where, is the real space that makes it possible? Where are the rest of the dimensions to be found? For Deleuze, the answer is intensive or virtual space, which, as I have mentioned, he calls depth. For Deleuze, the virtual is absolutely necessary if we are to account for the fact that, in three domains—space, time, and thought[16]—things are constantly becoming other than what they are, like an egg or an embryo. Deleuze's examiners were quite taken aback when, at the 1967 defense of his *doctorat d'état*, i.e., *Difference and Repetition*, Deleuze straightforwardly repeated the following claim from that text: "the entire world is an egg."[17] This statement certainly sounds strange, but it is actually part of the most succinct deduction of the virtual that Deleuze ever offered. Because Deleuze's dissertation defense, titled "The Method of Dramatization," is not so well known, and because it may also be the simplest *possible* deduction of the virtual, I will quote it at length. Deleuze begins with what appears to be a classic philosophical question: "What is a thing?" I say he "appears" to begin with a classical philosophical question because throughout the whole first part of his dissertation defense Deleuze takes pains to show that the question "what is?" is misguided insofar as it predisposes philosophy to seek essences, and of course, Deleuze's battle against philosophy's system of representation (that is to say, its "image of thought" based on recognition) is also a battle against essences (and in favor of multiplicities). So, it's a bit humorous when he then proceeds to ask: "What is a thing?" But it is also informative because, first of all, this is not some postmodernist denial of the question, for instance, by way of social construction, and, second, as we'll see, it is clear that Deleuze's answer to this question is not really an answer to the question "What?" but to the question "How?" In other words, it should become apparent that Deleuze wants to replace timeless essences with processes of morphogenesis and to replace identity with difference. Here, then, is the quote:

> What is the characteristic or distinctive trait of a thing in general? Such a trait is twofold: the quality or qualities which it possesses, the extension which it occupies . . . In a word, each thing is at the intersection of a twofold synthesis: a synthesis of qualification or specification, and of partition, composition, or organization. There is no quality without an extension underlying it, and in which the quality is diffused, no species without organic parts or points. These are the two correlative aspects of differentiation: species and parts, specification and organization. These constitute the conditions of the representation of things in general. But if differentiation thus has two complimentary forms, what is the agent of this distinction and this complimentarity? Beneath organization and specification, we discover nothing more than spatio-temporal dynamisms: that is to say, agitations of space, holes of time, pure syntheses of space, direction, and rhythms. The most general characteristics of branching, order, and class, right on up to generic and specific characteristics, already depend on such dynamisms or such directions of development. And simultaneously, beneath the partitioning phenomena of cellular division, we again find instances of dynamism: cellular migrations, foldings, invaginations, stretchings; these constitute a "dynamics of the egg." In this sense, the whole world is an egg. No concept could receive a logical division in representation, if this division were not determined already by sub-representational dynamisms. . . . These dynamisms always presuppose a field in which they are produced, outside of which they would not be produced. This field is intensive, that is, it implies differences of intensity distributed at different depths. Though experience always shows us intensities already developed in extensions, already covered over by qualities, we must conceive, precisely as a condition of experience, of pure intensities enveloped in a depth,

in an intensive *spatium* that preexists every quality and every extension. Depth is the power of pure unextended *spatium*; intensity is only the power of differentiation or the unequal in itself, and each intensity is already difference. . . . Such an intensive field constitutes an environment of individuation.[18]

We thus have three levels. All of them are real, even if they are not actual: actual individual things, spatio-temporal dynamisms, and a virtual field of intensities, which we may also refer to as the spatium or depth. Even if all we have is an egg. But if the egg is all we need, this is because Deleuze's larger claim is that the radical transformations and potentials of embryogenesis are determined, not merely by DNA or heredity, but primarily by the power of the virtual itself. Moreover, this mutability is characteristic of all life, be it organic or inorganic life. The intensities of the virtual do not just determine the stretchings, foldings, and strange topological behavior of embryos; they also condition, for instance, the bifurcations, dissipative structures, oscillations, and symmetry breaks of chemical reactions.[19] In other words, what Deleuze is really appealing to when he appeals to spatio-temporal dynamisms is the self-organizing capacities of the cosmos. If there is an empty space vacated by God, it is the space of virtual depth. It is the entire set of possible worlds, but, as opposed to the Leibnizian conception whereby this zone of infinite possibility would be subjected to the constraints that identity requires to ensure that ours is the best world, with Deleuze, difference, not identity, is motor of the genetic process, such that the space evacuated by God becomes the entire set of divergent virtual worlds that underlie any becoming whatsoever. In other words, the best world is the most multiple, where the empty space of God is substituted by the absolutely full spatium of infinite potentiality—the virtual.

Manuel DeLanda is in the habit of taking from science concrete examples of the virtual's ability to structure objects in space: for instance, the soap bubble and the salt crystal.[20] In both cases the spatio-temporal dynamism at work minimizes energy. Bubbles minimize surface tension, while salt crystals minimize bonding energy. *There is an intensity, a difference, itself conditioned by a singularity, or a topological form, i.e., the minimum,* that cannot be said to exist in extended space but which nevertheless determines the actual extended shapes produced by the virtual. This tells us three more things about Deleuze's virtual: (1) The virtual condition does not resemble the actual, or the conditioned. What we sense is "asymmetrically" determined by the virtual.[21] (2) The same virtual condition, or topological form, can determine two different outcomes. That is to say, intensities differ; crucially, they even differ from themselves, which is why the virtual can do what it claims to do—i.e., give a transcendental account of the genesis of the actual. And (3) the virtual condition tends to disappear in the conditioned. We cannot observe the condition, but only its effect. Deleuze calls this the objective transcendental illusion: "Difference is cancelled qualitatively and in extension. . . . Difference is intensive, indistinguishable from *depth* in the form of the non-extended and non-qualified *spatium*, the matrix of the different and unequal. Intensity is not the sensible but the being *of* the sensible, where different relates to different."[22]

The intensities in question, however, do not simply vanish into thin air. They are indeed canceled out in extension or qualities, such that we do not directly perceive them, just as we do not directly perceive forces such as gravity or electromagnetism (in Latin, force is *virtus*, the etymological root of virtual), but rather we feel their effects. They do, nevertheless, subsist as that which must be perceived, as Boundas writes, in "the production of new and heterogenous sensations."[23] Deleuze says: "intensity is simultaneously the imperceptible and that which can only be sensed."[24] The topological form of the minimum is not something we experience, but it can be determined on the basis of what we do perceive, and we would not perceive a soap bubble without it. Furthermore, singularities, like minima, do not determine phenomena, without being related to other intensities. The soap bubble is also subject to virtual conditioning by heat, gravity, pressure, etc.: in other words, by multiple differences of intensity, with their own thresholds and singularities. This combination is what Deleuze calls a multiplicity, and multiplicities correspond to the Idea, the virtual conditions of real experience. Multiplicities don't just account for the existence of eggs and soap bubbles. They also determine us. Our senses are determined by synesthesia, or intersensory perception. For instance, our eyes have a tactile function when we see texture. Or, to cite a more humorous example, one scientist, who happened to be an experienced pilot, went so far as to anesthetize his posterior, in order to demonstrate that without perception of his center of gravity (a virtual point, by the way) he could not orient himself by vision alone during high-altitude flight. Massumi says of him, "he scientifically demonstrated that we see by the seat of our pants."[25] Even more importantly, he demonstrated that our perception is riddled with traces of the virtual, as if the virtual and the actual were interlaced or superimposed on one another. This superimposition is what Deleuze means by the intensive spatium: there is an originary depth to all experience, which we cannot directly sense, though we are lost without it. In point of fact, the experimental mode of Deleuzean thought is a sort of askesis that is designed to train our sensibilities to be sensitive, like our pilot's, to the subsistence of the virtual in every actualized state of affairs, thereby opening ourselves up to new becomings that go beyond the kinds of experience that we have thus far represented to ourselves. To return to the Stoic archer, then: he intuits that his shot, which actually occurs, is subtended by a multiplicity of unrealized virtual conditions. These conditions superimpose a set of inexhaustible singularities on the event, and if the archer has a feel for them, it is possible for him to double, or counteractualize the event. Like an actor performing a role, he can act out the event's realization in an endless variety of ways. The original depth of the spatium means that the archer who actually manages to heighten his awareness of his shot's virtual ordinates actually hits the bull's-eye.

· II ·

Science and Process

CHAPTER 4

Interstitial Life: Remarks on Causality and Purpose in Biology

Steven Shaviro

THE QUESTION OF PURPOSE has long haunted biology. Darwin's explanation of evolution by means of natural selection was intended, among other things, to get rid of teleological explanations of living things. Darwin explicitly answered the "argument from design" invoked most prominently in the nineteenth century by William Paley in his once-famous book *Natural Theology* (1802). Recapitulating what was already an old argument, Paley said that living organisms were so intricately structured that they could not have arisen at random; they must have been produced by the deliberate actions of some Designer. The evident purposiveness and organized complexity of living things strongly suggests that they were purposefully created. This argument is so intuitively appealing that it is still being made today by creationists, or proponents of so-called "intelligent design."[1] Darwin was the first to explain the genesis of organic complexity in naturalistic terms without appealing to supernatural forces. It is important to understand how radical a move this was. In his exposition of the logic of natural selection, Richard Dawkins insists that, prior to Darwin, there was simply no satisfactory explanation for organic complexity, aside from the theistic one.[2] Even Hume, who ridiculed the argument from design in his *Dialogues Concerning Natural Religion* (1779), "did not offer any alternative explanation for apparent design, but left the question open" (6). For Dawkins, Darwinian natural selection is still the *only* theory that is able to account for the complex structures and properties exhibited by living things in cause-and-effect terms acceptable to modern science, without invoking prior intentions or purposes.

In other words, Darwin provides an immanent, nonteleological mechanism for the development of life. Given the theory of natural selection, it is no longer necessary to invoke such teleological agencies as the hand of God, or the Lamarckian acquisition of striven-after qualities, or the workings of some inner vitalistic force like Bergson's *élan vital.* Nonetheless, even the most dedicated evolutionists *are* unable to avoid reverting to the language of purpose. After all, living organisms, and their parts, are evidently purposive in terms of how they grow and how they relate to the world around them. It scarcely matters that this purposiveness, or intentionality, was never itself intended by any higher agency, but arose through the workings of natural selection. It still remains the case that biologists can only explain an organism by speaking *as if* its features (eyes, reproductive behaviors, or whatever) were purposive. When we study living organisms, we cannot get away from what Michael Ruse, in his book *Darwin and Design: Does Evolution Have a Purpose?*, calls "the metaphor of design."[3] Even as Ruse defends the reductionist program of orthodox neo-Darwinism against the more holistic approaches of such biologists as Stephen Jay Gould and Stuart Kauffman, he concedes that it is impossible to eliminate metaphors (274ff), teleological arguments (282ff), and pragmatic modes of evaluation (286ff) from biological discourse. "Darwinism does not have design built in as a premise," Ruse says, "but the design emerges as Darwinism does its work and some organisms get naturally selected over others" (269).

Ruse professes to find it unproblematic, and even "comforting" (270), that Darwinism both allows us and requires us to pursue an old-fashioned "understanding in terms of final causes" (289). I would like to suggest, however, that there is a greater tension than Ruse is willing to admit between the reductionist program of mainstream physical science and the implications of an unavoidable appeal to purpose, design, and final causality whenever living organisms are in question. Edward O. Wilson's principle of "consilience," for instance, maintains that "all tangible phenomena, from the birth of stars to the workings of social institutions, are based on material processes that are ultimately reducible, however long and tortuous the sequences, to the laws of physics."[4] But if this is correct, then biological explanations of human culture and society of the sort that Wilson copiously proffers are just as dubious a stopgap as are the sociocultural explanations that Wilson rejects. If we are to assert reductionism, then we need to follow it all the way down. We should focus upon the quantum interactions of subatomic particles, rather than upon genes and genomes, or upon organisms and their adaptations to their environment. In that way, we can banish all talk of purposes and final causes. If, on the other hand, we are willing to accord a certain relative autonomy to the biological realm and accept the existence of emergent properties on this level that have a certain explanatory power of their own, then we have admitted teleology and purpose. And by the same logic, we should be willing to accept the explanatory power of higher levels of emergent complexity (social, cultural, economic, political, aesthetic, etc.) as well. There is no good reason for us to stop at the biological and genetic level and assert against all evidence (as sociobiology and evolutionary psychology are wont to do) that everything having to do with human beings was somehow frozen in stone (or in DNA) at the end of the Pleistocene.

Elsewhere, I have criticized the way that devotees of evolutionary psychology,[5] in particular, tend to invoke "purpose," attributed to such reified agencies as "evolution," "natural selection," or "the genes," in much the same way that Paley invoked the intentions of the Deity, as a kind of catch-all principle of explanation. I will not rehearse that argument in detail here. Suffice it to say that if purpose and design have emerged in the course of biological evolution, then it is what Kant calls a *paralogism* to attribute such purpose to evolutionary processes themselves. Similarly, given that a certain biological trait came into being as a result of the fact that organisms possessing that trait were able to produce more viable offspring than competing organisms lacking it, it is a paralogism to infer that the purpose or goal of the trait, as it currently exists, is therefore to "maximiz[e] the number of copies of the genes that created it."[6] The given trait—whether it be a physiological organ or a pattern of behavior—is likely to be purposive in itself (e.g., the eye serves for seeing; the male bowerbird builds a bower in order to attract a mate). But such purposes cannot be equated with the alleged "purpose" of maximizing the transmission and inheritance of the organism's genes. The outcome of a process is not the same as the conditions that led to its existence in the first place. To equate the two is precisely to confuse the "efficient cause" that gave rise to the trait with the trait's concrete action as "final cause." I may well be more "reproductively fit" because I can see, but the purpose of my eyes is seeing and *not* reproductive fitness. Even when (as in the case of the bowerbird's building a bower) we find an actual purpose of a sexual and/or reproductive nature, it is a category error to translate this into overall evolutionary "purpose." However a trait was initially selected for, and even as it continues to be selected in contrast to alternative traits, its actual functionir may well turn out to involve exaptations, repurposings, autonomous strivings, and other violations of strict adaptationist logic.

Like so much in modern thought, this dilemma over biological explanation finds its clearest, and historically most crucial, exposition in the philosophy of Kant. In the second half of the *Critique of Judgment*, Kant formalizes this dilemma as what he calls the Antinomy of teleological judgment. On the one hand, Kant says, we *must* assume that the complex organization of living beings is "produced through the mere mechanism of nature"; indeed, no other explanation is possible. Ever since Galileo, Western science has proceeded on the basis of eliminating teleology and occult properties, and explaining everything in terms of efficient causes. And yet, on the other hand, mechanistic determinism "cannot provide our cognitive power with a basis on which we could explain the production of organized beings." When we try to establish such a basis, we are *compelled* "to think a causality distinct from mechanism—viz., the causality of an (intelligent) world cause that acts according to purposes."[7] For "we cannot even think [living things] as organized beings without also thinking that they were produced intentionally" (281). In other words, we are unable to avoid the idea of purposive design, even though "we make no claim that this idea has reality" (269), and even though such an idea goes against everything that we know and believe about the phenomenal universe. When biologists look at living organisms, they are forced to accept "the maxim that nothing in such a creature is *gratuitous* . . . Indeed, they can no more give up that teleological principle than they can this universal physical principle" (256).

Kant, of course, was writing long before Darwin. It is sometimes argued that Darwin's discovery of a naturalistic basis, or "physical principle," for the organized complexity of life—something that Kant considered to be impossible (282–83)—entirely obviates Kant's arguments about teleological judgment. Yet Kant's Antinomy still exists in contemporary biology, even though its location has been displaced. In the mainstream neo-Darwinian synthesis, any appeal to higher purposes is rejected; natural selection operates blindly, without foresight. At the same time, however, neo-Darwinian explanation depends entirely upon the maxim that "nothing in [living organisms] is *gratuitous*": even the most minute features of living beings are assumed to possess adaptive significance. Selection itself—defined by population genetics as the change over time in the distribution of allele frequencies for a given gene in a given population—operates mechanistically, as the statistical outcome of a multitude of small, contingent encounters. But selection is rendered intelligible, in retrospect, only by means of the "teleological principle" that particular traits have been selected for because they are adaptive. Thus, the theory of natural selection takes away teleology with one hand, but gives it back with the other. The "argument from design" is rejected as an appeal to a transcendent, external cause, but restored as an immanent principle of emergent order.

As Ruse summarizes the matter, even though "there is no reason to think that biology calls for special life forces over and above the usual processes of physics and chemistry," nonetheless "there does seem to be something distinctive about biological understanding—something having to do with purposes and ends in evolution . . . We have 'final causes.'"[8] Ruse concludes, almost in spite of himself, that "Kant was right in seeing that *we* do the science, and *we* try to make sense of the trilobite . . . looking at the trilobite as if it had intentions and interests. As if it had values" (288). For Ruse, as for Kant, the trilobite (even when it was alive) didn't actually have any intentions or interests or values of its own; we can only interpret it by looking at it as if it did and judging it by analogy with our own intentions, interests, and values. For "the aim of science" is not "to give an unvarnished report on reality," but rather "to make sense of reality" in our own terms (288). Starting from entirely naturalistic premises, Ruse ends up finding himself forced to reject positivist reductionism, and instead adopt a Kantian transcendental argument.

The problem with Ruse's argument is his unexamined assumption that only human beings have intentions, interests, and values, while organisms like trilobites (and presumably also cats, fruit flies, trees, slime molds, and bacteria) do not. "The trilobite has no interests," he writes; "it just is" (288). It is only we who comprehend it *in terms of* interests. But isn't this a rather un-Darwinian reversion to the Cartesian idea that human beings alone have souls, while all other organisms are merely things, or machines? Of course, Kant himself already assumes this prejudice. I would like to suggest, however, that a post-Darwinian Kantianism can, and should, drop the anthropocentrism altogether. It should concentrate on the structure and consequences of Kant's transcendental argument, while rejecting the idea that the argument applies only to human (or "rational") minds.

In fact, this is precisely what Gilles Deleuze does, when he converts Kant's transcendental idealism into what he calls a "transcendental empiricism." For Deleuze, Kant's discovery of "the prodigious realm of the transcendental" is crucially important,[9] but limited by the way that Kant's own account of the transcendental "retains the form of the person, of personal consciousness, and of subjective identity."[10] In place of this, Deleuze posits an "impersonal and pre-individual transcendental field," which "can not be determined as that of a consciousness" (102). This amounts to saying that Kant's transcendental syntheses have a general ontological significance, instead of a merely epistemological and psychological one. They apply to the inner being of all entities, rather than just to the way that we apprehend those entities. Deleuze, in short, proposes a Kantianism for trilobites, fruit flies, and birds, as well as for human beings.

Read in this manner, Kant implicitly proposes what Deleuze explicitly develops in *The Logic of Sense* as a theory of "double causality" (94–99). On the one hand, Deleuze says, there is real, or physical, causality: causes relate to other causes in the depths of matter. This is the materialist realm of efficient causes, of "bodies penetrating other bodies . . . of passions-bodies and of the infernal mixtures which they organize or submit to" (131). On the other hand, there is the idealized, or transcendental, "quasi-causality" of effects relating solely to other effects, on the surfaces of bodies or of things (6). This is the realm of intentions or final causes. Quasi-causality is "incorporeal . . . ideational or 'fictive,'" rather than actual and effective; it works, not to constrain things to a predetermined destiny, but to "assur[e] the full autonomy of the effect" (94–95). And this autonomy, this splitting of the causal relation, "preserve[s]" or "grounds freedom," liberating events from the destiny that weighs down upon them (6). An act is free, even though it is *also* causally determined, to the extent that the agent or actor is able "to be the mime of what effectively occurs, to double the actualization with a counter-actualization, the identification with a distance" (161). That is to say, Deleuze's counter-actualizing "dancer" makes a *decision* that supplements causal efficacy and remains irreducible to it, without actually violating it.

Deleuze is really just giving a de-anthropomorphized, nonrationalist account of something that already happens in both Kant's Second and Third Critiques. In the *Critique of Practical Reason*, everything turns upon the gap between the rational subject and the empirical subject, and the corresponding distinction between "causality as freedom" and "causality as natural mechanism."[11] This distinction also takes the form of an Antinomy: "The determination of the causality of beings . . . can never be unconditioned, and yet for every series of conditions there must necessarily be something unconditioned, and hence there must be a causality that determines itself entirely on its own" (69). Kant's solution to this Antinomy is that physical, efficient causality always obtains in the phenomenal world, but "a freely acting cause" can be conceived as operating *at the same time,* to the extent that the phenomenal being who wills and acts is "also regarded as a noumenon" (67).

Kant thus insists that linear, mechanistic causality is universally valid for all phenomena. But at the same time, he *also* proposes a second kind of causality, one that is purposive and freely willed. This second causality does not negate the first and does not offer any exceptions to it. Rather, "freedom" and "purpose" exist *alongside* "natural mechanism": Derrida would say that they are *supplementary* to it. According to the Second Critique, "nothing corresponding to [the morally good] can be found in any sensible intuition" (90); this is precisely why the moral law, or "causality as freedom," can only be a pure, empty form. The *content* of an action is *always* "pathological" or empirically determined, "dependen[t] on the natural law of following some impulse or inclination" (49). The second sort of causality, a free determination that operates according to moral law rather than natural law, may coexist with this "pathological" determination but cannot suspend it. This is why Kant incessantly qualifies his affirmations of freedom, reminding us that "there is no intuition and hence no schema that can be laid at its basis for the sake of an application *in concreto*" (91), and that it is an "empty" concept, theoretically speaking, that can be justified "for the sake not of the theoretical but merely of the practical use of reason" (75).

In the *Critique of Judgment,* Kant similarly solves the Antinomy of teleological judgment by differentiating between the claims made by the two sorts of causality. As normative physical science insists, mechanistic causality is the law of the phenomenal world, the way that things must necessarily appear to us. Purposive (teleological) causality is not altogether eliminated, but it can only be accorded a ghostly, supplemental status. Kant says that "we do not actually *observe* purposes in nature as intentional ones, but merely add this concept [to nature's products] in our *thought,* as a guide for judgment in reflecting on these products" (282). Purpose is "a universal *regulative* principle" for coping with the universe (287); but we cannot apply it constitutively. The idea of "natural purpose" is only "a principle of reason for the power of judgment, not for the understanding"

(289). That is to say, when we regard a given being as something that is alive, as an *organism,* we are rightly judging it to be an effectively purposive unity; but we do not thereby actually understand what impels it, or how it came to be.

The understanding has to do with a one-way "descending series" of "efficient causes," or "real causes." But judgment in terms of purposes invokes a nonlinear (both ascending and descending) series of "final causes," or "ideal causes" (251–52). The idea of purpose, or of final cause, involves a circular relation between parts and whole. The whole precedes the parts, in the sense that "the possibility of [a thing's] parts (as concerns both their existence and their form) must depend on their relation to the whole." But the parts also precede and produce the whole, insofar as they mutually determine, and adapt to, one another: "the parts of the thing combine into the unity of a whole because they are reciprocally cause and effect of their form" (252). An organism must therefore be regarded as "both an *organized* and a *self-organizing* being." It is both the passive effect of preceding, external causes, and something that is actively, immanently self-caused and self-generating. Only in this way can "the connection of *efficient causes* . . . at the same time be judged to be a *causation through final causes*" (253).

To what extent must final causality be taken into account, and to what extent can it safely be ignored? Kant's own answer to this, of course, is starkly binary. Morality applies to human beings as rational agents, and not to any other entities. Teleological design applies to living organisms, and not to anything in the inorganic world; however, if we follow Deleuze's revision of Kant, together with the general tendencies of modern science, we will tend rather to blur these boundaries. It is largely a matter of degree. In many inorganic physical processes, the scope of supplemental causality is vanishingly small; mechanistic, efficient causality accounts for everything that we are able to observe. But in cases of complexity, or of higher-order emergence, supplemental causality becomes far more important. "Deterministic chaos" is, like all empirical phenomena, entirely determined in principle (or, as Kant would say, "theoretically") by linear cause and effect. But because its development is sensitive to differences in initial conditions too slight to be measured, it is not actually determinable ahead of time pragmatically (or, as Kant would say, "practically"). In these cases, linear, mechanistic causality is inadequate for the purposes of our understanding, and an explanation in terms of purpose, or "causation through final causes" becomes unavoidable. This is why Kant's account of teleological circularity and "reciprocal cause and effect" is very much alive today: it lies at the heart of most versions of cybernetics, systems theory, and theories of self-organization.

The role of teleology and final causes is especially crucial when we get to those emergent processes of self-organization known as living things. Where Ruse and other defenders of the neo-Darwinian paradigm defend talk of purpose only as a rhetorical and epistemological necessity, other biological paradigms embrace this mode of explanation more wholeheartedly. In modern biology, Kantian "causation through final causes" can be traced in Bertalanffy's General System Theory, Humberto Maturana and Franciso Varela's theory of autopoiesis, Stuart Kauffman's explorations of emergence, Susan Oyama's Developmental Systems Theory, James Lovelock's Gaia hypothesis, and Lynn Margulis's accounts of symbiotic mergers.[12] None of these approaches dispute the central Darwinian claim that purposiveness in living organisms emerged through the efficient cause of natural selection. But they all insist that mechanistic causality must be supplemented by processes that involve reciprocity and feedback, autoproduction, and some sort of innovation or "decision." In this way, they all resonate with Deleuze's account of double causality, as well as with Deleuze's (and Guattari's) interest in such things as lateral gene transfer, "aparallel evolution," and symbiosis.[13]

Despite their considerable differences, all these approaches implicitly propose a common image of life, one that diverges sharply from that of the mainstream neo-Darwinist synthesis. According to the latter, genetic inheritance, when combined with occasional random mutation and the force of natural selection, is sufficient to account for biological variation. This is because life is essentially conservative, organized for the purposes of self-preservation and self-reproduction. Organisms strive to maintain homeostatic equilibrium in relation to their environment and to perpetuate themselves through reproduction. Innovation and change are not primary processes, but are forced adaptive reactions to environmental pressures. Life is not a vitalistic outpouring; it is better described as an inescapable compulsion. The image of a "life force" that we have today is not anything like Bergson's *élan vital*; rather, it is a virus, a mindlessly, relentlessly self-replicating bit of DNA or RNA. It would seem that organic beings only innovate when they are absolutely compelled to, and as if it were in spite of themselves.

In contrast to the neo-Darwinist synthesis, the alternative approaches that I have mentioned all suggest that, at least to a certain extent, "the nature of selective pressures is creative and active," rather than merely negative and eliminative.[14] Inheritance, modified by random mutation and selection, is a necessary condition for evolutionary change—but not an altogether sufficient one. A supplemental, emergent factor is also required. The best philosophical account of what this might entail comes not from Deleuze, but from Alfred North Whitehead's "philosophy of organism."[15] Whitehead, like Kant and Deleuze, posits a double causality: efficient and final. Efficient, or mechanistic, causality is universal; it applies to all events or "occasions" in the universe. Final causality, for its part, does not suspend or interrupt the action of efficient causality; it supervenes upon it, accompanies it, demands to be recognized alongside it. For Whitehead, the final cause is the "decision" (43) by means of which an actual entity "concresces," or becomes what it is. "However far the sphere of efficient causation be pushed in the determination of components of a concrescence . . . beyond the determination of these components there always remains the final reaction of the self-creative unity of the universe" (47). The point is "that 'decided' conditions are never such as to banish freedom. They only qualify it. There is always a contingency left open for immediate decision" (284). This contingency, this opening, is the point of every entity's self-determining activity: its creative self-actualization or "self-production" (224).

Whitehead defines "life" itself (to the extent that a concept with such fuzzy boundaries can be defined at all) as "the origination of conceptual novelty—novelty of appetition" (102). By "appetition," he means "a principle of unrest . . . an appetite towards a difference . . . something with a definite novelty" (32). Most broadly, "appetition" has to do with the fact that "all physical experience is accompanied by an appetite for, or against, its continuance: an example is the appetition of self-preservation" (32). But experience becomes more complex, aesthetically and conceptually, when the appetition pushes beyond itself and does not merely work toward the preservation and continuation of whatever already exists. This is precisely the case with living beings. When an entity displays "appetite towards a difference"—Whitehead gives the simple example of "thirst"— the initial physical experience is supplemented and expanded by a "novel conceptual prehension," an "envisagement" (34) of something that is not already given, not (yet) actual. Even "at a low level," such a process "shows the germ of a free imagination" (32). Thus, Whitehead's theory of life is also a theory of desire. It is insufficient to interpret something like an animal's thirst, and its consequent behavior of searching for water, as merely a mechanism for maintaining (or returning to) a state of homeostatic equilibrium. "Appetition towards a difference" seeks transformation, not preservation. For Whitehead, life cannot be adequately defined in terms of concepts like Spinoza's *conatus,* or even Maturana and Varela's *autopoiesis.* Rather, an entity is alive precisely to the extent that it envisions difference and thereby strives for something other than the mere continuation of what it already is. "'Life' means novelty . . . A single occasion is alive when the subjective aim which determines its process of concrescence has introduced a novelty of definiteness not to be found in the inherited data of its primary phase" (104). Appetition is what Whitehead calls a "conceptual prehension": it involves the grasping toward, and then the making-definite of, something that has no prior existence in the "inherited data" (i.e., something that, prior to the appetition, was merely potential).

If life is appetition, then it must be understood, not as a matter of continuity or endurance (for things like stones endure much longer, and more successfully, than living things do), nor even in terms of response to stimulus (for "the mere response to stimulus is characteristic of all societies whether inorganic or alive" [104]); but only in terms of "*originality* of response to stimulus" (emphasis added). Life is "a bid for freedom," and a process that "disturbs the inherited 'responsive' adjustment of subjective forms" (104). It happens "when there is intense experience without the shackle of reiteration from the past" (105). In sum, Whitehead maintains "the doctrine that an organism is 'alive' when in some measure its reactions are inexplicable by *any* tradition of pure physical inheritance" (104). Being alive means being able to make unforeseen decisions.

This evidently goes against the currently fashionable belief that everything an organism does is predetermined by its genes, but it fits rather well with some of the most interesting current research. When biologists actually look at the concrete behavior of living organisms, they continually discover the important role of "decision" in this behavior. This is not only the case for mammals and other "higher" animals. Even "bacteria are sensitive, communicative and decisive organisms . . . bacterial behaviour is highly flexible and involves complicated decision-making."[16] Slime molds can negotiate mazes and choose one path over another.[17] Plants do not have brains or central nervous systems, but "decisions are made continually as plants grow," concerning such matters as the placement of roots, shoots, and leaves, and orientation with regard to sunlight.[18] In the animal kingdom, even fruit flies exhibit "spontaneous behavior" that is nondeterministic, unpredictable, "nonlinear and unstable." This behavioral variability cannot be attributed to "residual deviations due to extrinsic random noise." Rather, it has an "intrinsic" origin: "spontaneity ('voluntariness') [is] a biological trait even in flies."[19] In sum, it would seem that *all* living organisms make decisions that are not causally programmed or predetermined. We must posit that "cognition is part of basic biological function, like respiration."[20] Indeed, there is good evidence that, in multicellular organisms, not only does the entire organism spontaneously generate novelty, but "each cell has a certain intelligence to make decisions on its own."[21]

Thus, biologists (Eshel Ben Jacob, Yoash Shapira, and Alfred I. Tauber) have come to see cognition, or "information processing," at work everywhere in the living world: "all organisms, including bacteria, the most primitive (fundamental) ones, must be able to sense the environment and perform internal information processing for thriving on latent information embedded in the complexity of their environment."[22] Organisms would then make decisions—which are "free," in the sense that they are not preprogrammed, mechanistically forced, or determined in advance—in accordance with this cognitive processing. This fits quite well with Whitehead's account of "conceptual prehension" as the "valuation"[23] of possibilities for change (33), the envisioning of "conditioned alternatives" that are then "reduced to coherence" (224). But it is getting things backward to see this whole process as the *result* of cognition or information processing. For "conceptual prehension" basically means "appetition" (33). It deals in abstract potentialities and not just concrete actualities; it is emotional, and desiring, before it is cognitive. Following Whitehead, we should say that it is the very act of *decision* (conceptual prehension, valuation in accordance with subjective aim, active selection) that makes cognition possible—rather than cognition providing the grounds for decision. And this applies all the way from bacteria to human beings, for whom, as Whitehead puts it, "the final decision . . . constituting the ultimate modification of subjective aim, is the foundation of our experience of responsibility, of approbation or of disapprobation, of self-approval or of self-reproach, of freedom, of emphasis" (47). We don't make decisions because we are free and responsible; rather, we are free and responsible because—and precisely to the extent that—we make decisions.

Life itself is characterized by indeterminacy, nonclosure, and what Whitehead calls "spontaneity of conceptual reaction" (105). It necessarily involves "a certain absoluteness of self-enjoyment," together with "self-creation," defined as "the transformation of the potential into the actual."[24] All this does not imply, however, any sort of mysticism or vitalism; it can be accounted for in wholly Darwinian terms. In fruit fly brains no less than in human ones, "the nonlinear processes underlying spontaneous behavior initiation have evolved to generate behavioral indeterminacy."[25] That is to say, strict determinism no longer applies to living things, or applies to them only to a limited extent because "freedom," or the ability to generate indeterminacy, has itself been developed and elaborated in the course of evolution. As Morse Peckham speculated long ago, "randomness has a survival value . . . The brain's potentiality for the production of random responses is evolutionarily selected for survival. As evolutionary development increases and more complex organisms come into existence, a result of that randomness, the brain's potentiality for randomness accumulates and increases with each emerging species."[26] The power of making an unguided and unforeseeable decision has proved to be evolutionarily adaptive. Some simple life processes can be regulated through preprogrammed behavior, but "more complex interactions require behavioral indeterminism" in order to be effective.[27] Organisms that remain inflexible tend to perish; the flexible ones survive, by transforming themselves instead of merely perpetuating themselves. In this way, the "appetition of self-preservation" itself creates a counter-appetition for transformation and difference. Life has evolved so as to crave, and to generate, novelty.

Of course, the sort of "freedom" described by Whitehead, and recently discovered in fruit flies, is not the same as Kant's notion of noumenal moral freedom, or other traditional notions of "free will." Whiteheadian freedom is not absolute in this sense. It is, however, self-generated, and irreducible either to external contingency or to internal predetermination. As the scientists involved in the fruit fly study put it, "free" behavior exists in "the middle ground between chance and necessity," or in "a brain function which appears evolutionarily designed to always spontaneously vary ongoing behavior."[28] Such spontaneous variation may be seen as an empiricist conversion, or a phenomenalization, of Kant's notion of freedom. In Whitehead's account of final cause and decision, as in Deleuze's account of quasi-causality and the virtual, the ghost, or the trace, that the noumenal leaves in the phenomenal world is more an absence than a presence, more a vacuum than a force. If life is a locus of appetition and decision, this can only be because, as Whitehead suggests, "life is a characteristic of 'empty space' . . . Life lurks in the interstices of each living cell, and in the interstices of the brain."[29] Life involves a kind of subtraction, a rupturing or emptying-out of the chains of physical causality. As a result of this de-linking, "the transmission of physical influence, through the empty space within [the animal body], has not been entirely in conformity with the physical laws holding for inorganic societies" (106). These empty spaces or interstices are the realm of the potential, or of the virtual. Interstitial life points toward a futurity that remains open: one that is not entirely determined by the present.

CHAPTER 5

Digital Ontology and Example

Aden Evens

T HIS CHAPTER PRESENTS two incompatible ontologies of the digital. The first ontology contrasts the digital with Deleuze's notion of the virtual: whereas the virtual is creative and fecund, the digital is sterile and hermetic, precluding creativity. The second ontology describes how the digital is (nevertheless) creative: by virtue of the *fold* in the digital, a subtle but crucial feature of digital ontology, the digital reaches beyond its flat plane to connect to the human world. In the fold, described by way of an extended example, the digital and the virtual thus overlap. These two ontologies are both essential to the digital and both are valid, but their simultaneous assertion is paradoxical. The paradox of these two ontologies is the political and aesthetic problem of the digital.

As against the frightening fantasy of a world become virtual, theorists since the late 1990s have asserted the digital's materiality: The digital is necessarily inaccessible in its pure abstraction and must always be apprehended through a material instance.[1] This materiality cannot be dismissed as incidental, for the digital generates a copious material history in its artifacts.[2] Any understanding of, any relation to these artifacts requires an embodied human being to mediate the passage from abstraction to sensation.[3] So to engage with the digital is never to leave this human body entirely behind.[4] The digital is inextricably material.

Though digital materiality is essential, it is also deeply problematic.[5] The digital is not just a particular case of the visual or cultural; it warrants its own ontology. Its distinction from other material histories lies in the abstraction at the heart of the digital, a constitutive abstraction that *persists in its material*. The abstraction of *form*, 0 and 1, remains at the surface, haunts the phenomenon of the pixel, the icon, the document. To use a word processor is to interact with a form of text; to use a graphic arts program is to interact with a form of image. To play a computer game is to conform to its logic, its formal structure. The digital may be material, but it is a material become digital.

The digital attempts to push the resistance of its material to a margin of irrelevance, to overcode the material via its apparatus of capture, the binary code. This erosion of the material constitutes much of the history of the digital. The computer chip shrinks, struggling against the material (heat, power, size, speed) that will not finally relinquish its hold over computer processing. An immaterial chip would be a node of pure programmable logic, not a Turing machine but *the* Turing machine, and this fantasy of the perfect computer drives the high-tech industry and shapes our relationships to digital technologies. Hard-drive storage grows larger and faster, the capacious and speedy a never-quite-adequate substitute for an infinite and immediate memory. The computer offers to the user a representation of a digital object, but ideally this representation would be fully equivalent to the data behind it, as if sensation could connect directly to form. The representation is a necessary prosthesis for the digital, an inconvenient and always too clumsy means to the end of access. What the user works *on*, after all, is the form of the digital object; every operation at the computer is a specific, determinate intervention to alter one or more formal characteristics of the data. What action do you take at your terminal that does not have a particular predetermined goal? This is the distinguishing feature of digital ontology: to work with the digital is to confront a material idealization, an abstraction made flesh.

If the digital is an abstraction incarnate, if it informs the material and gives it its sense, isn't this also true of Deleuze's *virtual*? This similarity is false, however. Like the digital, the virtual is abstract, and it persists in the actual to which it gives rise and in which it makes sense. But the power of abstraction (and the nature of sense) in these two ontologies are entirely different. In chapter 4 of *Difference and Repetition,*[6] Deleuze details an ontology of the virtual in terms of the problematic Idea: the virtual is the site of ontogenesis, where the Idea forms. In the beginning (and ever after), there is a throw of the dice, an aleatory point spewing difference; and the pips of the dice are differentials. Differentials (dx, dy), undetermined in themselves, obtain relations in which they are reciprocally determined (dx in relation to dy, or dy/dx). Reciprocal determination motivates a complete determination of the ordinary and the singular, a specification of an actual solution concretized in time and space. From the indeterminate to the reciprocally determined to the complete determination, the Idea proceeds from problem to solution, making sense of the world. Though he borrows the symbols of differential calculus, Deleuze emphasizes (179) that problems are not themselves mathematical (or biological, economic, sexual, etc.); only solutions are so categorized. Problems are genuinely undetermined, belonging to no particular domain. They are pure dynamism that refuses determinacy, escapes category, and defies character. This dynamism persists in the actual, the concrete forms of solution to which the Idea gives rise. Negation, proposes Deleuze, occurs only when the actual is cut off from the virtual, when its genesis ceases to propel it, so that it appears static, positive, and definite.

The digital, too, is a source of the actual, which begins to explain the popular confusion of digital and virtual, but these terms are, rather, opposed and complementary, each one emphasizing what the other most excludes. The virtual denies all form, all representation; it is an action, a production but not a product. For its part, the digital is entirely form, maximally determinate. Every 0 is precisely 0 and every 1 is precisely 1, and the digital has nothing but these 0s and 1s, no ambiguity, no indeterminacy. Even the dynamism of action in the digital is reduced to a static representation, a procedure captured and codified by the formal tools of the digital; code (program) is just more data, more information. Where the digital is entirely *formal,* the virtual is inscrutably *problematic.* Problematic, the virtual is neither fixed nor final. Problematic, the virtual is urgent and insistent. Problematic, the virtual is creative. Table 5.1 summarizes the complementarity of virtual and digital. Attention to these contrasts (in the subsequent paragraphs of this section) will help to define the unique ontology of the digital.

Deleuze takes pains to distinguish the virtual from the possible. A possibility is a fixed and definite state of things that differs from the actual only in that it does not (yet) actually happen to be the case, a modal or formal distinction. When the possible becomes actual, it is like flipping a switch: what was before possible is now also actual. But in changing from possible to actual, it does not otherwise change. In stark contrast, the virtual is the site of a progressive generation, from problem to solution, from undetermined to determination. Problems and solutions do not resemble each other, and a problem develops as it takes shape. Deleuze marks numerous stages of this ontogenesis: differentiation, differenciation, dramatization, individuation, etc.[7] The actual is a surface effect, the impression that remains when the generative process has passed along. Nothing like the

virtual	digital
problematic	formal
accidental/aleatory	possible/necessary
undetermined	determinate
universal/singular	general/specific
dynamic	static
insistent	indifferent
decentered	deliberate
genital	sterile
confused	discrete
perplicated	hermetic
creative	representational

Table 5.1. *The virtual and the digital are opposed but complementary.*

possible, the virtual is different in kind (and not just in mode) from the actual. Deleuze emphasizes that the virtual, as opposed to the possible, is *accidental* and *aleatory*.

That the virtual is aleatory means that it is irreducibly, originally creative. The aleatory defies any rule, for it is the source of rules, a fount of spontaneity. Similarly, the virtual is accidental because it follows no plan and expresses no essence or internal image. There is no preconception in the virtual, only a working out, a working through. It proceeds by accident, its essence is accident in that each step, each choice is without prejudice, without forethought, spontaneous.

The digital could hardly be more different. The exclusion of accident is a top priority in the digital, and this project is remarkably successful. Digital technologies operate with astonishing regularity and accuracy, doing exactly what they are programmed to do nearly one hundred percent of the time. (This is not to say that computers always do what the user desires; rather they follow the rules established in their hardware and software architectures, which may sometimes be flawed.) A digital process is never aleatory, and it is in fact impossible to simulate the aleatory. Rather, the digital is *determinate*. Every operation proceeds according to a fixed rule calculable in advance. The data themselves are also wholly determinate; every 0 is exactly 0, every 1 exactly 1. There is no excess, no character, no "kind" of bit.

Determined in advance by a rigid and calculable logic, the digital displaces accident or contingency in favor of *possibility* and *necessity*. Every result on the computer is a necessary one, logically implicit in the design of the hardware and software and the values of the particular inputs. Every input has been anticipated in advance, and every output is a necessary consequence of some input. To draw a line is only to designate a line already drawn; to choose which key next to press is to pick from a menu of key-presses. In other words, every move the computer makes is already there as a possibility; the inputs to the program (and thus the necessary outputs) are determinate before they become actual. As Brian Massumi puts it, "Digital coding is possibilistic to the limit."[8]

That the computer is the dominion of possibility is not only a technical claim about its hardware and software design. Even from the user's perspective, the computer presents itself as a field of determinate possibilities. To use a computer is to engage *deliberately* and explicitly with form, to actualize possibilities. Every operation is performed with a specific goal in mind. Success is generally measured by a criterion decided in advance of the operation. A programmer writes a line of code in order to accomplish something predefined and determinate; as she codes, she could describe precisely what she intends each line to do. Though she may be uncertain while coding whether a given algorithm will function properly, the criteria that measure proper functioning are fully determined before the code is written. For "end users" as well, actions at a computer are entirely deliberate; each click, each gesture intends a predictable response. Digital technologies are the ultimate instruments; the computer aims to render transparent the passage from means to ends.

But the virtual is never an instrument. One cannot employ the virtual, for conversely *the virtual seizes you,* exercising a power of *fiat* (197), striking like a bolt of lightning.[9] To engage the virtual is not to predict but to experiment; the indeterminate at the ontogenesis of the virtual poses a problem, an urge. Those who play a role in the problem are changed thereby, but the reward for this subjection is a stake in the creative. Where would the digital, with its perfect predictability, leave room for a genuine creativity?

It may be objected that while os and 1s are ideal, discrete, and perfect, their material instantiations can never be so perfect. Materially, in a running computer, a 0 or 1 is not an absolute and discrete value but a range of voltages or magnetic field strengths or the length of a pit carved into a reflective surface. While it is true that material instantiations of the digital are not precise in the way that digital bits are, these materials are designed to minimize their fuzzy or accidental aspects, so as to allow the abstraction to manifest. The range of voltages across which a given reading will count as a 0 is designed to ensure that in practice almost all material os in the computer chip will fall squarely within that range, curtailing ambiguity.[10] Materials and their properties are selected precisely for their abilities to yield to abstraction, to give themselves over to the idealization that overcodes the concrete material.

So extensive is the overcoding of the material by the digital that common characteristics of the objective or actual world lose their traction in the digital. The digital leaves intact neither copy nor original, offering instead only the generic, the reproducible. (The law is thus ill-prepared to accommodate the ontology of the digital. Legal and economic systems, developed around singular objects, now struggle to incorporate the digital, rearranging their own codes accordingly.) Can one even identify such a thing as a "single" digital object? A single material instance of a digital object is not the same as a single digital object, for the object is in principle abstract. Thus, there is no question of singularity in the digital. If we take seriously the abstraction of an object, then there is no count, no cardinal reckoning of the digital. One digital image is as good as ten or a billion such images. A copy of your file is functionally identical to the "original" file. It is not just that the material costs of reproduction are so low. This plenitude of digital resources and ease of digital labor are only the symptoms of the prior ontological condition: existence is incidental in the digital. (How many number 4s are there?)

A digital object lacks the singularity that stubbornly adheres to any material object, pinning it to a history and a location. With no material uniqueness, no history, a digital object is not particular but *generic* and *specific*. A digital object is a representation, impeccably reproducible as many times as one desires. Each reproduction, each representation will be an ideal copy, an instance that perfectly satisfies the code that gave rise to it; the digital is wholly general. It is also perfectly specific because it is one pattern of os and 1s and not any other pattern. Indeed, it is the exemplar of specificity, differing demonstrably and definitively from every other digital object.[11]

Again the virtual differs point by point from the digital. Perhaps we aren't even tempted to ask after a virtual object, as the notion of an object is already antipathetic to the process that is the virtual. The virtual answers to the digital's generality and specificity with *singularity* and *universality*. The virtual is singular inasmuch as its process advances in relation to singular points whose reciprocal interactions work out a problem and so constitute a solution. This aleatory process is radically undetermined; anything might happen. Thus the virtual cannot be mapped, cannot be specified in advance of its production, for singularity does not admit the specificity or generality of maps or conceptual structures.

And the virtual is also universal. All of the actual derives from the same, univocal perplication of differentials: "In a certain sense all Ideas coexist."[12] A given problem might distinguish itself from others by raising itself to a level of clarity, generating a solution that establishes its integrity, drawing together certain singular points and placing others at a greater distance. But the articulation of this problem/solution necessarily includes these other differentials, a perplex of them, such that the only origin is the univocity of difference, the aleatory point or dice throw. That is, every problem includes other problems, and this complex entanglement is the condition of the problem, it is what is problematic about the problem.

In describing ontogenesis as problematic, Deleuze is not referring only to the dynamic or processual nature of ontogenesis; after all, even a computer program runs through a process to get from beginning to end, and this logical sequence of fixed states is not necessarily problematic. Distinguishing the virtual from a mechanistic process, Deleuze explicitly gainsays the notion that the problem contains a solution as its complement. A problem in the virtual is not the shadow of its solution, a solution rearranged into an interrogative form; instead, it is a problem because *there is no solution* as yet, because the problem isn't clear enough, isn't articulate enough to generate its solution. The problem does not lack its solution, as though the solution were a possibility, predetermined and ready to be switched on. Rather, the problem has an urgency, it *is* this urgency, a nonspecific demand. Two dancers, strangers to each other, meet on the dance floor; how will they relate, where will they touch, where will they move? There is not enough energy in the batteries to support the typical demands of life-support systems and engines in the earth orbiter—what to switch off, what trajectory to follow, what to jettison? These problems have no solution but must generate solutions in themselves, determining their own boundaries through a play of tension and release, conservation and expenditure, despair and elation. Thus, the problem is desiring, but it is a desire without object, a desire that eventually gives rise to its own object. The virtual is thus willful, it insists, and this *insistence* drives onto-genesis, producing sense in things.

Complementary yet again, the digital bypasses will in favor of an utter *indifference*. The digital plods along, one calculation at a time, and it proceeds robotically, without urgency, without desire, arriving mechanically at whatever result is logically implicit in its structure. Each output is either a 0 or 1, and these have no substantive value in the digital. The digital will lend itself to any calculation, serving an evil master as heedlessly as a beneficent one, mindlessly performing mind-bogglingly dull operations for countless years, off or on without so much as a shrug. Where the virtual produces a variegated or textured time and space by generating the sense of the world, the digital proceeds without regard for time and space. It will calculate just as well and just as readily for a thousand years as for a second, and its component parts can be spread out across the universe or compressed into a square millimeter of silicon surface. In short, the digital offers no *resistance,* does not push back or exert any sort of desire. It simply obeys.[13]

In Deleuze, the virtual is fecund or *genital,* an autochthonous source of novelty; it has no being but the becoming of the new. Every event in the virtual generates not just a new object (one more) but a whole new dimension, a new kind. Virtual production is unpredictable not because it is random or without pattern, but because it produces what was inconceivable hitherto, a new thought in every moment. By contrast, the digital, working from only the neutered difference of 0 and 1, produces nothing outside of itself. Indifferent and pliable, lacking a will of its own, the digital is effectively *sterile.* Thus, it is also *hermetic,* failing to reach beyond its own surface. Instead of a new dimension at each production, the digital gives rise only to one more bit, extending in the same direction with the same difference every time.

And this points to perhaps the starkest contrast of all: whereas the virtual is *creative*, is nothing but the fact of creativity, the digital appears incapable of even the smallest creation. Creativity relies essentially on accident, for the calculable can only produce the already known. Creativity depends on resistance, for what bends entirely to the will becomes only what the will has already conceived. Desire would never be creative if it were entirely fulfilled. Where would creativity find its moment in a system where the output is necessarily and logically implicit in the input?

Consider again the relationship between the digital and accident. The abstraction constitutive of the digital code repels accident in two ways. First, by eliminating every ambiguity, the digital erases the context for accident. Mechanisms ensure that a straying bit will be brought back into the flock, defining and enforcing tolerances that interpret a *range* of voltages as a 1 (or 0), and anticipating even wider strays with a system that flags voltage readings outside the nominal ranges of 0 or 1 and triggering a fallback plan (such as a reread of the data or an error message) for such an event. With absolute 0 and absolute 1 as the only possible values, accident doesn't have much territory to work with.

Second, to the extent that the computer deviates from its expected or intended behavior, the results are usually pretty uninteresting. The material resistance of the hardware substrate that actualizes the digital occasionally interferes with the proper functioning of this material, as when a broken cable or a speck of dust on the drive platter prevents a clear reading of the data. But these deviations from expected behavior typically just result in a computer that fails to operate at all. While a program crash or system crash is not without consequence, it is not generally a productive or creative consequence.

Why is it that a material accident in the computer so rarely produces an interesting result? Because the digital code must anticipate every possible input, a faulty input "does not compute." The digital as an abstraction remains essential to the operation of the machine throughout the various subsystems that constitute the computer. When the digital lapses, when it collapses into the actual, the machine is simply unable to continue its process. The resting state of the analog or material equipment is blank or chaotic. The state that the computer enters when it malfunctions is completely disconnected from the meaning or sense of the data that its correct or normal process would operate on. Any small departure from anticipated form immediately becomes nonsense. "A single error in the components, wiring, or programming will typically lead to complete gibberish in the output."[14] Thus, when the computer malfunctions, it rarely produces an interesting accident.

A material accident is materially related to the sense of its context. A painter who spills paint does so in a location near the area of canvas she was working on and using colors from the palette she was holding. A wrong note on an instrument is in the same temporal position as the intended note, and may have many of the intended timbral characteristics and dynamics as well. Mistaking milliliters for centiliters in a chemistry experiment is likely to produce only an inert concoction, but the material conjunction of substances in unintended ratios may possibly reveal undiscovered properties of those substances. Even a car accident brings together people, metal, plastic, insurance companies, and other entities that otherwise would not have met, and the rich context of any human life ensures that this encounter could produce resonances, connections that are certain however unlikely they may seem.

Accident on a computer does not invoke such rich contexts, fields of connection. The digital code connects only to what it is designed to connect to. Despite the proliferation of standards (busses, protocols, etc.), the very specificity of these standards assures that a deviation in the data is likely to violate the standard itself, rendering the data unreadable, uninterpretable. (One might expect that common standards would mean a greater possibility of connection by accident or connection without too much prearrangement, and this is true. But at the level of a typical computer malfunction, an accident tends to undermine those standards rather than just altering the data they constrain. Given the flat or hermetic nature of the digital, the standard becomes part of the data themselves; there is no definitive separation between administrative information and "real" content, making the standard just as vulnerable to catastrophic and unproductive accident as are the data that it informs.)

Importantly, this cleavage of the digital from the virtual, this point-by-point complementarity and opposition between digital and virtual demonstrates that the challenge to creativity in the digital is not a *de facto* but a *de jure* limitation. According to its popular conception, the shortcoming of the digital is a technical inadequacy, a resolution insufficiently fine to capture the subtlety of the real world: it is thought that as digital tools gain additional memory, become faster, divide time, space, and quality into ever smaller samples, eventually the gap between the actual and the digital will close (or at least become insignificant). The foregoing analysis of the ontology of the digital shows this optimism to be mistaken. The digital falls short of the actual in principle, not because it lacks an adequate resolution but because unlike the digital the actual retains the virtual insistent within it. The virtual lends to the actual its indeterminacy, a leading edge that is unspecified and unspecifiable, for it is undetermined in itself and not just in need of a closer examination. The actual is surrounded by a halo of indeterminacy, it floats over roiling depths of creative production that cannot in principle be captured by a formal code. This rumbling beneath the actual has no dominion in the digital, does not get a foothold there.[15]

If these contrasts with the virtual jeopardize the digital's relationship to creativity, they cannot simply be strategically removed or bypassed, for these same facets of digital ontology are the wellsprings of the digital's extraordinary and unique capability. Only a total abstraction imparts to the digital such an unflagging consistency, for the abstract does not operate by accident. To introduce or amplify ambiguity in the operation of the digital might soften the sterile rigidity of possibility and necessity, but only at the cost of the digital's powerful reliability; no longer confined to the realm of the possible, the digital would also fail to be transportable, for it would behave differently according to context, to climate, to contingency. Only as maximally abstract can the digital strip itself of all character, inviting the impression or inscription of any material whatever. If the digital is a universal machine, capable of simulating any other medium, this is only because it gets out of the way, imposing no character but the pure abstract opposition between 0 and 1, the ultimatum of form. The phenomenal breadth of application of digital technologies is possible only because the digital minimizes the resistance of the material, allowing it to function across a broad range of media and to accommodate spatial and temporal constraints that other technologies cannot stand up to.

This leaves creativity as a significant challenge in the digital. Determinate, without accident, lacking resistance, generic in principle, the digital does not exactly invite novelty; nevertheless, the digital, at least on the surface of things, is the most fecund medium in history. Its ubiquity, born of its abstraction, rivals writing, speech, and money, and it is involved in countless areas of human creative endeavor. An increasing amount of human desire takes the form of the digital somewhere along its path of expression. (Think not just of art-making but of communication, commerce, leisure, etc.) But, maximally abstract, the digital is moribund, hermetic. Its "existence" is only a rarefied idealization, a set of numbers that are impotent without an extensive material apparatus of transduction, rendering that rarefied code into sights and sounds.

To this point, my description of the ontology of the digital is patently flawed, as it fails to explain (and indeed outright denies) the manifest creativity of digital technologies; this account of digital ontology must therefore be missing something fundamental. Massumi proposes that the digital is open to creativity but only when it is no longer digital: "The digital, a form of inactuality, must be actualized. That is its openness."[16] While Massumi's polemic against the impotence of the digital prior to its actualization is accurate and perceptive, it nevertheless moves too quickly, for it treats the transition from digital to actual much like a transition from possible to actual, as though the digital were suddenly actual, as though the form captured in the digital code appeared to the user in a fell swoop. In fact, the actualization of the digital is a process, not an unaccountable change of state, and only a close examination of this process can reveal how the digital might become effective. How does the digital reach beyond itself to make a contact with a human world? The digital is creative, abundantly so, and must therefore overlap at some point with the virtual. Where do digital and virtual meet? It is not enough to point to the materiality of the digital, even though this materiality is essential; for digital materiality is overcoded, dominated by the abstraction that squeezes out creativity. It is also inadequate to note that the digital is employed by human beings who are themselves seized by the virtual; the question of creativity stands, for we must discover how a realm of pure form can accommodate a desire that is rooted partly in the virtual. Form does not seem to make room for the expression of desire. There must be some transitory event, some ontological structure by which the digital reaches beyond itself to avail itself to human and material ends. If the digital in isolation is hermetic, sealed up into its own ideality, laminated onto a plane that renders everything level, then the structure that describes its contact with the actual would be a *fold* or pleat in that plane, a wrinkle that gives the surface of the digital enough texture to engender a friction with the actual, to perturb the human world. *Anywhere the digital meets the human, anywhere these worlds touch, there must be a fold.*

Example: the interrupt. The interrupt exists to address a problem of computer engineering: computers pay no attention; hermetic, the digital has no context, no permeable border surrounding the logic that constitutes its processes.[17] A computer's operation is based primarily in its central processing unit (CPU), its *chip*. The CPU processes instructions in sequence, one after another. It has only these commands in sequence, its entirety is the execution of these commands.[18] Thus, the computer is effectively single-minded (or null-minded). It is wholly occupied with the task at hand, entirely consumed by the operation being executed. Centered in its CPU, the computer is equivalent to the execution of a given command, followed by the execution of the next command, and so on. Commands are the basic units of a computer's process, the lowest layer of abstraction prior to the point at which the operation of the computer can only be understood in terms of hardware, electrical impulses passing along material pathways and the junctions at which they meet.

At this low level, an individual command is a number, an ordered sequence of bits that enters the chip as electrical impulses on small wires specifically devoted to that purpose. Different chips are distinguished by different sets of commands that they can accept, a given set of commands constituting a *machine language*. Each machine language designates a model of computer chip, and the name of the language is generally also the name of the chip that it runs on: 8088, 6502, G5, Xeon, etc. (There are also families of chips where each member of the family works with a similar but not identical language.) Generally, when programmers write in machine language (which is increasingly rare), they use alphanumeric mnemonic codes (assembly language) that stand for the strictly numeric codes of the machine language, just to make the code easier to write and read. (For example, if the number 01100100 is the numeric code for a multiplication operation, the programmer might type into the computer the more easily remembered alphanumeric code MULT, which prior to execution gets translated into the binary number 01100100 by a piece of software called an *assembler*.)

When a given command is fed into the CPU, it enters the CPU as a binary number, a sequence of 0s and 1s that are the binary representation of the numeric machine code. (Even when a programmer works directly in machine language, she would typically read and write the numeric codes not in binary [base-2] but in hexadecimal [base-16], which is much easier to scan visually: 01100100 would be represented as 0x64, where the *zero-ex* is one of many conventions to indicate a hexadecimal representation.) These os and 1s are flows of electricity that trigger the opening and closing of junctions on the circuitry of the chip, thus allowing other flows of electricity to flow (or not).[19] The binary number (command) imposes a logic on the chip, a series of gateways that, admitting or blocking the flow of electricity, eventually produces another binary number at the other "end" of the chip, the output. This process of reading in a number that controls the flow or stoppage of electrical impulses is the entirety of the CPU's process. There is no place in this scheme for something like attention. Utterly occupied by the execution of the current command, the computer does not attend to its surroundings.

It is largely by not paying attention that the computer gains the speed and consistency that constitute much of its fantastic utility. The commands that the computer executes are not themselves complicated; they are things like "multiply two small numbers together," or "shift a row of os and 1s one place to the right," or "copy the integer in one memory location into another memory location." These commands, the basic operations of the CPU, are tasks that any ten-year-old could perform. But a ten-year-old would surely become quickly distracted, as her horizon of attention extends well beyond the task at hand. (Just having a body ensures that one is already distracted by one's immediate surroundings. In fact, one is already distracted by one's body, with its various needs and its autonomous and autonomic desires and habits.) Lacking a body, lacking a world, the computer's strength is its ability to perform simple tasks with astonishing rapidity and perfect accuracy. Millions of times per second the CPU can multiply two integers together or copy a number from one location in memory to another. And it (practically) never makes a mistake; instead, the computer does *exactly* what it is programmed to do, for years at a time, billions and trillions of operations in a row without error.

The commands that the computer executes are organized in sequence and packaged into discrete functional units or *algorithms*. And algorithms usually take the form of loops, sequences of commands that are executed in order, to begin again when the end of the sequence is reached. The loop does not cease to run until its termination condition is reached, which might be the achievement of the desired result (for example, an algorithm to find the prime factors of a given number would not terminate until it had discovered all of these factors), or it might be some fixed number of iterations (for example, an algorithm that processes calendar data might execute twelve times, once for each month), or it might be some other condition (for example, a loop might continue to execute until a given time or until an error occurs).

Locked into its loop, the computer has no mechanism by which to recognize that the user wants its attention. Looping indifferently and incessantly, how would the CPU undertake a new project, head in a different direction? The CPU just executes one command after another. What happens when one needs the computer to perform a task other than the one currently being executed? There are a number of solutions to this dilemma.

First, one could simply allow the computer to finish the task at hand, reach the termination condition of its currently executing loop, only then directing it toward the next task to be performed. Computers are very fast, so it probably won't be too long of a wait. This effective strategy is frequently employed in software design. But it leaves something to be desired. For instance, what if the algorithm currently being executed tends to repeat many times before reaching its termination condition? While some loops terminate in milliseconds, others may continue for seconds, hours, or even days. In fact, there is no guarantee that an algorithm need ever terminate.

A second solution compensates for the inadequacy of the first: write the code so that each pass through the loop checks whether the user is pressing a key or holding down the mouse button. If so, read in and act on the user's new command, suspending indefinitely the operation of the currently executing loop. When the user presses a key on the computer keyboard, this closes an electronic circuit, allowing electricity to flow. This flow of current passes through the USB chip inside the keyboard,[20] which reacts by sending a digital message (usually in the form of electricity) through the wire that connects the keyboard to the computer. The USB chip on the computer's motherboard receives this message and appropriately interprets the message as a key-press. The USB subsystem on the motherboard then adds this key (or, rather, the numeric code that represents this key-press) to a buffer, a digital queue of keystrokes held in memory, waiting for the CPU to access these data. In addition, the USB subsystem sets a flag, that is, it changes a value stored in some prearranged memory location (from 0 to 1, for example), to indicate that there are data waiting in the queue.

If the loop executing in the CPU includes code to check whether the user is pressing a key, this is tantamount to code that checks whether the flag has been set. A set flag triggers a branch to some other algorithm that reads in the data from the queue and acts accordingly. This is what happens in the principal mode of operation of a word processing program, for instance, where the default loop checks whether a key has been pressed and, if so, reads in the data and updates the display to reflect this new input (by making the appropriate character appear at the insertion point on the screen, for instance).

But this method, too, though frequently employed, has a significant problem: remember that while many loops execute only a few times before reaching their termination conditions, others only terminate after millions of executions. And some algorithms perform such basic operations that they are triggered all the time, thousands of times per second. If each algorithm included code to check whether the user-input flag has been set, the computer would waste a lot of processor cycles checking for input that, for the most part, is not there.[21] The result would be a computer that feels slower, possibly a lot slower, because the computer would be wasting resources to check for nonexistent user input instead of using those resources to work on the task at hand.

The interrupt is a third solution that trumps the other two. It is generally engineered into the system at a low level, inscribed in the hardware of the machine. It acts as a direct line of transmission, a *hotline* to the CPU of the computer. When the user triggers an interrupt (say, by pressing a key combination or even a special key specifically designed to trigger an interrupt), the CPU, no matter what operation is currently executing, abruptly stops what it is doing and begins executing a new algorithm determined by the interrupt.

Your computer has a number of interrupt mechanisms. Pressing CTRL-ALT-DEL on a Windows machine or OPT-CMD-ESC on a Mac triggers an interrupt, which is why these key combinations tend to function even when the running program has become unresponsive. (Built into the hardware, the interrupt does not rely on the proper operation of the currently executing software.) Though it is an important part of the design of many aspects of modern computers, the interrupt is not itself technically remarkable. But this brute technique, a jolt to the CPU, demonstrates a significant feature of the ontology of the digital, showing how the digital escapes its own flatness to make a contact with the actual world. Consider these four traits of the interrupt that mark its status as example.

(1) An increase in the number of dimensions.

Topologically, the interrupt represents an increase in the number of dimensions of the computing context. Whereas the algorithm operating within the CPU is effectively a line or a closed loop, the interrupt implies a perpendicular or transverse direction that cuts into this loop, redirecting the computer onto a different path unrelated and indifferent to the one the computer has been treading.

(2) Hierarchy.

But the implicit additional dimension is not just a mirror or copy of the plane that the CPU was already traversing. A second characteristic is that the interrupt, carving out another dimension, instates a hierarchical distinction, asserting the priority of the new dimension over the old one. The operation executing in software is curtailed and the CPU redirected based on a message sent through hardware, and thus the hardware takes priority over the software, at least for the moment of the interrupt. No matter what process was executing at the CPU, no matter what algorithm is running, the interrupt trumps it, takes priority, asserts an absolute, unassailable dominance.[22]

(3) Inside/outside distinction.

The difference between a process taking place in software and a redirection command sent through hardware implies a specific topology: inside/outside. The interrupt itself comes from an outside of the process, a human world beyond the self-contained hermetic world of the CPU. A process operating within the material-ideal inside of the machine gets terminated by a message sent from without. As long as the CPU plods from one command to the next, there may as well be no outside, no world beyond the computer itself. But the interrupt implies an outside, a realm of contingency that disrupts the hermetic indifference of the linear process of the CPU.

(4) Enfolding.

Most dramatically, the interrupt enfolds logically, temporally, and spatially distinct domains, so that they overlap or connect to each other. Logical folding is evident when we examine the operation of an interrupt as compared to the normal termination of a given algorithm. When an algorithm terminates normally, it generally "cleans up" after itself, zeroing out certain registers, deallocating used memory, setting flags to indicate successful termination, saving files that have been altered, and returning a value to the subroutine that called it. But when an algorithm gets interrupted, these housekeeping measures are preempted; the interrupt is comparatively brutal, wresting control of the CPU from the process underway and not even giving it time to grab its stuff on the way out. This has the effect of destabilizing the entire system to some degree because the bookkeeping chores ordinarily performed when an algorithm terminates maintain the computer in a predictable and normal state, the condition anticipated by the various subsystems and subroutines that operate within the machine. If memory is not deallocated and no value returned to the triggering subroutine, the overall system may be more likely to crash in the wake of an interrupt.[23]

Moreover, the interrupt folds the human time of the operator into the inhuman time of the computer. For the computer, time is effectively indifferent, undifferentiated. The computer operates according to a clock that doesn't so much keep time as mark off successive clicks. A process might take twenty milliseconds or two centuries; the computer addresses it exactly the same way in either case and doesn't distinguish between a short time and a long time, between morning and evening, between summer and winter. Human beings have a rather different experience of time. For us, it often matters whether a given event happens now or now. Important, even life-altering events may depend on whether a task is accomplished in two seconds or two minutes. (Thirty seconds late for a bus may change the course of a career. A fiftieth of a second may decide the outcome of a sporting event. On eBay, the action happens in the last five seconds of the auction.) The interrupt represents the imposition of human time on the undifferentiated time of the machine. By triggering an interrupt we have an opportunity to coax the computer into adhering more closely to our time, and we guide the computer to behave according to the vicissitudes of human existence.

(5) Impermanence.

There is a fifth characteristic, distinguished from the other four because it refers to something different in kind. The interrupt, like other folds in the digital, is momentary or fleeting. It is not a constant or abiding event in the digital but a threshold, a boundary between two other states. As soon as the interrupt takes place, the computer returns to its "normal" operation, the CPU simply begins to execute a new algorithm (as determined by the interrupt), and the system continues as though nothing has happened. The interrupt does not last, does not cause an evolution of the system, but only a hiccup, an aberration, a flash in which the system makes a contact with its outside. (Massumi refers to "the fleeting of the virtual.")[24]

I have chosen to describe the interrupt using the example of a human operator getting the "attention" of the CPU when it is executing an operation that does not include a provision to pay attention. In fact, this is only one example of an interrupt and, though illustrative, probably not the most typical. Interrupts are frequently used internally, within the machine itself, as when a particular subsystem needs to alert the CPU to an emergent condition. For example, when an I/O subsystem (a disk driver, for instance) has a buffer full of data, it lets the CPU know about this condition using an interrupt, so that the CPU can deal with the full buffer before going on to process further data. Systems with multiple CPUs (or multiple CPU cores) may employ interrupts as a mechanism of queuing: when one CPU wants access to a shared resource, it triggers an interrupt so that the other CPU knows to relinquish this resource, making sure that the resource is in a safe condition to be handed over. (This sort of interrupt is not so brutal as the ones I have been describing to this point because the interrupted CPU gets a chance to tidy up before passing control of the resource over to the interrupting CPU.) In another interesting application, interrupts can be used to maintain security. Access to certain data may be restricted to subroutines that operate at a given "layer" of security; while a higher (less secure) layer may not be able to access these data directly, this higher layer can trigger an interrupt so that a lower layer algorithm gets called to process the data without anything (but the interrupt signal) passing back and forth across the layer boundary. Note that these interrupts, internal to the computer, don't directly involve a human user, but still exemplify the four (or five) key characteristics described above. At bottom, these interrupts still constitute a crease in the topology of the digital; they cut into the flat plane of the operation of the digital, giving it a texture or fold that reaches beyond the digital to afford a contact with the actual.

Increasing dimensions, hierarchical distinction, the auto-organization of an inside, and an adjunction of disjoint times and places, these are the symptoms of rupture in the digital plane, the ontologically subtle and potent tools that allow the digital to exceed its own borders. These are the features of the fold. The chief claim of this chapter is that the fold (exemplified by the interrupt) is *the* mechanism by which the digital connects to the human and the actual; only by virtue of the fold does the digital become creative. And the fold is not altogether outside the digital; as the example of the interrupt demonstrates, it bridges the gap between digital and actual. This process of connection is critical if we are to understand and augment the ways in which the digital is creative. The ontology presented previously that contrasts digital and virtual makes the mistake of dividing so sharply between them that one cannot investigate their overlap. If creativity arises within the digital only by leaving the digital entirely behind, then we are dealing once again with the impotent ontogenesis of the possible, a flick of a switch that offers no account and no understanding of the genesis of creativity in the digital. Instead, we must discover how even in the midst of its rigid determinism, its seizure by possibility, the digital opens itself to some sort of virtuality, ambiguates or blurs at its borders to engender a friction with the actual that serves virtual creativity. The identification of a fold as an element of digital ontology also thus imposes an imperative: to make the digital creative, one must seek out and ramify its folds.

As part of the structure of the digital itself, the fold is often tied to technical complexities of the operation of digital technologies; but it is not esoteric or rare. The fold is not a singular sign of the meeting of digital and actual, as though the fold appears, like the obelisk in *2001*, to mark sublime occasions of evolutionary advance. The obelisk is a symbol of positive, irreversible advance, a phallic progression that marks a claim over territory. By contrast, the fold is neither a phallic propulsion, 1, nor a vaginal enclosure, 0, but a genital, virtual process of invagination that enfolds both penis and vagina, producing 0 and 1 as complements, artifacts of the originary creative drive. This is why becoming-woman is so essential (as discussed in *A Thousand Plateaus*), for the feminine generates difference while the masculine asserts only opposition; invagination produces not only penis and vagina but also vulva, a whole series of folds within folds. The fold distinguishes 0 and 1 only after the fact, only eventually, and connects them to their origins as their principle of differentiation.

Rather it is necessary to discover the fold at every scale, the fold as prior to and constitutive of the flat digital plane that it populates. Everywhere the digital meets the actual, one discovers a fold. Even the simple binary code is folded, as the difference between 0 and 1 implies an asymmetry that does not obtain between 0 and 1 themselves; that is, there must be a perspective, another dimension from which this difference can be recognized or put to use. Thus, the operation of the binary depends not only on 0 and 1 but also on the linear sequence of these bits. Each bit has a particular value (0 or 1) but also a place in that sequence, and this linear organization of bits is spread across a dimension perpendicular to the polarity of values, 0 or 1, for a given bit.

The ambiguous distinction between software and hardware also locates its junction at a fold in the plane of the digital. Software represents a generalization from the material specificity (and resistance) of hardware, an abstraction that already implies another dimension. Software's principle is an independence from a particular machine, so that the same software might run on more than one machine and different pieces of software might take turns animating the material of a single piece of hardware. But they are also interdependent, as software is ineffective, purely ideal without a realization in the materiality of a computer. Hierarchy is biunivocal between hardware and software, as hardware trumps software at one level (such as the interrupt), while software overcodes and dominates hardware at another. Above all, software is ephemeral, as any attempt to pin it down ultimately pursues it into the hardware itself, revealing software as a phantasm that is no less effective for being fantastical.

Two ontologies, incompatible but simultaneously valid: the digital is Deleuzian after all. The fold at the juncture of the digital and the virtual draws lines that connect the pure form of the digital to the material, actual, and human worlds. The fold textures the plane of the digital, breaking up its flat contours to make it available for creative and original ends. But the cardinal rule of the digital is its purity of form, and this fundamental basis asserts itself not only as a limitation of the digital but also as a source of astonishing capability; by maximizing its abstraction, by confining itself to the hermetic world of 0 and 1, the digital achieves an unprecedented extension, powers of simulation, transportability, interoperability, speed, and accuracy. In short, by reducing its reach to only the forms of things, the digital all but abandons material reality, forgoing at one stroke both material singularity and material resistance. The struggle between these two ontologies, the moribund and the productive, is not only a matter of philosophy; it is a real struggle in the real world and the stakes are high. The digital constantly overcodes the domain of the fold, redrawing the fold as yet another formal distinction, pulling the fold back into the plane of the digital. The question—which in another context is also Deleuze's question—is how to cultivate the second ontology, to fold, if only for a moment, the flat surface of the digital whose lack of texture eludes the grip of the human. Only an always-new fold will ensure that the digital remains creative instead of falling into the aesthetic and political homogeneity that results from the ascendance of abstraction.

CHAPTER 6

Virtual Architecture

Manola Antonioli
Translated by Julie-Françoise Kruidenier and Peter Gaffney

What is ironic in a time of unprecedented advancement in scientific and technological inventions is the reactionary and superficial appropriation of historical forms. The problem here is not just one of form, but of the tendency for this architecture to be acquiescent to the day-to-day demands of utility and economics. . . . This romanticising of an earlier time as 'simpler,' fails to grasp that it is in the realisation of complexity and contradiction that we begin to find our way out of the psychological malaise we're currently suffering.

—Thomas Mayne, "Connected Isolation"

COMPARED TO THE WEALTH AND COMPLEXITY OF ANALYSES that Deleuze and Guattari devote to painting, cinema, literature, theater, and music, the place accorded to architecture seems extremely meager, indeed "minor." But we know that the "minor," in all its forms, plays an essential role in their philosophy, which is why we will not be overly surprised to read in *What Is Philosophy?* that "Art begins not with flesh but with the house. That is why architecture is the first of the arts."[1] Architecture makes its first appearance in the last chapter of *What Is Philosophy?* following long analyses devoted to literature and painting; however, this reference is anticipated at the beginning of the chapter by passages that define art as a "monument," "house," or "territory," terms that bring us back to the notions of dwelling and home, and whose appearance here may cause some surprise, coming from two philosophers who have not ceased to think in terms of deterritorialization and nomadism. We must try then to make sense of this at least apparent process of "sedentarization" in their discourse on art.

If the monument has long been the model-form of the work of art in Western culture, this chapter in *What Is Philosophy?* seems to inscribe itself in a traditional approach, at the same time radically modifying it:

> It is true that every work of art is a *monument,* but here the monument is not something commemorating a past, it is a bloc of present sensations that owe their preservation only to themselves and that provide the event with the compound that celebrates it.[2]
>
> A monument does not commemorate or celebrate something that happened but confides to the ear of the future the persistent sensations that embody the event.[3]

The art-monument, therefore, is not simply that which takes place and endures in its perfected being, not that which takes place as the unchanging repetition of an essence, as the last resort of transcendence and the stability of sense. Rather, its "taking place" constitutes a unique experience of duration as the duration of an event, or as the place of an encounter. A work that only effectuates itself once, one that is nonrepeatable, could nonetheless create a new duration and a new "spacing" of sense. The artistic compound is not given once and for all, but inscribes itself in duration because the new percepts and affects that it creates lead us in turn into new becomings.

From this point of view, architecture (as an applied science and as "the first of the arts") will also be concealed from the memorial dimension of "monumentality" in order to inscribe itself in a field of immanence that is not only spatial or spatialized, but also temporal. It will be an "architecture event" that configures space as well as all the dimensions of time.

The Fold and the Virtual

The work by Deleuze that has exerted the greatest influence on contemporary architecture is almost certainly *The Fold,* even if it seems at first to have less to do with space than with Leibniz's philosophy and its relation to all manifestations of the Baroque. The concept of the fold, which Deleuze bases on Leibniz, establishes a certain unity among the artistic manifestations of the Baroque, "the age of the fold which goes to infinity." Moreover, and more importantly, it permits Deleuze to redefine the natures of subject, concept, object and the reciprocal relations among them. We are dealing then with yet one more variation on the idea of the multiple, where multiplicity is not that which has many parts but "that which is folded in many ways."

The Baroque does not cease to make folds, to bend and rebend them, to push them to infinity and to differentiate them according to two directions or "levels" of the infinite: the pleats (*replis*) of matter and the folds (*plis*) in the soul.[4] Leibniz carries out a vast Baroque montage, never ceasing to affirm a correspondence between the two labyrinths, one marked by the pleats in matter and the other by folds in the soul, or to create a new fold between the two folds in the world. The extreme complexity of the notion of the fold is rooted in the apparently insoluble Leibnizian philosophy, in which:

(1) The monad is a mirror of and point of view on the world, but it is also a unity "without doors or windows" which draws all its perceptions from its own ground (*fond*).[5] This link between exterior and interior is seemingly paradoxical because the monad is a point of view on the world that only constitutes itself at the interior of the monad itself.

(2) The world of Leibniz and of the Baroque has two levels. The upper level is occupied by souls, each one distinct, each one expressing the world according to its unique point of view. These souls do not exert direct action on one another, but must draw everything from their own ground; furthermore, they are made up of infinite perceptions, of which very few ever achieve consciousness. On the lower level one finds matter, both organic and inorganic, subject to forces that give it a curvilinear movement. Organic and inorganic matter can be distinguished according to two different kinds of folding: organic matter is endowed with an internal originary fold, which is transformed alongside the development of the organism, while inorganic matter is folded exogenously, determined by the outside or by surrounding matter.

We must consider then how the fold cuts across these two levels and makes them communicate—a feature without which they would be destined to remain separate (interior and exterior, body and the soul, sensible and insensible). In this way, the curvilinear character of the Leibnizian and

Baroque universe implies a redefinition of the object, subject, and concept. In *The Fold*, Deleuze uses the term *objectile:* the objectile is not conceived in terms of the traditional relationship between matter and form, it does not have the structure of a spatial mold, but constitutes a temporal modulation that "implies as much the beginnings of continuous variation of matter as a continuous development of form."[6] Deleuze explicitly links this concept of modulation, which is based on the research of architect and designer Bernard Cache and on the philosophy of Simondon, to the domain of advanced technology where "industrial automation or serial machineries replace stamped forms" (19), and to practices of standardization that characterized the dawn of the Industrial Age. Already in the universe of Leibniz, and even more so in the actual world, the *essentializing* object becomes a *manneristic* object, one that is closely linked to the event and the virtual.

Relativism is not "a variation of truth according to the subject, but the condition in which the truth of variation appears to the subject. This is the very idea of Baroque perspective" (20). As in all of Deleuze's philosophy, the Baroque subject-point-of-view is never a given entity, but the product of a subjectivation that only takes place starting from a preindividual substrate. The subject is only the folding of an Outside, of something preexisting and exterior; it is the actualization of a virtuality that does not cease to persist at the surface of the actual (individual). In turn, "the whole world is only a virtuality that currently exists only in the folds of the soul that conveys it, the soul implementing inner pleats through which it endows itself with a representation of the enclosed world" (23), according to the dynamic of the Baroque universe, which goes from inflection to inclusion in a subject, from the virtual to the actual.

With Leibniz, the soul or the subject as metaphysical point is the monad. Because Leibniz wants to preserve the plurality of substances (or of individuals), a plurality composed of infinite individual beings, each one maintaining its own irreducible point of view, this means that every monad must express the entire world, but it does so by expressing more clearly a small region of the world, a finite and singular sequence. The world actually only exists in the monad, but the monads correspond to this world as to a virtuality that they actualize: "the soul is the expression of the world (actuality), but because the world is what the soul expresses (virtuality)" (26).

This complex relationship of reciprocal inclusion explains the torsion, Leibniz's folds in the world and in the soul, but it also involves the internal law that orients the relationships between the Outside and Inside in the world as Deleuze conceives it. The point of departure is always an Outside, a fabric starting from which the folds will be able to form and that is constituted by preindividual singularities. These singularities, intensive and virtual by nature, actualize themselves in an infinity of folds that correspond to points of individuation (points of inflections as things and states of things, points of inclusion in the sense of "subjects"). These infinite foldings exhaust Being, in the sense that nothing exists aside from them; at the same time, they are contingent and provisional in the sense that a virtual dimension always endures outside them, a dimension that may always actualize itself in new forms and new combinations that are also, in turn, contingent and provisional. Between that which one customarily calls the "subject" and that which is defined as the "object," there exists no causal link or direct action because subject and object are no more than the result, always provisional, of a process of individuation and actualization. The subject only constitutes itself by expressing a world, but the world has no existence except as the expression of a subject. This involves thinking of the relation, typical in the Baroque conception, between an interior without an exterior (the world actualized in the subject) and an exterior without an interior (as the subject is no more than the provisional site of consistency of a virtual surface that precedes it). Deleuze speaks of this relation as a strange "non-relation," which translates into characteristics proper to Baroque architecture: the split between the façade and the inside of a building, between interior and exterior, the autonomy of the interior and the independence of the exterior, the matter that reveals its texture and the matter that becomes material, the form that reveals its folds and the form that becomes force.

In such a universe, logic itself must be radically modified: to understand cannot signify explaining becomings by having recourse to the abstract forms or essences that are exterior to them. In a Leibnizian logic, the predicates are never simple attributes of a subject, but events, relations to existence, and time. Traditionally, the attribute expresses a quality and designates an essence; for example, in the proposition "I am thinking" ("*Je suis pensant*"), the thought appears as a universal and constant attribute of the human being. But Leibniz's grammar is a Baroque grammar, where the predicate defines a relation and an event according to the schema subject-verb-complement and not according to the schema subject-copula-attribute. If the predicate is not an attribute, the subject is not an essence but the contingent unity, the envelope of all its predicates; thus, a thinking nature will not suffice to make sense of thought, but thought will be comprised as the whole of all its events-predicates, as incessant passage from one thought to the other.

In this way, the world itself must be an event, which is included in each subject as a foundation, from which each subject extracts the manners that correspond to its point of view: Leibniz's thought (like that of Deleuze) is not an essentialism but a mannerism, a way of thinking manners and infinite variations. That which constitutes a principle of permanence (subject, object, or concept) is not endowed with a transcendent nature or eternal essence, but results always from a process of actualization or realization, and only has permanence within the limits of the flows that realize it. The universe is thus made of virtualities that actualize themselves and of possibilities that realize themselves in matter, without the actualization or realization ever being definitive or static.

The Baroque is the last attempt to reconstitute classical reason because it distributes divergences in so many worlds and because it transforms incompossibilities into frontiers between worlds. All the contrasts that emerge in a single world resolve themselves in accords, following the principles of preestablished harmony, while irreducible dissonance and contrasts without solution take place between different worlds. The contemporary Neo-Baroque is made on the contrary of dissonance or harmonically unresolved accords, of divergent series that coexist in the same universe. Divergent series trace, in a single chaotic world, paths that are always bifurcating: the cosmos becomes *chaosmos*. The infinite folds of world, matter, and spirit will no longer be submitted to a condition of closure but rather organized according to a mechanism of capture.

> We are all still Leibnizian, although accords no longer convey our world or our text. We are discovering new ways of folding, akin to new envelopments, but we all remain Leibnizian because what always matters is folding, unfolding, refolding. (189)

The Objectile and the Architecture-Interface

This complex and difficult text by Deleuze does not cease to exercise a profound and durable influence on new research, practices, and reflections in the field of architecture. This influence, exceeding to a large extent the simple phenomenon of "citation," manifests itself as a proliferation of folds and curvilinear elements in the design of contemporary buildings. The fold can, in a much more essential way, furnish architecture with the conceptual tools necessary to interpret and translate the progressive virtualization of buildings in space—spaces for living, working, leisure, and circulation. Architecture is no longer an art exclusively bound to matter or monuments, and it will increasingly separate itself from these designations because it may now conceive of inhabitable spaces in a world that is becoming more and more immaterial and virtual. A building is no longer simply a definitively constructed and fixed space; it is a creation of virtual spaces that permit the constructed "object" to enter into resonance with the subjective points of view that traverse or inhabit it and that must integrate territorial modifications of urban spaces with the transformation of individual and collective forms of life, new technologies of information and communication, and new chronotopes.

The Baroque and Neo-Baroque universe that Deleuze invents in *The Fold* thus furnishes several axes of possible reflection for new forms of architecture that have begun to emerge:

(1) A double dynamic or chiasmus between the dematerialization of matter and materials on the one side and, on the other, a transformation of form into forces that are indissociable from the affects and precepts that they produce.

(2) A transformation of the object into the *objectile,* product of an incessant modulation that is actualized from a fabric of preindividual and virtual singularities.

(3) A world transformed into event, which takes the place of the dimension of permanence that always characterized the monument.

Urban space is generated today out of activities that are at once physical and virtual, and its physiognomy is endlessly reconfigured by material and immaterial flows, as a function of new networks and systems of communication and transportation. The purely monumental dimension of architecture, from the building as an "object" that is more or less completed from the symbolic, aesthetic, or functional point of view, is no longer adequate to the task of understanding the multiple dimensions of a constructed space that has begun to function more and more like a *fold* between the rhizomatic fabric of the city and the multiple perspectives of its inhabitants, between the reality of materials and the virtuality of exchanges, contacts, functions, and experiences produced by and within urban space.[7] Architecture is no longer only a spatial practice; it is a practice that causes heterogeneous temporalities to coexist, to open on the urban fabric and its ceaseless transformations, to presuppose more and more transitory functions and successive moments of life.

The monument-form that traditionally characterized architecture is evolving toward multiple manifestations of events-processes, or becomings, which first manifested themselves in the field of design: the aim of the most contemporary design is no longer, in effect, to produce perfect forms, forms that are perfectly ergonomic or functional, but to modulate objects as a function of the new hybrid identities of their users and of new ways of living. Even the notion of the objectile that Deleuze introduces in *The Fold* is born of research by the architect Bernard Cache (who later called his architecture firm Objectile). Bernard Cache and his firm are working on the implementation of a production technique based on models derived from engineering, mathematics, technology, and philosophy, and which aims at nonstandardized industrially fabricated objects, "lines and surfaces of variable curvature like the folds of a Baroque sculpture."[8]

Thanks to the digital revolution, as affirmed by developments in architecture that began in the 1990s, architects now conceive projects with the help of computer programs that open a new field of investigation and creation, characterized by the variability and modulation of forms. The building is thus allowed to interact with its context, and unique and singular objects can be produced on an industrial scale. The generalization of digital technology, which accompanies the architectural process to its realization, profoundly modifies the creative process. The building, in turn, is more and more often conceived as an objectile, a new type of technological object that integrates the process of variability of forms. This implies a temporal dynamic, inaugurating an evolutional process in terms of the potential to react to environment and to climate, and to interact with changes in the environment.

The presence of a virtual dimension in architecture is not new: the architectural project has always been a virtuality seeking to translate itself into the real. If the ensemble of graphic documents that make up a project have always constituted a virtual dimension in architecture, this ceased to be virtual with the realization of the project, even if there exist projects that remain unrealizable for technical or economic reasons. But virtual architecture as it appears today at the intersection of space and cyberspace is not only an ensemble of potentialities seeking to become real; rather, it has to do with a process of destabilization of architectural form that inscribes itself into the dynamic of continuous transformation, presenting itself as a totality of transitory configurations that capture forces and flows. If, therefore, the traditional conception of the project inscribes itself in the possible-real dyad, digital architecture that utilizes mathematical or algorithmic methods orients itself rather, in Deleuzian terms, toward the virtual-actual dyad. Deleuze, in effect, does not oppose the virtual to the real but to the actual. The virtual is not actual, but it possesses a reality that is proper to it. The possible can realize itself or not realize itself; realization is a form of limitation of the possible, by which certain possibles are rejected or blocked, while others "pass" into the real. The virtual is not realized but actualized, following lines of divergence and creation. If there is a resemblance between the possible and the real, the actual does not resemble the virtuality that it incarnates. The course that leads from the virtual to the actual is one of selection, of a creation of differences, a route that must be invented: the virtual is never simply a possible that is realized.

Architects who use the Internet as a new tool and/or a new construction space are seeking a radically more dynamic approach to architecture. For example, former students of the French architect Bernard Tschumi at Columbia University's Paperless Studio consider how computer programs designed to generate cinematic special effects might also generate new strategies for research into the fluidity and manipulation of architectural form. Other architects study the capacity of materials to integrate computer chips and bio- and nanotechnologies to produce light and heat or react to the movement of their occupants.

It is difficult at this time to evaluate the real reach of such an evolution, its capacity to actualize itself in concrete and efficient forms in the public space, to reverse the principles of permanence, monumentality, fixity that have always characterized architecture; however, these experiments open a Neo-Baroque perspective on contemporary architecture:

> The play of the world has changed in a unique way, because now it has become the play that diverges. Beings are pushed apart, kept open through divergent series and incompossible totalities that

pull them outside, instead of being closed upon the compossible
and convergent world that they express from within. Modern
mathematics has been able to develop a fibered conception according to which "monads" test the paths in the universe and enter
in syntheses associated with each path. It is a world of captures
instead of closures.[9]

Connected Isolation

The influence of Deleuze's philosophy has been equally consequen-
tial for the work and thought of the American Thom Mayne and his
interdisciplinary firm Morphosis, whose experimental realizations, at first
firmly encamped in avant-garde studies and art galleries, have begun to
profoundly modify the American architectural landscape. The architecture
of Morphosis, which takes its name from the Greek word *morphosis*
("to form" and "to be in formation") is fueled by the proliferating fabric
of big cities, where the Baroque folds of the world, matter, and spirit are
transformed into Neo-Baroque rhizomes, producing an assemblage out
of natural environment, materials and techniques, flows and networks.
The rhizome is the principle of interconnection between one multiplicity
and another—multiplicities that are never more than provisional assemblages in the process of becoming. It thus excludes all exterior organizing principles because the elements and forces that compose it ceaselessly
create new configurations. The rhizome, as the modern-day megalopolis,
does not proliferate in an arborescent, binary fashion and does not possess
a structure in which one could easily recognize a beginning or an end. The
"nonstandardized" character of the new architecture thus opposes itself
to rationalism, to the imposition of norms, to the deadening standardization imposed in the past by modernist and functionalist architecture. The
right angle gives way to a new repertory of forms (helicoid lines, ribbons,
imprints, etc.). Architecture, design, and city planning become blurred,
interactive, and dynamic systems, which are situated in the continuity
between the living and the artificial—art, technology, and the industrial
mode of production.

In 1993, Mayne published a text, *Morphosis: Connected Isolation,*[10] which
has quickly become the manifesto of an Event-Architecture inscribed in
urban space and technological mutation. Mayne affirms the necessity of
redefining the way in which architects and city planners intervene, during

a period of crisis, in the relation between architecture and the globalized political-economic context. Strategies that are merely regressive (citation of historical precedents, study of the urban past, superficial procedures that aim to produce an ersatz cultural experience, "museification" of city centers) represent an inefficient and dangerous response to contemporary challenges: "It is necessary for architecture to be based in the present and aspire to that presence" (7).

The acceleration of telecommunication, as well as the mutation of lifestyles that this implies, have replaced traditional communities founded on the physical proximity by way of multiple interactions in a network. In the urban space, it is more and more difficult to find a satisfying articulation between a "public" and "private" sphere, like that between city and country, center and periphery. To overcome this crisis, Mayne affirms the necessity of abandoning conventional ideas about urbanism, which tend toward a simple and homogenous order, and to take account of the complexity of the actual urban experience, which can only be understood in terms of the relations between heterogeneous experiences.

Rather than considering the city as a chaotic system, one that is simply deprived of order, we will have to recognize it as a complex system. The alternative to the organized *cosmos* of the traditional city is not necessarily the evolution toward an undifferentiated *chaos* (an evolution that is familiar in numerous megalopolises, and always with catastrophic consequences): architects and city planners can work toward a *chaosmos* made of folds and multiple assemblages, and of dynamic and evolving forms of organization. Rather than seeking in vain to return to a fixed or stable space, Mayne considers how to "utilise the tremendous energy of the city," to put to work organizing strategies that attain a "high degree of differentiation in an orderly and continuous framework" (13). It is in this sense that the 2006 exposition at the Centre Pompidou that introduced Morphosis to the French public was called "Continuités de l'inachèvement" ("Continuities of the Incomplete").

The architects at Morphosis work toward the creation of an architecture-landscape of a fragmentary and fractured nature, conceiving and realizing projects that have a perpetual open-endedness and unfinished quality. Their interventions in the urban context, for example, strive to create architectural elements that are at once in harmony and in tension with their natural setting, in order to "challenge the traditional dominance of the man-made object in favour of a more dispersed, fragmentary, and integrated relationship of the built environment and its site" (13). This is a relational architecture, one that aims at a continuous exchange between exterior and interior worlds, between subjective and objective experience and therefore, according to the Baroque tradition analyzed by Deleuze, to continuously fold, unfold, and refold the spaces and temporalities of experience.

The most innovative construction techniques and materials allow us to bring these conceptual initiatives to life: physical components (bricks, cement, steel, and glass) exist together with immaterial ones in the same spatial construct. Each building is destined to live in continuous osmosis with the modern and dystopian city, which overlays differences rather than segmenting them:

> A city is a living organism, a work-in-progress, an impasto of forms made by successive waves of habitation. One should continue to choose to do only projects which offer hope of a complex, inte-grated, contradictory and meaningful future. (17)

In turn, this text by Thom Mayne has become one of the principal sources of inspiration for the *Spheres* project of philosopher Peter Sloterdijk, who for several years has been developing an original theory based on spaces and technological mutations. In the cycle of the *Spheres*,[11] Sloterdijk imagines a universe where each individual is already tuned by rhythms, melodies, and collective projects with millions of other people, so that the individual forms a dyad with a place, is nested in a sphere, globe, or foam of which the dimensions and connections are modified in the course of individual and collective history:

> To every social form corresponds its respective world-house (*Welthaus*), a bell of meaning (*Sinn-Glocke*), under which human beings first commune, relate, grow, protect, and set themselves apart from others. Hordes, tribes, peoples, and to a much greater extent empires are, in their respective formations, the psycho-sociospherical dimensions which organize themselves, establish the right climate, and contain themselves. In any given moment of their existence, they find it necessary to build above themselves, with their typical means, a semiotic sky from which flow the com-mon inspirations that form their character.[12]

For Sloterdijk, the diverse forms of space and architecture become a key to interpreting changes in the social and political spheres that mark the passage from modernity to the postmodern era of globalization. The malaise that affects world civilization appears to him first of all as a crisis in the political form of the world, as a symptom of the passage from a culture of rootedness ("spirit of the agrarian era") characteristic of classical politics toward the new global game of the industrial era, for which we have yet to determine a set of rules.

From Baroque architecture and philosophy as reinterpreted by Deleuze to contemporary avant-garde architecture and the recent philosophical endeavors of Peter Sloterdijk, we thus see a constellation or rhizome of meditations and practices that establish novel lines and connections, new folds between architecture, urbanism, philosophy, and new technology. We cannot speak of a temporal succession or of a linear derivation that would run from Leibniz to Deleuze, and on to Mayne and Sloterdijk, but rather of a genealogy in the Nietzschean sense, or of a system of resonances that joins together art, technology, philosophy, and politics, and that aims at understanding and developing the virtualities of the present.

Architectural Enunciation and Ecosophy

In *A Thousand Plateaus*, Deleuze and Guattari develop a complex theory of territoriality, deterritorialization, and reterritorialization. Similarly, Guattari has developed a theory of ecology that includes a meditation on architecture. The resulting "ecosophy," outlined in Guattari's 1989 work *The Three Ecologies*,[13] articulates an ethico-philosophical complex "between the three ecological registers (the environment, social relations and human subjectivity)" (28). The common principle among the three ecologies is in the need to give form to new existential Territories, in concordance with the Greek root of the word ecology (*oïkos*), which refers to the house, habitat, and natural environment, to the domestic and living space; however, these Territories are never closed or static realities, but finite and singular entities, susceptible at any moment to bifurcate toward processes of stratification or mortification, or, conversely, toward processes of opening that make them amenable to human undertakings. If Deleuze and Guattari first borrow the notion from ethology, they proceed to use it in a very broad sense that exceeds the false opposition or the dangerous superpositions between animal territorializations and human territorializations. From a perspective that is transversal to the notions of nature and culture, territory becomes a form of organization that delimits the vital space of a being while articulating it with other beings, with the Earth and Cosmos. Territory is a form of appropriation of time-space, which implies existential, social, cultural, aesthetic, and cognitive investments. Yet territories that appear to be fully constituted and closed on themselves do not cease to open and to be deterritorialized, comprising an immense movement of deterritorialization that marks the entire history of the human species and has accelerated considerably with capitalism and advances in technology.

Ecosophy also comprises forms of praxis that are oriented toward the re-creation of existential Territories, concerning not only the sensibility and intelligence of each person, the body, and the environment, but also, on a larger scale, those assemblages relative to ethnicity, nation, humanity, and the planet. Thus, the problems of the future will all be ecological problems, to the extent that the advance of capitalism and ensuing Westernization of the world have swept away and destroyed all manners of rootedness in identity, culture, territory, and natural resources as cultural territories and resources. Though it is customary to associate Deleuze and Guattari with a critique of all forms of sedentarization and territorialization, to see their work as an affirmation of deterritorialization, these authors have also given much thought to the notion of territory in its ethical, political, and philosophical aspects, notably in the chapter of *A Thousand Plateaus* that elaborates the concept of the refrain.[14] Guattari, too, has done much to elaborate the notion of territory in *The Machinic Unconscious*.[15]

In *A Thousand Plateaus*, beginning with their study of the territorializing function of the refrain, Deleuze and Guattari present territory as the product of an *act* of territorialization that affects milieus and rhythms. A territory borrows from all milieus, is constructed out of aspects or parts of milieus, and is thus marked by multiple indices: positions and colors of bodies, melodies and cries, leaves and blades of grass. Territory is presented first of all as the site of a passage, the place where a transversal assemblage conjugates forces of the earth and milieus, as well as functional and expressive rhythms. When a territory takes root at a center of intensity (the unity of earth as receptacle or pedestal, for example, which permitted the emergence of the sacred and of religious practice), it gives rise to the illusion of a unique territory, of a singular ground (*fondement*)[16] identified with the Home (*le Natal*). But the center of intensity can be situated at the

point of convergence between two or more territories that are very different or distant from one another. A territory is always on the way to being
deterritorialized; the assemblages that constitute it undergo a continuous
process of transformation: "deterritorialization, in all its forms, 'precedes'
the existence of strata and territories."[17]

Thus, there exists a constant tension in A Thousand Plateaus between
the forces of territorialization and those of deterritorialization, and the
pages devoted to the concept of the refrain also contribute to a redefinition of the link between territory and identity, a link that constitutes one
of the essential issues in the current debate regarding political ecology. If
territory may function as a basis for the identity of a group or for each
member of a group, if it has historically functioned as such, then the coexistence of territorial assemblage with the dynamics of deterritorialization
prevents us from thinking the connection to a territory (or territoriality)
according to the vegetal paradigm of a root or of taking root. We must
avoid reducing the relationship to territory to the false concept of "nature"
because the territory inscribes itself in a series of becomings and is always
the product of a creation (among all animals, not only among human
beings). Territory is always differentiated, composite, complex, and traversed by variable coefficients of deterritorialization.

Still thinking of the refrain, Guattari develops a political interpretation of the territory in The Machinic Unconscious. He affirms that, to
the extent that the deterritorialized flows of late capitalism have done
away with territorialized assemblages of people, cultures, languages, and
arts, subjectivities have a tendency of clinging to various attempts at
reterritorialization, attempts that are sometimes pathetic but that have
become more and more dangerous (myths regarding the blood, the
earth, national identity, etc.). Guattari's aim is not to applaud a return
to archaic territories, dispersed long ago by the formidable deterritorializing capacity of capitalism, but to work toward the construction of new
existential territories. It is in this way that philosophy as well becomes an
ongoing meditation on territoriality, on the relationship of a subject to
a body, to time and to space. These subjective territories are implicated
not only by way of new technologies (post-media), a problem of urban
space, and the evolution of forms of habitation, but also more far-reach-
ing problems, as for example those revealed by environmental ecology
that involve the relationship between humanity with the totality of the
planet and nature.

Above all, the concept of territoriality concerns connections between subjectivity and bodies inhabiting space. Already in 1976, Paul Virilio—one of the central points of reference for Deleuze and Guattari's thinking on space, the city, and architecture—affirmed the necessity for architects to abandon a purely technical approach to living space, which he attributes to traditional modernist and rationalist architecture (*The Insecurity of the Territory*). Virilio is in favor, rather, of an approach that emphasizes the way a space is used and inhabited. The monumental conception of architecture forgets that the building is inscribed in a duration and in a complex fabric of social relations. The work of an architect does not end when actual construction is completed; the building continues to act within its environment and in regard to the various ways it is used by a multitude of people, each one carrying out a different strategy. Virilio thus affirms that, to the extent that ecology involves the study of relations carried out by living beings vis-à-vis their environment, it will be necessary to engage a way of thinking that comprises a true ecology of the building. Indeed, the milieu of architecture, urbanism, and living space constitutes the milieu and environment of the human being, in exactly the same way as a natural environment properly so-called. The education received by architects and city planners has a tendency to valorize technical performance in the construction of buildings, but it, too, often neglects the importance of the concrete ways in which spaces will be used. The tendency to reduce architecture to its monumental function or to its immediate—i.e., technological or economic—effectiveness without thinking of its effects on the environment or on future users is, for Virilio, the principal cause of the current crisis in cities and in land management. Modernist architects have too often forgotten that to inhabit a space means first of all to besiege it and to appropriate it; as a consequence, they have produced spaces that are uninhabitable because they do not allow for an individual appropriation of space.

Seeing the problem from a similar point of view (that of the "ecology of the constructed domain"), Guattari has shown a great interest in architecture and is particularly interested in the works of the Japanese architect Shin Takamatsu.[18] Japanese architecture on the whole provides fertile ground for Guattari's thinking because of its system of correspondences between macrocosm and microcosm, between interior and exterior; because of its capacity to transfigure the ancient connection between nature and culture that are inscribed in the history of Japanese art; and because of its striking invention of "architectural machines" that succeed in creating a new *nature* in the hypertechnological fabric of modern day cities. Guattari's "Architectural Enunciation" is a veritable manifesto of this aspect in his ecosophy.[19] When Guattari speaks of enunciation, of discourse or of discursiveness, he never refers exclusively to the written text or the spoken language: text and language are only two components among others belonging to a transversal journey that binds together elements of expression and content, words but also percepts and affects. The article opens with a strange passage, once again combining the traditionally separate domains of nature and culture: "For several millennia, and perhaps in imitation of crustaceans and termites, human beings have had the habit of surrounding themselves in carapaces of all kinds."[20] These carapaces are not only made of buildings, but also of clothes, automobiles, images, or messages. The construction, occupation, or destruction of symbolic buildings (the construction of zigurrats or pyramids, the demolition of the Bastille, the capture of the Winter Palace) has often accompanied the delineation of social and/or political assemblages. But in the current era, the importance of stone has been supplanted by that of steel, concrete, and glass; simultaneously, power has been founded on speed of communication and control of information.

The role of the architect has also been put in question and even, more often than not, deprived of all sense: "What use would it be today, in Mexico City, for example, which charges deliriously towards a population of 40 million, to invoke Le Corbusier! Even Baron Haussmann, in this case, would be useless."[21] The power of politicians, technocrats, engineers, and money has long been incommensurably greater than that of the architect, who retains control only over the smallest niche in the domain of extravagant construction projects, which imply the acceptance of heavy politico-financial risks. The object of architecture seems in this way to have been exploded, torn as it is along political lines, by demographic and ethnic tensions, shaken by technological and industrial mutations, "pulled and rent apart in all directions."[22]

The positive consequence of this state of things is that architects will no longer be able to take refuge, one by one, in a nostalgia for the past, in utopia, in art for art's sake or in pure science; they will be confronted more and more by the urgency of ethical and political choices that are related to the practice of architecture. To reinvent architecture today can no longer mean to launch a style, to found a school, or to elaborate a theory, but to recompose architectural enunciation and the work of the architect in all its complexity. If the architect can no longer aim to be a plastic artist of constructed forms, he must aspire to reveal virtual desires of space, sites, and territories. He must undertake an analysis of the transformations carried out within the individual and collective body, and of the ways in which it occupies space; he must singularize his approaches to architecture; he must mediate between the emergence of these desires and the interests that oppose them; and he must become "an artist and an artisan of the sensibly lived and relational."[23] Architecture will thus become an essential component in the project of ecosophy, not only in the sense of safeguarding and reconstituting the natural environment together with its material resources, but also in the sense of safeguarding and reconstituting the constructed environment and its inhabited territories, as well as the economic, social, and political dynamics that transform it.

Like all domains of existence, the structure of architecture should be "polyphonic" and should imply percepts and affects, including both discursive and nondiscursive elements. In architectural polyphony, which comprises innumerable, heterogeneous voices, Guattari gives prominence to eight principal components, which are still today of extreme relevance:

(1) A geopolitical enunciation, which takes into account the data of terrain, climate, and demography, but also perspectives of the evolution of inhabited space as a function of fluctuations in the world economy.

(2) An urbanistic enunciation, which involves the laws and regulations that affect construction, forming connections between the bodies of the State or the state of public opinion.

(3) An economic enunciation, implying the requisite evolution of costs, profits, public utility, prestige, and political impact.

(4) A functional enunciation, which involves maximizing the suitability of constructed spaces in regard to their specific uses.

(5) A technical enunciation, taking into account equipment and materials and implying an exchange not only among building engineers but also within the totality of technical and scientific disciplines.

(6) A signifying enunciation, which assigns to each constructed form a meaningful content shared by the human community

that must make use of and inhabit it, but also integrated into the totality of communities that do not share the same content.

(7) An enunciation regarding existential territorialization, where the architectural space would become a concrete operator of the metabolism between external objects and the internal intensities that constitute a territory.

(8) A representative enunciation that integrates and specifies the totality of other components, in particular the ethico-aesthetic aspects of the constructed object.

The ethico-aesthetic component must not be considered as a simple "supplement of the soul" or superficial personalization added after the fact to the architectural object; rather, it is an essential dimension that gives that object its most intrinsic consistency. If the architectural object produces spatialized effects, it must avoid giving way to the dream or to the imaginary; it must avoid causing such aspects as otherness and desire to disappear, or else interest in the constructed space wanes and the desire to inhabit it collapses. The art of the architect can be defined in the last instance by its capacity to perceive the affects of a spatialized enunciation and to give life to paradoxical objects that must not be approached exclusively from the point of view of scientific or technological rationality, but in a more roundabout way, via aesthetics and mythology. The successful architectural enunciation can only take place in a transversal manner, causing the most heterogeneous levels to coexist: "Architectural form is not called to function as a gestalt enclosed in itself, but as a catalytic operator triggering chain reactions at the heart of modes of semiotization that cause us to go beyond ourselves and that open us to as yet unimagined fields of possibility."[24] Even if the architect cannot, all by himself, compose or program in advance the totality of all the fragmentary components of subjectivation that constitute the spatialized affect or the aura of a building, he must strive at least not to compromise in advance the birth of these virtualities by creating fixed spaces, spaces that are anonymous and nonprocessual and that exclude (as is often the case) all possibility of becoming.

Guattari's reflections remind us that interrogating the virtual dimension of an architecture "to come" must not mean forgetting the experience of the bodies that inhabit and traverse constructed spaces, annulling affects and percepts: *virtual architecture* cannot be reduced to *digital architecture*, which strives rather to conceive new forms, but of a kind that is abstracted from any sensible, social, or political context. To explore the theoretical and practical perspectives of a virtual architecture would thus mean, paradoxically, to adapt oneself more and more to the real in order to integrate, within an expanded ecology, the spatial needs and desires of everyone, as well as technological flows, networks and the natural environment. Architecture cannot be a speculative or abstract activity, conceived by specialists who act on the basis of reason alone, but must become once again a practical and political activity, responding to the impossible challenge of reconciling the global-urban dimension with a properly ecological concern. It must take advantage of the resources of science, of the most advanced materials and technologies, without forgetting the body and its sensations. It must allow each person to inhabit a place and a territory and to create his or her own existential refrains, but also to have the capacity of safeguarding his or her relations with nature and the landscape, with flows and networks, with society and the world.

Such reflections that aspire to a virtual architecture, inscribed in the event-world, thus allow for a confluence of the Baroque perspectivism studied by Deleuze in *The Fold* and the ecosophical dimension developed by Guattari. The latter, in turn, is rooted in the philosophical tools developed by Deleuze and Guattari throughout their collaboration (the collective assemblages of enunciation, the production of subjectivity, the rhizome, the distinction between smooth and striated space).

· III ·

Science and Subjectivity

CHAPTER 7

The Subject of Chaos

Gregory Flaxman

The Daughters of Chaos

The relationship between science and philosophy constitutes one of the most difficult and perplexing aspects of Gilles Deleuze's work. Both with and without Félix Guattari, Deleuze develops a notion of philosophy that draws upon the domain of science (as well as that of art) at the same time that he seems to draw abiding distinction between them. In *What Is Philosophy?* where the eponymous question demands their most explicit and enduring consideration of these three domains, Deleuze and Guattari insist that philosophy, science, and art achieve self-consistency by virtue of their respective problems and the modes of thought to which those problems give rise. Each discipline designates a separate sphere with separable concerns: science concerns *variables* that "enter into determinable relations in a function"; art concerns *varieties* that "no longer constitute a reproduction of the sensory in the organ but set up a being of the sensory, a being of sensation"; and, finally, philosophy concerns *variations* that "are not associations of distinct ideas, but reconnections through a zone of indistinction in a concept."[1] As a result, Deleuze and Guattari contend, each discipline claims its own creations: while science creates prospects and function, and art creates percepts and affects, philosophy creates concepts.

These divisions are justifiably the source of debate and disagreement—as if Deleuze, who once declared that the "only true criticism is comparative,"[2] had endeavored late in his life to define science, art, and philosophy as distinct and even disciplinary precincts. Is it really impossible to imagine science producing philosophical concepts, or philosophy producing aesthetic sensations (percepts or affects), or art forms producing scientific functions? No doubt these questions deserve careful reflection, but in this essay I want to forgo such specific concerns in order to consider what we might well take to be their cause or condition. We can only grasp Deleuze and Guattari's insistence that science, art, and philosophy are discrete domains—or, as they will say, "planes"—insofar as we understand that these domains traverse the same *topos*. This paradox effectively structures *What Is Philosophy?* which treats science, art, and philosophy as autonomous domains only to affirm, in its final pages, that this autonomy is based on a common brain that the authors call "chaos." Their respective autonomy notwithstanding, these domains are belied by a shared heritage, a genetic lineage, without which we cannot hope to grasp their subsequent divergences and relations. "In short," the authors write, "chaos has three daughters, depending on the plane that cuts through it: these are the *Chaoids*—art, science, and philosophy—as forms of thought or creation."[3] But what does it mean to say that these domains are born of chaos, and in what sense does this parentage reveal itself in and across these three siblings?

Even as Deleuze and Guattari explain that the variables of science induce the development of functions, that the varieties of art induce percepts and affects, and that the variations of philosophy induce concepts, they also insist that these distinctions bear witness to a mutual chaos. No matter which domain, science, art, and philosophy only exist insofar as they take a "cross-section" of chaos, as if each one drew a secant traversing a common brain, an infinitely complex gray matter. The brain is "the *junction* (not the unity) *of the three planes*,"[4] and it is with respect to this junction that we can begin to consider the liaison between philosophy and science. As we will see, philosophy and science unfurl completely distinct planes, or what we will call, respectively, "immanence" and "reference," but this distinction must always be grasped in relation to a *common nondenominator*—chaos. Each plane or domain consists in a relation to something that cannot be understood, grasped, or thought, but at the same time, Deleuze and Guattari insist, without this relation, or nonrelation, or "relation," thought as such would not exist. Indeed, Deleuze maintains that philosophical thought acquires its character and its consistency in relation to that which is nonphilosophical, and the same would have to be affirmed of chaos' other daughters. "*Philosophy needs a nonphilosophy that comprehends it; it needs a nonphilosophical comprehension just as art needs nonart and science needs nonscience.*"[5]

Therefore, the three daughters of chaos constitute distinct planes only on the condition that each one exists "in an essential relationship with the No that concerns it" (218), but what we must endeavor to understand here is that the relationship to the "No" or "Non" does not signify negation. Science, art, and philosophy do not negate or even sublate chaos so much as each affirms a relation to that which remains beyond its proper (*propre*) domain, which is to say, unthought. "The unthought is therefore not external to thought but lies at its very heart, as that impossibility of thinking which doubles or hollows out the inside,"[6] Deleuze writes. Following Blanchot and Foucault, Deleuze calls this thought "the outside" (*le dehors*), but it should already be clear that, by invoking this term, Deleuze is not interested in rendering a topography, or any such stratiography, that would lack the suppleness to undergo the torsions of thinking (folding). Rather, when we say that chaos is "outside," we lay down a *topological space* whereby the unthought consists in the delicate contours of a fold that is at once "deeper than any internal world" and "farther away than any external world."[7] Chaos is unthinkable, but this is why it lies at the very heart of thought, constituting the kernel that is at once the most intimate and most external to "what is called thinking" (Heidegger). It might be worth recalling Jacques Lacan's wonderful neologism, "*extimité*," in order to convey the intimate exteriority that characterizes the outside because the impulse to overreach ourselves that the outside demands invariably returns us to that which is folded into the unfathomable depths of the soul.

Only in this view can we understand Deleuze's avowal in *The Fold* that "[c]haos does not exist,"[8] which might otherwise strike us, once again, as a kind of negation. But in fact, Deleuze's claim ought to be regarded in the sense that the event that ostensibly marks our encounter with chaos is also the event that makes chaos itself emerge. Chaos does not exist apart from thinking and from being thought, even if this confrontation renders chaos unthought or outside. In each case, then, chaos consists in the impossible and unthinkable that does not exist apart from being filtered through the membrane (*crible*) of thought. There is no chaos apart from the screen, no chaos "in itself," but neither is there a screen apart from chaos. So long as we speak of chaos, we speak of a kind of quasi-chaos, of a little bit of sense that (however minimal) we extract because "we" are the filtration of chaos just as chaos is no less the expression of this filtration. Chaos "is an abstraction because it is inseparable from a screen that makes something—something rather than nothing—emerge from it," Deleuze writes. "Chaos would be a pure Many, a purely disjunctive diversity, while the something is a One, not a pregiven unity, but instead the indefinite article that designates a certain singularity."[9]

The Image of Thought

But what is it that "experiences" chaos? Not an "I think": the differential forces of thinking are expressed impersonally—"One thinks." In other words, the confrontation with chaos provokes a cerebral crystallization, the development of a brain that can only be called "subject" insofar as we immediately distinguish it from the determinate form ("I think") associated with the term. Typically, the subject precedes chaos and is defined, a priori, in advance of experience, but Deleuze submits thought to a becoming-anonymous (and this, after all, is the nature of all becoming) that will strip the mantle of any determinate identity or determinate experience. Deleuze lifts Whitehead's singular term, "superject," to describe this process: if the superject is thrown or throws itself (-ject, jacēre) into chaos, this is so as to surf along the fluid surface of the plane of immanence, to survey its variations, to make "something issue from chaos" without, for all that, detaching itself from chaos.[10] A "form in itself,"[11] the superject consists in an overflight of the plane of immanence: the emergence of a philosophical brain, of a faculty for creating concepts, the superject moves at absolute speed over this absolute surface.

The difficulty of this premise remains the substance of Deleuze's remarkable ruminations on the Baroque, in which the subject-superject corresponds to a kind of "Baroque line." In *The Fold*, Deleuze argues that Baroque thought in general and Leibniz's philosophy in particular evade the simple ascription of subject and object because these terms cease to refer to distinct poles whose adequation (*adequatio*) traditionally defines the condition of "Truth." The subject cannot be reduced to a vantage, or series of vantages, on a self-same object, just as the object cannot be referred to the self-same subject according to the concepts of representation ("conditions of possibility"). Deleuze's sense of this paradox is expressed in his enigmatic declaration that "all perception is hallucinatory," for the very coordinates of perception—the perceived and the perceiver—no longer anchor experience so much as they derive from its differential oscillations. The Baroque line synthesizes a variable perspective on variation: while the ostensible object is modulated (or, as Deleuze would say, "mannerized") by the event of its perception, the ostensible perception no less constitutes an inflection of variation. The "brain is the screen," Deleuze declares of cinema, but the cinema only reveals the nature of the brain itself. Indeed, the supple tissue undergoes a folding that, tracing the vector between chaos and thought (*membrain?*), will precipitate the creation of both.[12] "The transformation of the object refers to a correlative transformation of the subject," Deleuze explains, adding that "the subject is not a sub-ject but, as Whitehead says, a 'superject.'"[13]

But if Deleuze initially unfolds the fold of the superject in the context of a broadly Baroque spirit, his subsequent articulations of the concept with Guattari mark a significant transformation. No longer a general principle of thought with no particular domain, the superject undergoes a metamorphosis in *What Is Philosophy?* which divides thinking between three domains or chaoids and, thereby, renders the superject the province of philosophy and philosophy alone. As we have seen, Deleuze and Guattari suggest that each of these domains formulates its own plane—a plane of immanence (philosophy), a plane of reference (science), and a plane of sensation (art). Each of these planes describes a milieu and a corresponding force of thought. "Philosophy, science, and art are not the mental objects of an objectified brain," they write, "but the three aspects under which the brain becomes subject."[14] Philosophy, science, and art produce distinct cerebral crystallizations, each brain corresponding to the ontogenetic forces that it claims by right (*quid juris*). While chaos consists in a kind of gray matter, the daughters of chaos develop a distinct brain in relation to their respective planes: imagine Siamese triplets, cephalically linked, all three sharing an infinitely complex neuronal mass, but each bound to undertake thinking on its own.

In terms of philosophy, then, Deleuze and Guattari define the plane of immanence as a formless *topos* on which concepts—even and including the lamentable concept of the subject—are composed. The milieu of philosophy itself, the plane of immanence consists in a field of intensities moving at infinite speed. "Thought demands 'only' movement that can be carried to infinity," Deleuze and Guattari write. "What thought claims by right, what it selects, is infinite movement or movement of the infinite."[15] If thinking is better expressed by a weed than a tree, by a Pollack than a classical portrait, by an innumerable pack of wolves than any enumerable ensemble, this is because thinking consists in the deterritorialization of all forms of classical unity into a multiplicity of points (singularities) and lines (vectors) in perpetual variation. But what is movement or, should we say, what is it that moves on the plane? No doubt it is thinking that moves at infinite speed, but thinking does not exist on the plane of immanence as in a self-contained space of logic operations or a transcendent crypt ("unity of apperception") sealed off from experience. The movement of thought remains only one aspect of a plane that always has two facets. Even as we define the plane of immanence as the plane of thought, we must also understand that it does not exist in the absence of the material universe. "It is in this sense that thinking and being are said to be one and the same. Or rather, movement is not the image of thought without being also the substance of being."[16] The plane of immanence is Janus-faced, turned on one side to thought and on the obverse to nature. As Deleuze and Guattari write, the Pre-Socratics regarded these sides, *Nous* and *Physis,* along a single surface on which they communicate, and the Stoics refined this plane according to series along which mind and matter mutually vibrate. "The atom will traverse space with the speed of thought," Epicurus writes, because on the plane of immanence thought *literally* expresses its own materiality just as matter expresses its own thinking.

It is in this light that we can cash out the compromises and commonplaces into which philosophy lapses in relation to chaos. The problem with philosophy, at least insofar as Deleuze understands its traditional or majoritarian impulse, is that it tends to take leave of chaos, abandoning the delicacy and precision demanded by the plane of immanence in favor of regularity and predictability. We seem to gain purchase on chaos by means that, as Deleuze and Guattari suggest, go from "the elementary to the composite, or from the present to the future, or from the molecular to the molar."[17] Surely this is why Deleuze's treatment of the traditional subject (as opposed to the superject) seems to be both an attack and a lament, for he maintains that our subjective habitas perpetuates the repression of chaos. We are "slow beings" composed of fantastic speeds, Michaux remarks and Deleuze repeats, as if to mark the curious distance between our molecular becomings and the molar selves.[18] Hence, as Deleuze and Guattari remind us, the problem of philosophy concerns the process of laying down the plane of immanence without introducing the obstacles and limits, the presuppositions and opinions that invariably abort the creation of concepts. There is always a danger that we will botch the relation to chaos, that we will do too little or too much. On the one hand, when we refuse to recognize our need for sense, we supply too little order—as if we could simply except ourselves and thereby encounter chaos in a pure and untrammeled state. As Deleuze once commented, this eventuality would consist in the fantasy of "withdraw[ing] the pure sensible from representation" and, thereby, "determin[ing] it as that which remains once representation has been"—"a contradictory flux" or "rhapsody of sensations."[19] But on the other hand, and even more troublingly, when we go in the other direction, mustering the indelicate urge for order, we run the risk not of chaos but of the reaction (reactive) formation it induces. In the midst of the infinite variation of mind and matter, the development of the subject amounts to the static appearance of things, to the encrustation of identities, and to the assignment of values. Thus, for Deleuze, as for Nietzsche before him, the molarization of chaos is redoubled by its moralization whereby the formation of semistable extensionality and semistable intentionality reciprocally determine each other, the former projecting the latter, the latter introjecting the former.

Common Sense

Inasmuch as the task of treating chaos remains a matter of almost inexpressible difficulty, inasmuch as we naturally worry about lapsing into madness, we are coming to understand that this prospect also tends to lead to an even more worrisome reaction. "We require a little order to protect us from chaos," Deleuze and Guattari admit, but we almost always end up mustering too much; the confrontation with chaos tends to provoke a cure more serious than the disease. "It is as if the *struggle against chaos* does not take place without an affinity with the enemy," Deleuze and Guattari write, "because another struggle develops and takes on more importance—the struggle *against opinion,* which claims to protect us from chaos itself."[20] Why? As we have defined it, the plane of immanence consists in the infinite variability of thought moving at infinite speeds, which seem to "blend into the immobility of a colorless and silent nothingness" (201). Under purely chaotic conditions we cannot get ahold of our ideas, which slip through the sieve of our brain-plane without leaving any trace, any residue, any *sense* behind. "That is why we want to hang on to fixed opinions so much," they write. "We ask only that our ideas are linked together according to a minimum of constant rules."[21] It is not enough for us to muster ideas: rather, those ideas must be "linked together" according to "protective rules—resemblance, contiguity, causality—which enable us to put some order into ideas." In asking for such "rules," however, we

quickly divest ourselves of chaos altogether in favor of so many tacit presuppositions that provide the model of what it means to think (rightly, for
instance, according to transcendent values) and how thought proceeds
(properly, for instance, according to logical methods).

If our concern here revolves around the regime of opinion, this is
because Deleuze's philosophy and his philosophical commentary on science suggest the dangers of retreating into habits of thought, into all kinds
of fixed ideas. The paradox and problem of chaos, then, is that it should
give way to an even more troubled relationship with opinion (*doxa*), the
adoptive parent who threatens to alienate the daughters (chaoids) from
their natural-born mother—from their ontogenetic lineage. The children
fly from chaos to opinion to ensure the pretense of order, or regularity, and
of Truth, and this inclination extends, *mutatis mutandis,* from the domain
of philosophy to that of science. [22] In *Difference and Repetition,* as well as
elsewhere, Deleuze suggests that philosophy and science come together
to form a particular form of *Urdoxa* that we will have to elude if we are to
invent different means with which to approach the question of chaos. In
effect, this *Urdoxa* is composed of two distinct, though intimately related,
orthodoxies—namely, "common sense" and "good sense"—which we
will consider in turn. The difficulty of this consideration, as we will see,
concerns the respective tendencies (*dérives*) of each orthodoxy, for on the
face of it, common sense describes a broadly philosophical inclination and
good sense a broadly scientific one. Having said that, though, we would
have to admit that not only does philosophical common sense implicate
science and, conversely, scientific good sense implicate philosophy, but
that the general constituents of good and common sense reciprocally
demand each other. Common sense and good sense come together under
the auspices of philosophy and science, and vice versa, to formulate what
Deleuze calls our "image of thought."

Let us turn, then, to common sense. The great presupposition that
reigns over classical philosophy, serving as its first and fundamental law,
concerns the insistence upon the good will of thought and the thinker.
"The philosopher readily presupposes that the mind as mind, the thinker
as thinker, wants the truth, loves or desires the truth, naturally seeks out
the truth," Deleuze laments;[23] "He assumes in advance the good will of
thinking; all his investigation is based on a 'premeditated decision.'"[24] All
the more ubiquitous because it so often goes undetected, the convictions
of this *Eudoxus* constitute the enduring certainty with which philosophy

so often opens, assuring us about what "everybody knows . . ."[25] The philosopher presupposes that we share a common sense, the constituents of which define us as "thinking beings" (*Cogitas Cogitans*) and the ends of which are lodged in the overarching sphere of representation. Surely this is why Deleuze considers "beginnings" to be so troubling in the context of philosophy, for they almost always inspire the propadeutic innocence with which we claim to clean the slate and start afresh. "The philosopher takes the side of the idiot as though of a man without presuppositions,"[26] Deleuze says, but this ostensibly honest accounting forms the source of common sense itself.

If the implicit (and prephilosophical) presupposition of philosophy concerns what "everybody knows," the structure of this presupposition is always more important than its predication. This form

> consists only of the supposition that thought is the natural exercise of a faculty, of the presupposition that there is a natural capacity for thought endowed with a talent for truth or an affinity with the true, under the double aspect of a *good will on the part of the thinker* and an *upright nature on the part of thought*. It is because everybody naturally thinks that everybody is implicitly supposed to know what it means to think. The most general form of representation is thus found in the element of common sense understood as an upright nature and a good will (Eudoxus and orthodoxy). The implicit presupposition may be found in the idea of a common sense as *Cogitatio natura universalis*.[27]

Common sense comprises any number of points of departure, a series of countless different premises, but it always retains a kind of enduring logic concerning what it means to think. This is precisely because its premises proliferate among thinkers, forming the basis for what "everybody knows . . ." The image of thought is characterized by a model of common sense that circulates between the universal subject (*Cogitatio natura universalis*) and the universe of subjects. Indeed, common sense has two sides that are nonetheless intimately related. On one side, we discover the subject and the labor of recognition, namely, "the harmonious exercise of all faculties upon a supposed same object."[28] Not only does each faculty relate to the same object, but the identity of the object consists in the accord reached between faculties, which relate their particular "given" or "style" to each other such that the object is the basis for and result of their agreement. The accord between faculties, Deleuze suggests, relies on "a subjective principle of collaboration of the faculties for 'everybody'—in other words, a common sense as a *concordia facultatum.*"[29] Insofar as the common sense consists in the subjective agreement between faculties, its insistence that "everybody knows . . ." bears witness to the agreement between subjects themselves, who are compelled to overcome any such "conflict of the faculties" (Kant). Indeed, this is the other side of common sense, which is writ large in a *sensus communus,* just as this community of sense will be exercised in the instance of each and every individual subject.

For this reason, common sense prescribes not only what we know but, more importantly, the form of knowledge itself, which emerges as a consensus—both within the individual (as, say, the agreement of the legislated between faculties) and among individuals (as, say, an agreement legislated among the people). This double agreement can be clearly elucidated in terms of *doxa.* As we know, *doxa* generally pertains to a kind of abstraction from our lived situation: we immediately move from "this or that," "here and now," to a statement that pertains to conditions that derive from or purport to cover all manner of situations. But this procedure of abstraction would be inconceivable if the one who made the statement did not, in turn, identify himself or herself with "a generic subject experiencing a common affection."[30] If *doxa* takes its point of departure from the mode of recognition, this is because the latter likewise presupposes an object and a subject, an abstract enunciation that refers to an object and a general form of a subject capable of that enunciation.[31] *Doxa* prolongs the model of recognition into a discursive or communicative sphere because it "gives to the recognition of truth an extension and criteria that are naturally those of an 'orthodoxy': a true opinion will be the one that coincides with that of the group to which one belongs by expressing it."[32] The model of *doxa* is not just a kind of competition but a game show where we are openly or implicitly asked to reiterate what everyone knows—what the survey says, what the majority believes, what amounts to common sense.

Good Sense

Inasmuch as common sense constitutes the philosophical aspect of the orthodoxy that Deleuze calls the image of thought, we can now understand its reciprocal relation to the other half of this image—namely, the scientific aspect of good sense. Common sense appeals to the regularity of the subject, and thence all subjects, in the form of so many presuppositions. But if common sense concerns the universal determination of these presuppositions, how are these rules universally applied? In view of this question, Deleuze suggests, common sense looks to good sense as the other half of its orthodoxy, the principle that distributes its universal organization universally, over time and throughout space, providing the individuated and determinate sense of the abstract subjects and objects. Indeed, the endurance of objects is reciprocally insured by the self-consciousness of one's own endurance as habit and habitas. Subjected to this regime, the subject of representation projects this structure from one moment to the next as the continuity of its own experience. Deleuze terms this repetition our "passive synthesis," the first series of (brute) repetition, insofar as consciousness is a repetition of oneself, even as a *méconnaissance*, prolonged from one moment to the next. The modulation of objects in time, or movement, is thereby determined in accordance with the habit of the subject itself, a mobile milieu, which anticipates itself in the prolepsis of its own common sense.

Needless to say, the brain-plane of science is no less liable—and, arguably, even more—to lapse into molarization than philosophy, which circumscribes and regulates chaos out of existence. Indeed, science undergoes its own "parallel process" with respect to chaos and opinion: whereas philosophy engages variations on its plane of immanence, science derives variabilities on its plane of reference, and this provides the most fundamental basis to mark the distinction between these two brains. Recall that philosophy makes its plane a matter of infinite movement, the variations—i.e., intensities or differences—of which it surveys at infinite speed. Science also concerns intensities, which Deleuze defines in terms of inequality. "Every phenomenon refers to an inequality by which it is conditioned,"[33] he writes, but in our so doing, we are already bound to distinguish between inequality, or difference, and the surface effect, or phenomenon, it produces. Beneath the algorithmic pretense of diversity, or of any ordered, identifiable, and determinate world, Deleuze says, we glimpse inexact and unjust calculation—the "inequality"—that "forms the condition of our world." Deleuze's God is not a Cartesian, a Newtonian, or even a Laplacian, for the God who calculates never ultimately balances accounts. He is an imprecise mathematician, Deleuze says, who always leaves a remainder—namely, the world. Intensities are differential insofar as they emerge from differences, whether these are differences of "level, temperature, pressure, tension, potential" etc., but by the same token these measures already indicate the sense in which intensity will be covered up or over by surface effects. In other words, "we know intensity only as already developed within an extensity, and as covered over by qualities."[34]

"Difference is that by which the given is given as diverse,"[35] Deleuze writes, but good sense makes difference into diversity by introducing variables that agglomerate into values and probabilities. Good sense enacts the taming of chance. Indeed, when Deleuze and Guattari say that science extracts variable from chaos, they mean that the struggle that science conducts against chaos prompts the creation of problems and methods that allow us to slow down chaos by introducing "constants or limits," by creating "centers of equilibrium." In effect, the eject of science typically imposes a degree of order on chaos by: (1) selecting "only a small number of independent variables within coordinate axes"; and (2) considering the relationships between these variables "whose future can be determined on the basis of the present (determinist calculus)" or when we introduce "so many variables at once that the state of affairs is only statistical (calculus of probabilities)."[36] Whereas the philosophical brain surveys the surface of the plane of immanence, ordinating intensities (singularities) in the drama of creating concepts, science creates an entirely different relation to chaos. Science does not superject a plane of immanence, as philosophy does, nor does it inject itself onto a plane of sensation, as art does; rather, science develops functions so as to eject elements onto a plane of reference.

This formulation, surfacing as it does in *What Is Philosophy?* can be understood in light of Deleuze's much earlier critique of science in *Difference and Repetition,* where these predilections are grasped, in their more extreme and ultimately doxological manifestations, according to what he calls "good sense." The complement to common sense and, thence, the other half of the image of thought, good sense concerns the "faculty of distribution."[37] Whereas common sense imposes the presuppositions of the subject and object upon difference, good sense sustains these presuppositions (or opinions) in time by distributing difference according to a progressive homogenization. The nature of good sense is such that, when we begin with a field of differences, even an ostensible chaos of differences, this state of affairs is immediately posited as the "origin = x of the diverse." As such, difference only exists as the precedent for its own equalization. "Perhaps good sense even presupposes madness in order to come after and correct what madness there is an any prior distribution," Deleuze writes, but if this is the case then "madness"—or, let us say, the chaotic distributions of difference—only exists in order to subject intensity to new distributions or to "repartitions."[38] Good sense is God's other side, the careful divinity who watches over the endless series of sloppy calculations with an even greater measure of care. He is a sedentary figure with all the time in the world because his time is commensurate with the world he corrects, slowly but inevitably eliminating difference with entropic assurance. Neither inherently optimistic nor pessimistic, good sense assumes its patient posture on the basis of an enduring certainty that, "given world enough and time," we inexorably move from the proliferation of differences to their reduction. In this regard, Deleuze suggests, the equalization or homogenization of difference ("good sense") can be called "thermodynamic":

> Intensity defines an objective sense for a series of irreversible states which pass, like an "arrow of time," from more to less differenciated, from a productive to a reduced difference, and ultimately to a cancelled difference. We know how these themes of a reduction of difference, a uniformalization of identity, and an equalization of inequality stitched together for the last time a strange alliance at the end if the Nineteenth Century between science, good sense, and philosophy. Thermodynamics was the powerful furnace of that alloy.[39]

Notably, the genetic power of difference that is covered up by its very creation suggests a kind of structure that threatens, under the pressure of less delicate operations, to lapse into "grand dualisms" that include intensity (and its avatars) only within the framework of an opposition. Of course, we have already dealt with this kind of dualism in the oppositions between chaos and opinion, between molecular and molar, etc. And, no doubt, these oppositions seem to suggest the essential structure and perform the essential dynamic of Deleuze's work, providing the poles from which his philosophical tapestry is ostensibly hung and woven. As Fredric Jameson has written, it is impossible to read certain of Deleuze's writings (and perhaps, according to this logic, all of them) "without being stunned by the ceaseless flood of references that tirelessly nourish these texts, and which are processed into content and organized into dualisms."[40] At first glance, of course, we can hardly disagree with this diagnosis because

Deleuze's work both alone and with Guattari seems to generate countless opportunities for eager dualists. And yet, as Deleuze warns of the dialectic, "There is a false profundity in conflict," for the arrangement of oppositions consists in the superficial effect of difference—or, rather, the effect that difference inspires when we absolve ourselves of discerning its depths.[41] Deleuze's own ostensible dualisms should never be understood in the spirit of pure oppositions because they invariably exist along a dynamic continuum where we pass from one to the other along a field of qualitative differences.

But good sense accomplishes precisely this opposition and organization, and this is the point of Deleuze's critique. Inasmuch as good sense rests upon an irreversible synthesis of time, its arrow always shoots from the differences of the past to the homogenized diversity ("differenciation") of the future, from "the particular to the general." What does this mean? If we take the past to be that which was ruled by unruly differences, given over to mad distributions and "crowned anarchy,"[42] then we must understand time itself to consist in a transformation of difference from the improbable to the probabilistic. "In effect, since every partial system has its origin in a difference which individualizes its domain, how would an observer situated within the system grasp this difference except as past and highly 'improbable,' given that it is behind him?" Deleuze asks. "On the other hand, at the heart of the same system, the future, the probable and the cancellation of difference are identified in the direction indicated by the arrow of time—in other words, the right direction."[43]

The Series

While common sense distinguishes the "double identity" of subject and object, those universals would remain perpetually indeterminate unless they are actualized from one moment to the next. Thus, while common sense gave us the subject among determinable and recognizable objects, good sense represents the habitual synthesis whereby those constituents are prolonged into a future that is increasingly predictable. Given the predilections of common sense and good sense, no less the contract between philosophy and science that they tacitly underwrite, we might well ask: how can we elude the constraints of the image of thought and begin, however modestly, to introduce chaos into thinking? Our answer and our conclusion consists in turning to what Deleuze calls a "series," for the philosophical procedures he associates with seriality are no less grounded in the domain of science. On both planes, the series emerges as a means to order and organize thinking—and yet, the serial form also allows us to entertain the possibility of thinking in the absence of an image and, thereby, the possibility of foreclosing the regularity (common sense) and predictability (good sense) to which thought has been subjected. Prima facie, we are liable to regard this possibility with skepticism because we traditionally define the nature of any series by virtue of its order, whether we are talking about a succession of numbers, letters, or events. Typically defined, a series is a sequence in which a single rule relates each term (or member) to one or more of its predecessors. The rule of a series governs the distribution of its term and, by extension, its overarching pattern. There is no series without a rule, but as we will see, the series can induce a variation within those rules or can make variation its rule.

In any case, consider the series in its most basic and generative sense, namely, as an orderly sequence. The ancient Greeks spoke of a series as *kathexes*, an adverb meaning "thereafter," which, in the absence of a noun, assumes the sense of "that which is subsequent" (*hexes* alone signifies an "echo"). Indeed, among the Greeks the word was often associated with acts of selling and exhibition, such that we might take as the primary image of a series to be the ledger or audit, the publication of an orderly sequence of numbers. Consider a simple row of numbers—100, 106, 112.36, 119.1016, etc.—and let us say that these are financial entries in a ledger. Following the sequence, we quickly find a pattern: to derive the next term, we calculate the initial sum plus 6 percent interest that is compounded each year, ad infinitum. But what does it mean to determine a series, and why is this process of interest to what Deleuze once called chaos? Just as a broker looks for patterns in the fluctuations of the stock market, or SETI searches for some organization in the radio waves bombarding the earth, or a mathematician tries to divine some arrangement in the distribution of prime numbers, so any series depends upon our capacity to grasp when it begins, however complexly, to repeat. A series can intervally approach infinity or it can infinitely approach a limit within an interval, but what matters is that we can successfully predict the next ordinate.

Once intuited, a series consists in a promise that a pattern will hold, but do we really know that this sequence will repeat forever, can we really be sure that the law holds for every term? In the famous Mertens conjecture, for instance, mathematicians found that a statement about natural numbers held for the first 7.8 billion terms and was assumed to hold infinitely, until a supercomputer proved it false in 1983. Thus, mathematicians will grudgingly admit that the continuation of a series ultimately demands an "act of faith,"[44] but what I want to underscore here is the *factum brutum* of the act itself, the resolution at whatever point to presume that a pattern holds beyond our means or desire to continue calculating. A series depends upon a decision, the selection of a limit beyond which we do not or cannot venture: it is the point at which we are compelled to surrender skepticism and to trust that the pattern holds. Thus, however justified, natural, or inevitable it may seem, the determination of a series necessarily relies on the arbitrary decision. The mathematical act of faith may surrender at one level to the unforgiving nature of infinity, but at another level it consists in an action, the designation of a boundary. The determination of a series and the establishment of its order, which amount to the same thing, are only brought about by some measure of capriciousness, and this mean truth should hardly surprise us because it is instilled in life and language at virtually every juncture. The appeal that every child makes to infinite causality, the repetition of "Why? Why? Why?" invariably runs up against the unimpeachably parental "because" that draws a border, however arbitrary, which we are forced to respect.

The final term in a series, even in an infinite series, is the terminus or *horos,* a boundary (literally, in Greek, a rock) that traditionally demarcates a piece of property. In marking a series, then, we dissociate ourselves from the hermetic theater of the succession, which seems to have always already existed in nature or number. The determination of a series, such as the totalization of a set, demands an external instance—the exceptionality of the one who stands apart—but our point here is that this exclusion is also an intervention. The border of any series is determined by a decision, the intervention of judgment into a sequence: the law is not always-already there but, rather, what we find there, and the border term is not an incidental stopping-point but, rather, the co-incidence of one series with another. As Deleuze writes, "*the serial form is necessarily realized in the simultaneity of at least two series,*"[45] and at a fundamental level we can regard this according to the structure of the subject that we have outlined to this point. A series acquires its order; it becomes a series only in relation to a subject.

In this regard, Deleuze treats the serial form according to a kind of "games theory," for what games almost invariably demand is a set of rules, of presuppositions, which define the parameters of our play and, thence, the sequence that "plays out." Not only do players typically alternate turns, one move following or determining the next, but the overarching design of so many games consists in a number of rules that condition the orderly progression of a series. Hence, what Deleuze says of these games—and, indeed, we could add, of the series that they produce—is that they are doubly "partial." In the first place, they deal with "only one part of human activity," the remainder of which remains out of play, excluded from the game.[46] While we can make our move, these activities are circumscribed by rules that dictate not only the meaning of what we do, but the invariable limits: having completed our turn, and in the absence of whatever stipulations might redouble that move ("roll again"), we cede control to another player, etc. In the second place, and more importantly, these games are partial insofar as, "even if they are pushed to the absolute, they *retain chance only at certain points,* leaving the remainder to the mechanical development of consequences or to skill, understood as the art of causality."[47] The roll of this dice is permitted, but only at preset and privileged moments when we integrate chance into the series.

But if the proliferation of series seems to give way to an implicit or explicit invocation of regularity and predictability—in short, the subject—Deleuze pursues a sense of the serial form that, reverberating with particular trends in science, suggests a wholly different relationship to chaos. Whence Deleuze's invocation, in *The Logic of Sense,* of a new kind of game—an "ideal game." Inasmuch as it is "not enough to oppose a 'major' game to a minor game," Deleuze writes, what makes a game ideal, among other things, is that it will create the conditions to experience and ramify chance beyond any set of preestablished rules.[48] What if chance, instead of being determined within an invariable (and, should we not say, metaphysical-moral?) structure, were unleashed as the very principle upon which the variability of the structure, of a fluid and self-synthesizing structure, was contingent? In short, "it is necessary to imagine other principles, even those which appear inapplicable, by means of which the game would become pure"—which is to say, by means of which we can make life into an artful game whereby we affirm chance, and the game of chance into an art of living. The ideal game, Deleuze says, makes the world a "work of art." But what are the principles of this work?

1) There are no preexisting rules, each move [or throw] invents its own rules; it bears upon its own rule. 2) Far from dividing

and apportioning chance in a really distinct number of throws,
all throws affirm chance and endlessly ramify it with each throw.
3) The throws therefore are not really numerically distinct. They
are qualitatively distinct, but are the qualitative forms of a single
cast which is ontologically one. 4) Such a game—without rules,
with neither winner nor loser, without responsibility, a game of
innocence, a caucus-race, in which skill and chance are no longer
distinguishable—seems to have no reality. Besides, it would amuse
no one. (*Logic of Sense*, 59)

But is it possible to understand the world according to such a game? Perhaps, in light of our discussion to this point, we can say that the turn to
such a game culminates an implicit trajectory, for the movement from
common sense to good sense to seriality correlates with the transformation of scientific thought that certain theorists ("chaosticians") have analyzed in the development of modern sciences. For instance, Ilya Prigogine
and Isabelle Stengers, whose work notably influenced Deleuze, describe
the lineage of modern scientific paradigms as the transfomation from
mechanism to dynamism to, finally, self-organization (chaos theory).[49]
In other words, this sequence describes a movement from the static and
certain reversibility of mechanism, which we have aligned with common
sense, to the irreversible distribution of thermodynamics, which we have
aligned with good sense, to an emergent sense or science of complexity.
"Science turns against opinion, which lends to it a religious taste for unity
or unification," Deleuze and Guattari explain this transformation, but this
must be understood in the sense that science "turns within itself against
properly scientific opinion as Urdoxa, which consists sometimes in a
determinist prediction (Laplace's God) and sometimes in probabilistic
evaluation (Maxwell's demon)."[50]

Beyond determinism and probability, beyond common sense and good
sense, philosophy and science develop a sense of seriality that embraces
chaos rather than effacing it. "Chaos is not an inert or stationary state, nor
is it a chance mixture," Deleuze and Guattari argue, because chaos does
not describe a deterministic state of affairs, which would exist reversibly;
rather, chaos consists in a ceaselessly modulating field of differences within
which complexity occasionally and unpredictably precipitates itself. As
Prigogine and Stengers stressed, and as Deleuze and Guattari do in turn,
even the most dissipative ("far from equilibrium") systems are capable of

spontaneously producing self-organization.[51] In analyzing these systems, then, Deleuze looks to the domain of science to develop the conceptual means with which to approach chaos without, for all that, tipping over into its flux completely. Across the sciences of complexity, ranging from biology and chemistry to physics and astronomy, we discover that what appear to be the most irredeemably unpredictable systems may well conceal, or even induce, profound patterns and structures.

It is in the latter sense that Deleuze turns to the model of a Markov chain, providing as it does the expression of a "semi-accidental series," of a future that emerges independently of the past states or prior events. Neither determined nor arbitrary, this series approximates what we have called the ideal game, which is "like a series of draws in a lottery, each one operating at random but under extrinsic conditions laid down by the previous draw."[52] The Markov chain is "a mixture of the aleatory and the dependent"[53] in which elements are linked along a line that, with each cast of the dice, with the introduction of each new element, reformulates its own rules based on the newly constituted series. "There are no preexisting rules, each move [or throw] invents its own rules; it bears upon its own rule."[54] In the feedback loop between what has been and what is to come, the calculation of each new term retroactively changes the significance of the series, so that the rules are variable—or, rather, the only consistent rule is variability, variation, variety. "Far from dividing and apportioning chance in a really distinct number of throws, all throws affirm chance and endlessly ramify it with each throw." As Nietzsche explains, as Deleuze repeats, and as we can conclude, the series as such comes to consist in the "iron hand of necessity throwing the dice of chance."[55]

CHAPTER 8

Elemental Complexity and Relational Vitality: The Relevance of Nomadic Thought for Contemporary Science

Rosi Braidotti

The "Knowing" Subject as Multiplicity, Process, and Becoming

The theoretical core of nomadic thought consists in the rejection of the unitary vision of the subject as a self-regulating rationalist entity and of the traditional image of thought and of the scientific practices that rest upon it. These are traditionally expected to implement a number of Laws that discipline the practice of scientific research and police the borders of what counts as respectable, acceptable, and fundable science. In so doing, the Laws of scientific practice regulate what a mind is allowed to do, and thus they control the structures of our thinking. Foucault's *Archaeology of Knowledge* (1966) is a foundational text and a crucial point of origin for a critique of this intrinsically normative image of thought at work within the allegedly "objective" practice of science.

The "knowing" subject of European philosophical humanism has historically claimed to be structured along the royal axes of self-reflexive individualism and self-evident scientific rationality. They are indexed on a linear notion of time and a teleological vision of the purpose of scientific thought. The deeply entrenched anti-humanism of poststructuralist thought becomes radicalized in Deleuze and Guattari's conceptual redefinition of the practice of thinking and hence also of scientific reason. Nomadic subjectivity moves beyond the mere critique of both the identitarian category of a sovereign self and dominant subject position on the one hand and the image of thought that equates subjectivity with rational consciousness on the other. The linear, Enlightenment-based vision of human progress as the effect of a deployment of scientific reason upon the theater of the world historical experience of humans is accordingly abandoned. An alternative vision is proposed of both the thinking subject, of his or her evolution, and of the structure of thinking.

I will develop this insight in two parallel directions: the first is a sociopolitical critique of the identity politics of the allegedly universal subject of knowledge. The second is a more conceptual critique of the rationalist takes of subjectivity but also of what it means to think at all.

As for the former, social criticism of science: following the insights of feminist (Lloyd, Irigaray, Harding, Haraway), postcolonial (Spivak), and race theorists (Gilroy), I take the universalistic claim of scientific practice to task and expose the cluster of vested interests and particularities that actually sustain its claims. A binary logic of self-other opposition is at work in this falsely universalistic model, which results in reducing "difference" to pejoration, disqualification, and exclusion.[1] Subjectivity is postulated on the basis of sameness, i.e., as coinciding with the dominant image of thought and representation of the subject as a rational essence. Deleuze and Guattari offer the perfect synthesis of this dominant image of the subject as masculine/white/heterosexual/speaking a standard language/property-owning/urbanized. This paradigm equates the subject with rationality, consciousness, moral and cognitive universalism. This vision of the "knowing subject"—or the "Man" of humanism—constructs itself as much by what it includes within the circle of his entitlements as in what it excludes. Otherness is excluded by definition, which makes the others into structural and constitutive elements of the subject, albeit by negation. Throughout Western philosophy, Otherness has been constructed with distressing regularity along intertwined axes of sexualization, racialization, and naturalization.[2] The others—women or sexual minorities; natives, indigenous and non-Europeans, and earth or animal others—have been marginalized, excluded, exploited, and disposed of accordingly. The epistemic and world-historical violence engendered by the claim to universalism and by the oppositional view of consciousness lies at the heart of the conceptual Euro-centrism that Deleuze and Guattari are attacking.

In so far as rhizomatic subjectivity and nomadic thought challenge the methodological Euro-centrism of philosophy, they also critique the complicity between this discipline of thought and nationalism. After Deleuze and Guattari, it becomes not only feasible but even imperative to question the habit of thought that reiterates the Euro-centric character of philosophy. The question of what is European about Continental philosophy, for instance, can and should be raised as a way of suspending the assimilation of philosophy into a hegemonic vision of European consciousness.[3] In this respect I concur with Foucault's assessment that Deleuze brings to completion the denazification of European philosophy and lays to rest the belligerent vision of the philosopher's tasks as guardian of the *status quo*.

As a vintage Deleuzian, I am very keen to stress this aspect of Deleuze and Guattari's work and to add a note of concern at the relative neglect suffered by the methodological implications of nomadic thought for contemporary science and for philosophy. Especially in the last few years, there has been increasing compartmentalization and abstraction in the reception of Deleuze's thought. Thus, scholarship about the "political" Deleuze is often distinct from that on the "cultural" or the "epistemological" aspects of Deleuze and Guattari's complex *corpus*. This is a problematic tendency, which often projects a spurious veneer of scientism over a thinker whose rigor is—or should be—beyond dispute. Contrary to the ongoing recompartmentalization of Deleuze's thought, I would want to stress the practical implications of his philosophy for today's world as well as its methodological innovations. Deleuze's critique of capitalist power relations, for instance, is an integral part of his reconceptualization of the specific domain and responsibility of science. I would want to argue, in other words, for a multilayered unity of thought within the Deleuzian rhizomatic universe and to call accordingly for a more multifaceted reception of his work, by his admirers and his critics alike. The point of Deleuze and Guattari's work is to empower us to think differently about the analytic and historical preconditions for new forms of materialist and complex subjectivity. This transformative ethics provides the inner cohesion of their work.

The second direction in which I deploy Deleuze's insights is conceptual. Deleuze and Guattari's defense of the parallelism between philosophy, science, and the arts is not to be mistaken for a flattening out of the differences between these different genres of intellectual pursuit. There is no easy isomorphism, but rather an ontological unity among the three branches of knowledge. Deleuze and Guattari take care to stress the differences between the distinctive styles of intelligence that these practices embody, but these qualitative differentiations are possible only because they are indexed on a common plane of intensive self-transforming Life energy.

This continuum sustains the ontology of becoming that is the conceptual motor of nomadic thought. In so far as science has to come to terms with the real physical processes of an actualized and defined world, it is less open to the processes of becoming or differentiation that characterize Deleuze's monistic ontology. Philosophy is a subtler tool for the probing intellect, one that is more attuned to the virtual plane of immanence, to the generative force of a generative universe, or "chaosmosis," which is nonhuman and in constant flux.

Deleuze calls the radical alterity of a mind-independent reality "Chaos" and defines it positively as the virtual formation of all possible forms. The generative force of "Chaos" is the source of its vital elemental powers of renewal and transformation—through endless processes of actualization of determinate forms.

The key elements of this conceptual operation are: firstly, Deleuze and Guattari stress the notion of a deep vitalist interrelation between ourselves and the world, in an ecophilosophical move that binds us to the living organism that is the cosmos as a whole. By extension, for philosophers, this leads to a redefinition of the activity of thinking away from the rationalist paradigm to a more intensive and empathic mode. Thinking is the conceptual counterpart of the ability to enter modes of relation, to affect and be affected, sustaining qualitative shifts and creative tensions accordingly.

Secondly, there is the shift away from an epistemological theory of representation to ontology of becoming. By way of comparison, Lacan— and Derrida with him—defines Chaos epistemologically as that which precedes form, structure, and language. Confined to the unrepresentable, this post-Hegelian vision reduces "Chaos" to that which is incomprehensible. For Deleuze, however, following Spinoza, Bergson, and Leibniz, Chaos en-/unfolds the virtual copresence of any forms. This produces a number of significant shifts: from negative dialectics to affirmative affects; from entropic to generative notions of desire; from a focus on the constitutive outsides to a geometry of affects that require mutual actualization and synchronization; from an oppositional and split to an open-ended, relational vision of the subject; from the epistemological to the ontological turn in philosophy.

As a consequence, one can venture the conclusion that the main implication for the practice of science is that the scientific Laws need to be retuned according to a view of the subject of knowledge as a complex singularity, an affective assemblage, and a relational vitalist entity. This could also be described as a metamethodological shift.

The Decentering of Anthropo-Centrism

One of the great innovations of Deleuze's philosophy is the rigorous brand of methodological pacifism that animates it. The monistic ontology that he adapts from Spinoza, to which he adds the Bergsonian time-continuum, situates the researcher—be it the philosopher, the scientist, or the artist—in a situation of great intimacy with the world. There is no violent rupture or separation between the subject and the object of her inquiry, no predatory gaze of the cold clinician intent upon unveiling the secrets of nature.[4] An elemental ontological unity structures the debate. This nonessentialist vitalist position calls for more complexity and diversity in defining the processes of scientific inquiry.

As a result of the conceptual shifts introduced by Deleuze and Guattari, the burden of responsibility is placed on us to develop new tools of analysis for the subtler degrees of differentiation and variations of intensity that characterize the formation of the subject. The nomadic vision of the subject as a time continuum and a collective assemblage implies a double commitment, on the one hand to processes of change and on the other to a strong sense of community—of our being in *this* together. Our copresence, that is to say the simultaneity of our being in the world together, sets the tune for the ethics of our interaction. Our ethical relation requires us to synchronize the perception and anticipation of our shared, common condition. A collectively distributed consciousness emerges from this—i.e., a transversal form of nonsynthetic understanding of the relational bond that connects us. This places the concepts of relation and affect at the center of both the ethics and the epistemic structures and strategies of the subject.

The decentering of anthropocentrism is one of the effects of the scientific advances of today—from biogenetics to evolutionary theories. This means that the naturalized, animals, or "earth-others"—in fact, the planet as a whole—have ceased to be the boundary-markers of the metaphysical uniqueness of the human subject. They have consequently stopped acting as one of the privileged terms that indexes the European subject's relationship to otherness. Otherness or pejorative difference has a long and established history in scientific practice. As I suggested earlier, scientific inquiry and exploration has been historically an outward-looking enterprise, framed by the dominant human masculine habit of taking for granted free access to and the consumption of the bodies of others. As a mode of relation negative difference is Oedipalized, in that it is both hierarchical, and hence structurally violent, and saturated with projections, identifications, and fantasies. These are centered on the dyad: fear and desire, which is the trademark of the Western subject's relation to his "others." They are also the expression of his sense of entitlement to knowledge—that systematic "curiosity" that, from Odysseus on, has been the emblem of applied intelligence in our culture. Desire and fear are the motor of the quest for knowledge about, and control over, the others.

Take evolutionary theory as an example: fear and desire for the animal outside is echoed by fear for the animal within. The wild and passionate animal in us may be cheered as the trace of a primordial evolutionary trajectory or cherished as a repository of unconscious drives, but it also calls for containment and control for exactly the same reasons. The technologies to discipline these wild passions through specific practices, as Foucault teaches, are coextensive with the making of high scientific discourses and institutions. The technologies of control are both gender-ized and racialized to a very high degree, and historically they have harped with distressing regularity on the disposable bodies of "others."

Deleuze's work rests on the by-now proven hypothesis that this mode of Oedipal relation to one's object of inquiry is currently being restructured. As a result of the advances of our own scientific knowledge, a bioegalitarian turn is taking place that encourages us to engage in an animal relationship with animals—the ways hunters do and philosophers can only dream of.[5] The challenge today is how to transform, deterritorialize, or nomadize the human-nonhuman interaction in philosophical practice, so as to bypass the metaphysics of substance and its corollary, the dialectics of otherness, secularizing accordingly the concept of human nature and the life that animates it. With Deleuze and Guattari, I would speak of a generic becoming-minoritarian/animal as a figuration for the humanoid hybrids we are in the process of becoming. It is clear that our science can deal with this post-anthropocentric shift, but can philosophy rise to the occasion?

The answer lies in the ethical underpinnings of the nomadic vision of philosophical thinking. The displacement of anthropocentrism and the recognition of trans-species solidarity are based on the awareness of "our" being in *this* together; that is to say: environmentally based, embodied, and embedded and in symbiosis with each other. Biocentered egalitarianism is a philosophy of radical immanence and affirmative becoming, which activates a nomadic subject into sustainable processes of transformation. Becoming-animal/nonhuman consequently is a process of redefinition of one's sense of attachment and connection to a shared world, a territorial space. It expresses multiple ecologies of belonging, while it enacts the transformation of one's sensorial and perceptual coordinates, in order to acknowledge the collective nature and outward-bound direction of what we call the self. The subject is fully immersed in and immanent to a network of nonhuman (animal, vegetable, viral) relations. My code word for this relentless elemental vitality of Life itself is *zoe*. The *zoe*-centered embodied subject is shot through with relational linkages of the symbiotic, contaminating/viral kind that interconnect it to a variety of others, starting from the environmental or eco-others. This nonessentialist brand of vitalism reduces the hubris of rational consciousness, which far from being an act of vertical transcendence, is rather recast as a downward push, a grounding exercise. It is an act of unfolding of the self onto the world and the enfolding within of the world.

Methodological Implications

Transpositions

The postanthropocentric shift entails a number of important theoretical and methodological implications for the practice of science. Deleuze's nomadic vision of the subject does not necessarily preclude the position in which the subject is placed by scientific methods of inquiry, but displaces it in a number of structural and productive ways. The more obvious innovations are methodological: nomadic thought requires less linearity and more rhizomatic and dynamic thinking processes. A commitment to process ontology and to tracking the qualitative variations in the actualization of forces, forms, and relations forces some creativity on the usually sedate and conformist community of academic philosophers and institutional scientists.

The basis for this practical method is that of affirmative differences, or creative repetitions, i.e.: retelling, reconfiguring, and revisiting the concept, phenomenon, event, or location from different angles. This is the application of the key concept of Spinoza's perspectivism, but it also infuses it with a nomadic tendency that establishes multiple connections and lines of interaction. Central to this is the notion of repetition as the internal return of difference, not of sameness. It is creative mimesis, not static repetition. Revisiting the same idea or project or location from different angles is therefore not merely a quantitative multiplication of options, but rather a qualitative leap of perspective. This leap takes the form of a hybrid mixture of codes, genres, or modes of apprehension of the idea, event, or phenomenon in question.

This shift calls for an intensive form of interdisciplinarity, transversality, and boundary-crossings among a range of discourses. Deleuze's wide-ranging reading habits offer a perfect example of this approach: references to modernist literature and music coexist peacefully alongside comments on contemporary mathematics and physics. This transdisciplinary approach enacts a rhizomatic embrace of diversity in scholarship that can only be sustained by a double talent: enormous erudition and a rigorous structure of thought. No wonder most academics flee from the challenge of Deleuze's texts, arguing that they are either too complex or too "unfocused" for their liking. Nomadic texts are not written for those who confuse thinking with the mere exercise of sedentary protocols of institutional reason. Deleuze brings transdisciplinarity to bear in the actual methods of thought, thus making diversity into a core issue.

I have also defined this methodological approach as "transpositions." This is a situated method of tracking the qualitative shifts or ontological leaps from generative chaos or indeterminate forms to actualized or determinate forms, while avoiding the pitfalls of subjectivism and individualism. Theoretically, a transposition has a double genealogical source: from music and genetics. In both cases it indicates an intertextual, cross-boundary or transversal transfer of codes, in the sense of a leap from one code, field, or axis into another. These leaps are not to be understood merely in the quantitative mode of plural multiplications, but rather in the qualitative sense of complex multiplicities. In other words, it is not just a matter of weaving together different strands, variations on a theme (textual or musical), but rather of playing the positivity of difference as an ontological force and of setting up adequate frames of resonance for their specific rhythms of becoming. As a term in music, transposition indicates variations and shifts of scale in a discontinuous but harmonious pattern. It is thus created as an in-between space of zigzagging and of crossing: nonlinear and chaotic, but in the productive sense of unfolding virtual spaces. Nomadic, yet accountable and committed; creative and hence affective, relational and cognitively driven; discursive and also materially embedded—it is coherent without falling into the logocentric inflexibility of instrumental rationality.

In genetics "transposition" refers to processes of mutation, or the transferral of genetic information, that occur in a nonlinear manner, which is nonetheless neither random nor arbitrary.[6] This is set in opposition to the mainstream scientific vision that tends to define the gene as a steady entity that transmits fixed units of heredity in an autonomous and self-sufficient manner and genetic variation as random events. Transposable moves appear to proceed by leaps and bounds and are ruled by chance, but they are not deprived of their logic or coherence. Central to genetic transpositions is the notion of material embodiment and the decisive role played by the organism in framing and affecting the rate and the frequency of the mutations. Transpositions occur by a carefully regulated dissociation of the bonds that would normally maintain cohesiveness between the genes, which are laid out in a linear manner on the chromosome. Nobel Prize–winning geneticist Barbara McClintock shows that as a result of the dissociative impact, a mutation occurs that splits the chromosome into two detached segments. The rate of the mutation of these "jumping genes" is internally determined by the elements of the cell itself, and thus is not prewritten in the gene. The notion of transposition emphasizes the flexibility of the genome itself. This implies that the key to understanding genetics is the process itself, the sequence of the organized system. This can be traced a posteriori as the effect of the dissociative shifts or leaps, but these controlling agents remain immanent to the process itself and are contingent upon the rearrangements of the elements. In other words, genetics information is contained in the sequence of the elements, which in turn means that the function and the organization of the genetic elements are mutable and interdependent.

In other words, our genetic system does not operate under the law of evolution defined as selection and aggressive struggle for survival. Rather, it proceeds by variations and adaptations—that is to say by qualitative changes and structural transformations of the nonlinear and anti-teleological kind. Consequently, as Hilary Rose put it ever so wittily: "DNA, far from being the stable macho molecule of the 1962 Watson-Crick prize story, becomes a structure of complex dynamic equilibrium."[7] Nobody and no particle of matter is independent and self-propelled, in nature as in the social. Ultimately, genetic changes are under the control of the organisms, which, under the influence of environmental factors, are capable of influencing the reprogramming of the genetic sequence itself.

As if it were capable of "learning from experience," the organism defined as the host environment of the genetic sequence plays an interactive and determining role in the transmission of genetic information. Haraway sums it up brilliantly: "A gene is not a thing, much less a master molecule, or a self-contained code. Instead, the term 'gene' signifies a mode of durable action where many actors, human and non-human meet."[8] In other words, genetic evolution is about sustained changes and unpredictable variations, not about a neotranscendental discourse of survival of the fittest.

Resting on the assumption of a fundamental and necessary unity between subject and object, the method of transpositions offers a contemplative and creative stance that respects the visible and hidden complexities of the very phenomena it attempts to study. This makes it a paradigmatic model for scientific knowledge as a whole, which becomes indexed on a definition of "Life" as *zoe*, that is to say a dynamic entity, and not as an entropic force aiming at homeostatic stability. It also shows affinity with spiritual practices like Buddhism, not in a mystical mood, but in a cognitive mode.

Further, the notion of transposition describes the connection between the text and its social and historical context, in the material and discursive sense of the term. The passion that animates all scientific and philosophical endeavors for nomadic thought is a concern for our historical situation, in so-called advanced, postindustrial cultures at the start of the third millennium. In my work, this has become an emphasis on *amor fati*, not as fatalism, but rather in the pragmatic mode of the cartographer. In other words, my working definition of a nomadic scientific method in the human and social sciences (the "subtle sciences"), as well as in genetics, molecular biology, and evolutionary theory (the "hard sciences"), cannot be dissociated from an ethics of inquiry that is adequate to and respectful of the complexities of the real-life world we are living in. I am committed to start my critical work from this complexity, not from a nostalgic reinvention of an all-inclusive holistic ideal. I want to think from here and now, from Dolly my sister and OncoMouse as my totemic divinities; from missing seeds and dying species. But also, simultaneously and without contradiction, from the staggering, unexpected, and relentlessly generative ways in which Life, as *bios* (human) and as *Zoe* (nonhuman), is fighting back. This is the kind of materialism that makes me an anti-humanist nomadic subject at heart and a joyful member of multiple companion species in practice.[9]

Nonlinear Time

Linearity is especially problematic on the methodological front for radical epistemologies and marginal discourses. The question is how to implement a coherent but nonhierarchical system of knowledge transfer and the transmission of the cultural and political memory of a past that is often not recognized by official institutional culture. Foucault's early work on genealogies as counter-memories of resistance is again foundational. Deleuze expands this pioneering effort into a conceptual critique of the powers of historical discourses over the human and social sciences. Deleuze's favorite example for this is the ravages accomplished by the teaching of the history of philosophy as a normative canonical discipline.

A more poignant example of the same methodological issue—how to intervene creatively upon a canonized *corpus* of texts and a fixed idea of historical time—is the transmission of the cultural and political capital of a centuries-old political movement such as socialism, pacifism, or feminism. Linearity is the dominant time of *Chronos*, not the dynamic time of becoming or *Aion*, and as such it is a very inadequate way of accounting for intergenerational relations among political subjects of a countercultural movement: for instance, women who belong to different historical phases of the women's movement or youth that were born after the end of Communism. Nowadays, with a third feminist wave in full swing,[10] it is difficult to avoid both the hierarchical Oedipal narrative of mothers and daughters of the feminist revolution and the negative passions that inevitably accompany such narratives. The best antidote is an anti-Oedipal approach to the question of intergenerational ethics. It results in the need to find adequate accounts for the zigzagging nature of feminist intellectual and cultural memories, as well as their respective political genealogies.

This raises methodological issues of how to account for a different notion of time, focused on *Aion*, the dynamic and internally contradictory or circular time of becoming. Thus, instead of deference to the authority of the past, we have the fleeting copresence of multiple time zones, in a continuum that activates and deterritorializes stable identities. This dynamic vision of the subject enlists the creative resources of the imagination to the task of enacting transformative relations and actions in the present. This ontological nonlinearity rests on Spinoza's ethics of affirmation and becoming that predicates the positivity of difference. A nomadic methodology posits active processes of becoming: we need flows of empowering desire that mobilize the scientific subject and activate him or her out of the gravitational pull of envy, rivalry, and ego-indexed claims to recognition. This project requires a serious critique of institutional structures and modes of Oedipalized, competitive, and negative interaction.

Remembering in the nomadic mode is the active reinvention of a self that is joyfully discontinuous, as opposed to being mournfully consistent, as programmed by phallogocentric culture. It destabilizes the sanctity of the past and the authority of experience. This is the tense of a virtual sense of potential. Memories need the imagination to empower the actualization of virtual possibilities in the subject. They allow the subject to differ from oneself as much as possible while remaining faithful to oneself, or in other words: enduring. Becoming is molecular, in that it requires singular overthrowing of the internalized simulacra of the self, consolidated by habits and flat repetitions. The dynamic vision of the subject as assemblage is central to a vitalist, yet anti-essentialist theory of desire, which also prompts a new practice of ethics.

Desire is the propelling and compelling force that is driven by self-affirmation or the transformation of negative into positive passions. This is a desire not to preserve, but to change: it is a deep yearning for transformation or a process of affirmation. Empathy and compassion are key features of this nomadic yearning for in-depth transformation. Proximity, attraction, or intellectual sympathy is both a topological and qualitative notion: it is a question of ethical temperature. It calls for an affective framing for the becoming of subjects as sensible or intelligent matter. The affectivity of the imagination is the motor for these encounters and of the conceptual creativity they trigger.

One of the ways in which this can be accounted for is through an intensive or affective mapping of how each of us relates to and interacts with the ideas/events/codes as processes. I shall return to the affective element later. Ethically, each researcher or writer has to negotiate the often dramatic shifts of perspective and location that are required for the implementation of a process-oriented—as opposed to concept-based and system-driven—thought. In other words, we need to rise to the challenge of more conceptual creativity.

Defamiliarization: Toward an Anti-Oedipal Science

On the methodological front, de-Oedipalizing the relationship to the nonhuman others is a form of radical pacifism that sets strong ethical requirements upon the philosophical subject. It requires for instance a form of disidentification from a century-old habit of anthropocentric thought and humanist arrogance.

A few words about the method or strategy of disidentification first. Disidentification, estrangement, or defamiliarization from certain established views entails a radical repositioning on the part of the subject. In poststructuralist feminism, for instance, this discursive strategy has also been discussed in terms of disidentifying ourselves from familiar and hence comforting values and identities, such as the dominant institutions and representations of femininity and masculinity, so as to move sexual difference toward the process of becoming-minoritarian.[11] Disidentification involves the loss of familiar habits of thought and representation. Spinozist feminist political thinkers like Genevieve Lloyd and Moira Gatens (1999) argue that socially embedded and historically grounded changes require a qualitative shift of our "collective imaginings," or a shared desire for transformations. Race and postcolonial theories have resulted on the one hand in the critical reappraisal of blackness[12] and in the other on radical relocations of whiteness.[13] Critical studies of whiteness and transnational citizenship also produced a renewed sense of critical distance from set conventions about cultural identity. This has resulted among others in a postnationalistic redefinition of Europe as the site of mediation and transformation of its own history.[14]

Defamiliarization is a sobering process by which the knowing subject evolves from the normative vision of the self he or she had become accustomed to. The frame of reference becomes the open-ended, interrelational, multisexed, and transspecies flows of becoming by interaction with multiple others. A subject thus constituted explodes the boundaries of humanism at skin level.

For example, the Deleuzian unorganic body is delinked from the codes of phallologocentric functional identity.[15] The "body without organs" sings the praise of anomalies. It also introduces a sort of joyful insurrection of the senses, a vitalist and pan-erotic approach to the body. It is recomposed so as to induce creative disjunctions in this system, freeing organs from their indexation to certain prerequisite functions. This calls for a generalized recoding of the normative political anatomy, and its assigned bodily functions, as a way of scrambling the old metaphysical master code and loosening its power over the constitution of subjectivity. The subject is recast in the nomadic mode of collective assemblages. The aim of deterritorializing the norm also supports the process of becoming-animal/woman/minoritarian/nomadic.

Nonhuman others are no longer the signifying system that props up the humans' self-projections and moral aspirations. Nor are they the gatekeepers that trace the liminal positions in between species. They have rather started to function quite literally as a code system of their own. This neoliteral approach to otherness begins to appear with the masters of modernity. With Freud and Darwin's insights about the structures of subjectivity a profound inhumanity is opened up at the heart of the subject. Unconscious memories drill out timelines that stretch across generations and store the traces of events that may not have happened to any one single individual and yet endure in the generic imaginary of the community. Evolutionary theory acknowledges the cumulated and embodied memory of the species. It thus installs a timeline that connects us intergenerationally to the prehuman and prepersonal layers of our existence. From the angle of critical theory, psychoanalysis propels the instance of the unconscious into a critique of rationality and logocentrism. Evolutionary theory, on the other hand, pushes the line of inquiry outside the frame of anthropocentrism into a fast-moving field of sciences and technologies of "life." The politics of life itself is the end result of in-depth criticism of the subject of humanism.[16] Pushed even further with philosophical nomadology,[17] the metaphorical dimension of the human interaction with others is replaced by a literal approach based on the neovitalist immanence of life.

This deeply materialist approach has important ethical implications. In terms of the human-animal interaction, the ego-saturated familiarity of the past is replaced by the recognition of a deep bioegalitarianism, namely that "we" are in *this* together. The bond between "us" is a vital connection based on sharing *this* territory or environment on terms that are no longer hierarchical nor self-evident. They are rather fast evolving and need to be renegotiated accordingly. Gilles Deleuze and Félix Guattari's theory of "becoming animal" expresses this profound and vital interconnection by positing a qualitative shift of the relationship away from species-ism and toward an ethical appreciation of what bodies (human, animal, others) can do. An ethology of forces emerges as the ethical code that can reconnect humans and animals. As Deleuze put it: the workhorse is more different from the racehorse than it is from the ox. The animal is not classified according to scientific taxonomies, nor is it interpreted metaphorically. Rather, it is taken in its radical immanence as a body that can do a great deal, as a field of forces, a quantity of speed and intensity, and as a cluster of capabilities. This is posthuman bodily materialism laying the grounds for bioegalitarian ethics.[18]

Affirmative Ethics

What this means concretely is that Deleuze's vision of science cannot be dissociated from his Spinozist project of developing a science of ethics based on affects and an ethology of forces. Nietzsche is an important precedent. The eternal return in Nietzsche is the repetition of difference as positivity. Deleuzian-Nietzschean perspective ethics is essentially about transformation of negative into positive passions, i.e., moving beyond the pain. This does not mean denying the pain, but rather activating it, working it through. Again, the positivity here is not supposed to indicate a facile optimism or a careless dismissal of human suffering. In order to understand the kind of transmutation of values I am defending here, it is important to depsychologize this discussion about positivity, negativity, and affirmation and approach it instead in more conceptual terms. We can then see how common and familiar this transmutation of values actually is. The distinction between good and evil is replaced by that between affirmation and negation, or positive and negative affects.

What is positive in the ethics of affirmation is the belief that negative affects can be transformed. This implies a dynamic view of all affects, even those that freeze us in pain, horror, or mourning. The slightly depersonalizing effect of the negative or traumatic event involves a loss of ego-indexes perception, which allows for energetic forms of reaction. Clinical psychological research on trauma testifies to this, but I cannot pursue this angle here. Diasporic subjects of all kinds express the same insight. Multilocality is the affirmative translation of this negative sense of loss. Following Glissant, the becoming-nomadic marks the process of positive transformation of the pain of loss into the active production of multiple forms of belonging and complex allegiances.[19] Every event contains within it the potential for being overcome and overtaken—its negative charge can be transposed. The moment of the actualization is also the moment of its neutralization. The ethical subject is the one with the ability to grasp the freedom to depersonalize the event and transform its negative charge. Affirmative ethics puts the motion back into emotion and the active back into activism, introducing movement, process, becoming. This shift makes all the difference to the patterns of repetition of negative emotions. It also reopens the debate on secularity, in that it actually promotes an act of faith in our collective capacity to endure and to transform.

What is negative about negative affects is not a normative value judgment but rather the effect of arrest, blockage, rigidification, that comes as a result of a blow, a shock, an act of violence, betrayal, a trauma, or just intense boredom. Negative passions do not merely destroy the self, but they also harm the self's capacity to relate to others—both human and nonhuman others—and thus to grow in and through others. Negative affects diminish our capacity to express the high levels of interdependence, the vital reliance on others that is the key both to a nonunitary vision of the subject and to affirmative ethics. Again, the vitalist notion of Life as "zoe" is important here because it stresses that the Life I inhabit is not mine, it does not bear my name—it is a generative force of becoming, of individuation and differentiation: apersonal, indifferent, and generative. What is negated by negative passions is the power of Life itself—its *potentia*—as the dynamic force, vital flows of connections, and becoming. And this is why neither should they be encouraged nor should we be rewarded for lingering around them too long. Negative passions are black holes. In affirmative ethics, the harm you do to others is immediately reflected on the harm you do to yourself, in terms of loss of *potentia*, positivity, capacity to relate, and hence freedom. Affirmative ethics is not about the avoidance of pain, but rather about transcending the resignation and passivity that ensue from being hurt, lost, and dispossessed. One has to become ethical, as opposed to applying moral rules and protocols as a form of self-protection: one has to endure.

Endurance is the Spinozist code word for this process. Endurance has a spatial side to do with the space of the body as an "enfleshed" field of actualization of passions or forces. It evolves affectivity and joy, as in the capacity for being affected by these forces, to the point of pain or extreme pleasure. Endurance points to the struggle to sustain the pain without being annihilated by it. Endurance has also a temporal dimension, about duration in time. In my work, I have theoretically transposed endurance into a Deleuzian concept of sustainability. This has been my answer to the dilemma of how to combine an ontology of becoming (the task of philosophy) with a focus on actualized and definite objects of inquiry (the task of science).

There is, however, a conceptual difficulty at stake in this discussion, which has important implications for a Deleuzian practice of science. For philosophical nomadism, the problem with sustainability is that it has the feel of a qualitative (intensive) criterion, but, in fact, it is a quantitative one. Sustainability clashes with duration, which is not the same as pluralistic speed. Speed is a trajectory, it is spatialized and it deals with concepts like bodies or actualized entities. Duration, on the other hand, is an intensity, which deals with abstract diagrams or lines of becoming. Sustainability as a quantitative measure runs the risk of becoming effective and operational within the logic of advanced capitalism, which it aims to undermine, namely the liberal individual responsibility for one's well-being. This is an axiomatic system capable of considering all qualities as quantities and of instrumentalizing them in order to feed itself.

My response to this is that the concept of sustainability is particularly relevant, albeit it in an intriguing manner. It has two reference points: sustainability as a temporal notion (duration of Life as *zoe*) and sustainability as an intensive notion (the ability to sustain intensities). It brings together both the *durée* of life and the intensities of encounters. This is for me the most direct consequence of the relational vitality and elemental complexity that mark Deleuze's thought: life is not a teleological notion, and thus it does not seek or want to express itself. Life, simply by being life, expresses itself. This is why I defend the idea of *amor fati*. To accept *amor fati* is to change one's relation to life, and in doing so, perhaps change life itself— allow it expressive intensities it would not otherwise possess.

This complex task is facilitated by adopting a nonunitary vision of nomadic subjectivity, which coupled with the idea of desire as plenitude and not as lack, produces a more transformative approach to the ethics of thinking. The stated criteria for this new ethics include: nonprofit; emphasis on the collective; viral contaminations; and a link between theory and practice, including the importance of creation. They are not moral injunctions, but frames for an ongoing experiment. They need to be experimented with collectively, so as to produce effective cartographies of how much bodies can take, or thresholds of sustainability. They also aim to create collective bonds, a new affective community or polity.

This must include an evaluation of the costs involved in pursuing active processes of change and of recognition of the pain and the difficulty these entail. The problem of the costs within the schizoid logic of our times concerns mostly *potestas*, the quantitative, not *potentia*, or incorporeal intensities. Creation or the invention of the new can only emerge from the qualitative intensities and thus cannot apply to a notion that measures the tolerance of bodies as actualized systems. Hence again another aspect of the ethical question: if in the name of encouraging (prehuman or individual) life (*zoe*), we value the incorporeal invention of quality and primarily affect and precept; if (again, following Deleuze) we insist on the incorporeal insistence of affects and precepts or becoming (as distinguished from affected bodies and perceptions of entities), then how can we use a concept of sustainability to argue against the cost of fidelity to the concept or the precept? That would involve a corporeal criterion to the incorporeal. This is a conceptual double bind and a true ethical dilemma.

How can we combine sustainability with intensity? One line I would propose is to hold everyone, not only exceptional people like writers or thinkers but just anyone (*homo tantum*), accountable for the ethical effort to be worthy of the production of affect and precept. It is a noble ethics of overcoming the self and stretching the boundaries of how much a body can take; it also involves compassion for pain, but also an active desire to work through it and find a way across it. The ethical question would therefore emerge from the absolute difference (or *différend*) between incorporeal affects, or the capacity to experiment with thresholds of sustainability, and our corporeal fate as such and such an affected body. What ethical criterion can we invent in the context of this difference? How can one (simultaneously?) increase affectivities as the capacity to invent or capture affect and look after the affected bodies? What kind of synchronized effort could achieve this aim? In other words, what is the "cost" of the capacity to be affected that allows us to be the vehicle of creation? What would a qualitative concept of cost be? This is the core of the nomadic ethics agenda. It includes interrelationality and a relation to otherness, on the model of mutual specification and collective becoming.

CHAPTER 9

Numbers and Fractals: Neuroaesthetics and the Scientific Subject

Patricia Pisters

Scientific knowledge of the brain has evolved, and carried out a general arrangement. The situation is so complicated that we should not speak of a break, but rather of new orientations . . . It is obviously not through the influence of science that our relationship with the brain changed: perhaps it was the opposite, our relationship with the brain having changed first, obscurely guiding science. . . . The brain becomes our problem or our illness, our passion, rather than our mastery, our solution or decision. We are not copying Artaud, but Artaud lived and said something about the brain that concerns all of us: that "its antennae turned towards the invisible," that it has the capacity to "resume a resurrection from the death."

—Gilles Deleuze, *Cinema 2: The Time-Image*

THE POPULARITY of mathematics and scientific reasoning in contemporary culture is evident from popular television series such as *Numb3rs* (CBS, since 2005) and Hollywood films about mathematicians such as *Good Will Hunting* (Gus van Sant, 1997), *A Beautiful Mind* (Ron Howard, 2000), and *Proof* (John Madden, 2005). Besides a general fascination for mathematics as principle underlying all kind of phenomena in our world, these films also indicate a particular interest in the brain, the mind of the scientist in particular. It is a classic trope to feature the scientist as a mad mind, but contemporary cinema shows that something else is at stake as well. The mathematician in contemporary popular culture may be socially not adapted, even paranoid and schizophrenic, but what is going on in this particular mind is no longer considered as completely deranged and totally opposed to a normal functioning brain. Instead, the scientific and "mad" mind in popular culture seems to indicate deep metaphysical and ontological truths. In this essay I will propose the hypothesis that the popular obsession with mathematics and the mind of the scientist is related to a Deleuzian ontology of differences, repetitions, and folds that finds a full expression in films that not only deal with mathematics and madness in terms of their content (such as the Hollywood films indicated above) but also in terms of their particular "neuroaesthetic" style. The limitless beauty and power of numbers and geometric figures such as spirals and fractals that are at the basis of this style are related to the limitless powers of thought where madness and metaphysics fold and unfold in each other and point toward an "ungrounded ontology" of the virtual. Departing from the idea that "the brain is the screen,"[1] I will start by looking at the changing relationships between cinema and the (neuroscientific) brain, from the movement-image and the time-image to a contemporary "neuro-image." In the second part, I will develop this concept of the neuro-image further by looking at two films of Darren Aronofksy, *Pi* (1997) and *The Fountain* (2007), relating them to Deleuze's ideas on thought in *Difference and Repetition* and on Leibniz's Baroque mathematics in *The Fold*. I will argue that aesthetically both of these films give us, in two different ways, direct access to a scientific brain that reaches out to the universal questions of the genesis of the universe: life, death, and belief. In doing so, these films could be considered as the extreme poles of contemporary neuroaesthetics in cinema that reveal its profound relations to the forces of the virtual.

Cinema and the Brain

The Movement-Image: Thought, Tropes, and the Mad Scientist

Before entering the specific characteristics of contemporary cinema's relation to neuroscientific discoveries about the brain and the mind (and the importance of mathematics), it is useful to recall how in the past cinema dealt with these issues. As Deleuze reminds us in *The Time-Image,* cinema has always had a profound relation with thinking, the connection to the brain even being cinema's essence: "It is only when movement becomes automatic that the artistic essence of the image is realized: *producing a*

*shock to thought, communication vibrations to the cortex, touching the ner-
vous and cerebral system directly."*[2] Cinema produces "nooshocks" to
the brain; cinema and the brain enter into a circuit that produces new
thoughts. The cinema of Eisenstein, which combines emotional images
of attraction with intellectual montage, is for Deleuze the paradigmatic
example of the organic way in which the movement-image connects to
thought. In cinema of the movement-image, thinking proceeds by tropes,
metonymies, metaphors, inversions, oppositions, attractions, and so
forth. Deleuze calls this a form of action-thought,[3] where there is always
a relation between man and the world. Hence its organic qualities, always
relating to a synthetic Whole in which everything can be kept together.
Where classic American cinema operates mainly through metonymi-
cal principles of continuity editing, Eisenstein's films produce shocks to
thinking through metaphorical montage. The prime example here is the
intellectual montage in *October* (1927) where, for instance, images of the
commander-in-chief Kerensky entering a room in the Winter Palace are
dialectically intercut with a peacock, producing the synthetic thought of
his vanity (and eventual downfall).

Another way in which classical cinema or the movement-image is
related to the brain and to mental processes is its relation to memory and
to the imagination (dreams and fantasies). Here again the organic composition of the Whole is determining the place of memory and imagination.
Memories are always presented out of the necessity of a clearly defined
point in the present to which we always return. The flashbacks in *Daybreak*
(Marcel Carné, 1939), for instance, are motivated by the character's fate in
the present. Moving back and forth in time is always related to the organic
Whole of the tragic conditions of the present, and it explains how this
present has come about. Hitchcock's *Spellbound* (1945) most famously
shows how dreams figure in our unconscious minds. Here the main character suffers from amnesia and anxiety attacks whenever he sees black
stripes on a white surface (a fork scratching on a white table cloth, stripes
on pajamas, ski marks in the snow). The famous dream sequence designed
by Dalí is shown as an oneiric flashback that can be decoded by the psy-
choanalysts in the film to discover its significance and, again, the composition of a Whole that makes sense.

If we look at more literal images in which the brain and the mind
of the scientist feature in the movement-image, we find that classical
cinema presents us quite frequently with the trope of the mad scientist,

together with that of the brain, as metaphors for all kind of fears. In the fifties, a whole range of horror movies produced the B-genre of so-called "brain movies." A telling example is *Fiend without a Face* (Arthur Crabtree, 1958), in which a mad scientist secretly experiments in thought materializations to detach consciousness and give it an entity of its own. The experiments he performs on his own brain literally get a boost when his instruments are hit by lightening (another trope of mad science since Dr. Frankenstein), and he discovers that the atomic plant near his laboratory provides an even more powerful aid. Of course, the experiments soon become uncontrollable, and the scientist realizes that he has created an invisible fiend of expanded intelligence, a mental vampire that feeds on atomic power and the brains plus spinal cords of human beings. While the representation of materialized thought as literal disembodied brains is quite over the top, the metaphoric relations between the unleashed brain and the dangers of nuclear power during the Cold War are still striking. And again we see here how in the movement-image thought, tropes and the brain are connected in an organic way. The mad scientist soon regrets the effects of his thoughts and experiments when they disturb the Whole.

The Time-Image: Belief, Theorem/Problem, and Schizophrenia

With the arrival of the time-image, cinema's relationship to the brain takes on a different form. Deleuze now refers to Artaud, who argued that cinema can be brought together with the innermost reality of the brain. "But this innermost reality is not the Whole, but on the contrary a fissure, a crack," Deleuze adds.[4] This crack is quite literally related to a break with the organic sensory-motor link of man with the world, so that the time-image produces seers who find themselves struck by something intolerable in the world and confronted with something unthinkable in thought. So the "task" of cinema is now no longer to produce thought in showing the connections to the Whole, but to produce "the psychic situation of the seer, who sees better and further than he can react, that is, think."[5] When the sensory-motor link of man with the world is broken and we can no longer be sure of the exact relationships between man and world, of the great organic links between what he sees and hears and the world, it is *belief* that becomes the ontological basis of the image. Belief becomes a power of thought that replaces the model of knowledge. And, Deleuze adds, this also changes the nature of belief: "Whether we are Christians or atheists, in our universal schizophrenia, *we need reasons to believe in this world*."[6]

I will return to the schizophrenic nature of this belief in cinema, but first it is important to recall Deleuze's observation that thought in the time-image no longer operates through figures and tropes but becomes theorematic and problematic. The cinematographic image no longer only gives us an association of images, but "it also has the mental effect of a theorem, it makes the unrolling of the film a theorem . . . it makes thought immanent to the images."[7] In this description of the image as theorematic thought, Deleuze refers to mathematics:

> In fact there are two mathematical instances which constantly refer to each other, one enveloping the second, the second sliding into the first, but both very different in spite of their union: these are the theorem and the problem. A problem lives in the theorem, and gives it life, even when removing its power. The problematic is distinguished from the theorematic (or constructivism from the axiomatic) in that the theorem develops internal relationships from principle to consequences, while the problem introduces an event from the outside—removal, addition, cutting—which constitutes its own conditions and determines the "case" or cases: hence the ellipse, hyperbola, parabola, straight lines and the point are cases of projection of the circle on its secant planes, in relation to the apex of a cone. This outside of the problem is not reducible to the exteriority of the physical world any more than to the psychological interiority of a thinking ego. . . . There is a decision on which everything depends, deeper than all the explanations that can be given for it. . . . As Kierkegaard says, "the profound movements of the soul disarm psychology," precisely because they do not come from within.[8]

In my discussion of *Pi* and *The Fountain* I will return to the theorem and the problem. For now, it is important to see how thought in the time-image is related to the exteriority of a belief, a choice that has to be made outside any mode of knowledge.

Another important characteristic of the time-image is that it no longer refers to a Whole defined as an organic open totality as in the movement-image, defined by montage of associations or attractions between parts, the set, and the changing Whole. In the time-image, the Whole is the outside, which means that what is important now is what happens in between images, a spacing that according to Deleuze means that "each image is plucked from the void and falls back to it. . . . Given one image, another image has to be chosen which will induce an interstice *between* the two. This is not an operation of association, but of differentiation, as mathematicians say, or of disappearance, as physicists say: given one potential, another one has to be chosen, not any whatever, but in such a way that a difference of potential is established between the two, which will be productive of a third or something new."[9] Thought becomes irrational and not necessarily organic. It is known that for the time-image Deleuze has demonstrated how this inorganic power of thought and belief is related to the unsummonable of Welles, the inexplicable of Robbe-Grillet, the undecidable of Resnais, the impossible of Duras, and the incommensurable of Godard. In all these types of time-images, the power of thought is related to a confusing and confused experience of time and the reality of the virtual of the past and future.[10]

A final aspect of the time-image that has to be addressed is its schizophrenic nature. Deleuze never explicitly related the cinema book to his work on schizoanalysis that he developed with Guattari; however a schizoanalysis of cinema seems to be called for when reading the cinema books together with the two volumes on capitalism and schizophrenia, *Anti-Oedipus* and *A Thousand Plateaus*. In "Is a schizoanalysis of cinema possible?" Ian Buchanan refers to crucial passages in *A Thousand Plateaus* where Deleuze and Guattari turn to cinema to explain the schizoanalytical implications of the delirium and the delirious nature of cinema. He furthermore argues that the cinematographic apparatus is, in essence, schizoanalytic: the frame, the shot, and montage corresponding to the Body without Organs, the assemblage, and the abstract machine. Buchanan thus suggests that "if cinema is delirium we need a theory of delirium to form the basis of a schizoanalysis of cinema."[11] In "Delirium Cinema or Machines of the Invisible?" I have elaborated this suggestion by looking at the relationship between clinical and critical forms of schizophrenia, and its implications for cinema.[12] I here argue that a schizoanalysis of cinema entails a shift in the conception of the cinema as a "machine of the visible" (the images rendering the visible world) to cinema as a "machine of the invisible" (images making the invisible—thought, the virtual—visible).

Although from its beginnings cinema has had the potentiality to render thought visible, the time-image fully developed this possibility of giving the image the delirious and psychotic potentiality to giving us direct access to the mind, to consciousness, to the invisible: a camera-consciousness.

These schizoanalytic implications of the time-image become increasingly evident and important. Besides their fascination for mathematics and the (mad) mind of the scientist, contemporary cinema and popular culture at large are also populated with schizos, delirious and delusional characters, characters that suffer from amnesia and other brain disorders. Contemporary cinema has quite literally entered the mind of its characters, playing all kind of tricks with the mind of the spectators as well. Mind-game movies such as *The Game* (David Fincher, 1997) and *Minority Report* (Steven Spielberg, 2002) present complex narratives that play with the spectators' expectations.[13] In *Tierra* (Julio Medem, 1996), the main character is schizophrenic or perhaps even dead; *The Butterfly Effect* (Eric Bress, 2004) deals with blackouts and schizophrenic hallucinations; in *Eternal Sunshine of the Spotless Mind* (Michel Gondry, 2004), the classic screwball theme of remarriage is literally played out in the mind of the two main characters, who have their memories of one another erased by a company called Lacuna; *The Machinist* (Brad Anderson, 2004) presents events from the traumatized mind of its protagonist; in *Fight Club* we enter the movie quite literally on a ride through the brain's neural network, only to find out at the end that the two protagonists are actually one, a "crystal character," so to speak, whose virtual and actual sides are both real. Many more examples can be given of contemporary film characters who seem to have lost their minds or are, as Anna Powell indicates, in "altered states."[14] In any case, the wondering and wandering character of the time-image described by Deleuze seems to be replaced by a "delusional" character of what could be called the neuro-image (or schizo-image). The wondering characters of the time-image after World War II are paralyzed by something intolerable they see in the world. Such "schizophrenic" characters, whose brains we literally enter in the neuro-image, are not so much traumatized by something intolerable, but lost in the vortex of screens, data, and information of contemporary globalized media culture.

The Neuro-Image: Brains, Chaos, Interdisciplinarity

If we consider contemporary cinema as belonging to a new type of image, this always has to be seen in a continuum with the other types of image.

Just as a number of different relationships between the movement-image and the time-image can be distinguished,[15] neuro-images are distinct from the other two types of images, but they are also profoundly related to them. As Gregg Lambert and Gregory Flaxman argue in "Ten Propositions on the Brain," the future of the cinematic brain lies particularly in the development of the crystalline image.[16] The neuro-image is, in any case, a development of the time-image. In the conclusion of *The Time-Image*, Deleuze has already suggested several characteristics of video and digital images that came into being at the time of the publication of the cinema books. Here, Deleuze argues that cinema will change, but by no means is it meant to die, as long as it is produced from a will to art. So, on the one hand, the contemporary neuro-images are not at all dependent on new technologies of the digital age. On the other hand, contemporary digital and media culture seem to form an intrinsic part of the new image because it makes the chaos into which all images plunge very palpable and sensible. Neuro-images relate to chaos and complexity theory and to all kinds of neuroscientific findings on the workings of the brain. Increasingly, neuroscience demonstrates that aberrations of the brain tell us something about the normal functioning of the brain, and that the differences between madness and metaphysics can be very subtle, perhaps only a matter of differentiation: "cerebral creation or deficiency of the cerebellum."[17] Furthermore, Deleuze has indicated that already in the time-image a specific "brain-cinema" emerged (for instance the cinema of Kubrick and Resnais) that connects the inside and outside: "Between the two sides of the absolute, between the two deaths—death from the inside or past, death from the outside or future—the internal sheets of memory and the external sheets of reality will be mixed up, extended, short-circuited, and form a whole moving life, which is at once that of the cosmos and the brain, which sends out flashes from one pole to the other."[18]

It is worth recalling the three characteristics of the new image mentioned by Deleuze because they point toward this paradoxical nondependency and dependency to new (visualization) technologies. First, the organization of space is different. Instead of privileged directions, space has become omnidirectional, and there no longer seems to be an outside or out-of-field: "they have a right side and a reverse, reversible and non-superimposable, like a power to turn back on themselves."[19] Second, the screen itself can no longer be considered as a window or a painting, but rather it constitutes a table of information, a surface inscribed with "data," where information replaces nature, the brain-city replaces the eye of nature: "the image is constantly being cut to another image, being printed through a visible mesh, sliding over other images in an 'incessant stream of messages,' and the shot itself is less like an eye than an overloaded brain endlessly absorbing information: it is the brain-information, brain-city couple that replaces that of eye-Nature."[20] Finally, the new image gives way to a new psychological automaton, already present in the time-image, where characters are no longer psychologically (and psychoanalytically) motivated but become the performance of a speech-act: Bresson's "models," Rohmer's puppets, Robbe-Grillet's hypnotized ones, and Resnais's zombies. In Resnais, "there are no more flashbacks, but feedbacks and failed feedbacks, which, however, need no special machinery."[21]

The neuro-image is related to chaos and complexity theory. Translated into mathematical terms, it is related to the fractal organization of many elements in nature, where self-same structures constitute infinite variations of "difference and repetition." In *What Is Philosophy?* Deleuze and Guattari refer to the fractal nature of the plane of immanence.[22] Aesthetically, fractals have mesmerizing power, rhythm, and beauty.[23] I will return to the mathematics of fractals and numbers in the second part. For now it is important to note the relation between chaos and complexity theory and the neuro-image trying to create some temporary order. This temporary order is fractal, reproducing self-same relations of macro and micro parts of Chaos or reproducing other basic but infinitely variable geometric patterns. These complex patterns are related to a profound connection between microcosmic and macrocosmic perspectives that are held together "mid-way" in our brains. Some neuroscientists even argue that the brain itself is fractally structured.[24] Finally, chaos theory and schizophrenia seem to be connected in their nonlinear dynamic, as is suggested in recent neuroscientific studies.[25]

Clearly, all these connections and fractal enfoldings, variations and patterns of different levels of existence, the relation to mathematics and neuroscience and the clinical and metaphysical implications of the neuro-image, asks for an interdisciplinary approach that remains to be developed more profoundly. For now I will further analyze some characteristics of the neuro-image from a film-philosophical perspective, taking the films of Darren Aronofsky as a central focus.

Neurocinema and the Forces of the Virtual

Pi: Visceral Qualities of the Brain of a Mathematician

π (pi) is the Greek symbol for 3.1415926535 . . . (on to infinity), the ratio of a circle's circumference to its diameter. *Pi*'s main character is a mathematician, Max Cohen (Sean Gulette), who is obsessed by finding a universal pattern in the numbers that pi represents. He searches for a way of predicting the fluctuations in the stock market and is chased by both a Wall Street company and a group of Hasidic Jews. The mathematical theories and numerological references to the Kabbalah (the Gematria) that Aronofsky makes to are true, but the film is not about mathematics. Rather, it is about cool math theories and the belief that mathematics is related to the divine.[26] When asked if *Pi* is a science fiction film, the director emphasizes that it is sci-fi in the tradition of Philip K. Dick, a tradition of inner exploration. "It's pushing science forward within the fiction realm, so I think ultimately it is a science fiction film."[27] So, the film gives us a kind of pop-mathematics that nevertheless relates to bigger underlying questions about the origin of the world and cosmic or divine spirituality that are typical for the neuro-image.

Pi is a subjective movie. The images are completely shot from the perceptions of Max Cohen; they render his mental space.[28] I will return to the ways in which this is done stylistically. For now, it is important to see that Max suffers from paranoid schizophrenia, the initial idea for the film being about paranoiac schizophrenia.[29] The expression of a subjective mental space and the references to schizophrenia are again typical for the neuro-image. *Pi* refers to the brain on three different levels. First, as already indicated, the image itself is completely mental (the brain is the screen). At the same time, there are metaphorical references to the brain. The brain no longer stands for dangers of nuclear power and mad scientists. The brain is now seen as a complex computer network that can go wrong. In one scene Max literally discovers a bug in Euclid (Max's homemade computer), which can be read in relation to the "bugs" in his mind. Finally, inside Max's delirious hallucinations, which always happen after heavy headache attacks, he also sees an actual brain on the floor in the underground and in the washbasin ("that was Rudolph Guiliani's brain that we borrowed," Aronofsky jokes in the 1998 Artisan Entertainment DVD edition commentary about this brain).

What is very striking about *Pi*'s different relations to the brain is its visceral qualities. First of all, the choice of the film stock is remarkable. *Pi* is

Figure 9.1 Max sees an actual brain on the floor in the underground and in the washbasin. From the film *Pi* (1998), by Darren Aronofsky.

shot in black and white reversal film, which is difficult to develop and has no gray tones, only sharply contrasted black and white. Furthermore, the camera angles and movements bring the camera into Max's head space by always staying close to him, or showing his (hallucinating) point of view. Sometimes, a little camera on his body (a Snorri-cam) gives the sense of agitated movement (for example, when he is chased in the underground). "We wanted the audience to experience how it was to be a renegade genius mathematician standing on the verge of insanity," Aronofsky said.[30] The soundtrack is another important element that affects the senses directly. Max's headaches are announced by an uncontrollable shaking of his thumb, followed by what Aronofsky calls a hip-hop montage of Max taking pills, where the images and music get into a fast rhythm. Then, as the pain kicks in, we (with Max) physically experience it through a sharp and penetrating sound that sharply penetrates our brain. When Max opens his computer and gets the bug (an ant) out, his fingers are sticky with a sort of slimy substance that Max first looks at, listens to, smells, and then tastes. The actual three pounds of brains in the hallucinations are touched with a pen (which causes the sharp sound again) and are finally literally attacked and smashed. Contrary to many of the brain-films in the time-image, where the mental landscape is more often expressed in a more distant way (even if that could be very violent or passionate), the mental space and brain in the neuro-image are very physical and sensuous.

Pi's Theorematic Nature and Geometric Style

If we take Deleuze's definition that a theorem develops internal relationships from principle to consequences, we can consider *Pi* as a theorematic film. At several points in the movie Max's voiceover states his assumptions: "(1) Mathematics is the language of nature. (2) Everything around us can be translated and understood through numbers. (3) If you graph the numbers of any system, patterns emerge. Therefore everywhere in nature there are patterns." Clearly this is the theorem the film proposes; the principle of mathematics, as underlying principles of everything, should then also make it possible to decode and predict the patterns of the stock market, which, according to Max is "a living organism, screaming with life." This is the theorem Max explores.

So what about the numbers and geometric figures in the film? In the 1998 Artisan Entertainment DVD extras (the director's commentary track, actor's commentary track, notes on π, music video, and behind-the-scenes montage), the attraction of the number π is indicated as the attraction for the circle: It is perhaps the fact that a circle is probably the most perfect and simple form known to man. And lying at the heart of it is a specific, unchanging number that also manages to appear everywhere in functions of geometry, statistics, and biology. It keeps popping its head up, reminding us that it is there and defying us to understand why. Pi is a nonrepeating decimal that reaches out into infinity, and the biggest challenge now is to compute the number farther than before, farther than the many billions it has reached now. Besides the circle and the number pi, the Fibonacci sequence and the spiral are other mathematical figures that return in the film. The Fibonacci sequence is a sequence of numbers in which each succeeding number in the sequence is the sum of the two preceding ones (1, 1, 2, 3, 5, 8, 13, . . .). It appears that many phenomena in the world reproduce Fibonacci sequences (flowerhead arrangements, the human body, DNA, voting patterns). In *Pi*, Max also looks at the Fibonacci patterns of the stock market. Spiral logarithms are other frequent patterns in nature (seashells, whirlpools, hurricanes, an embryo, the galaxy). Many have argued that these patterns must have a meaning, perhaps a divine meaning.[31]

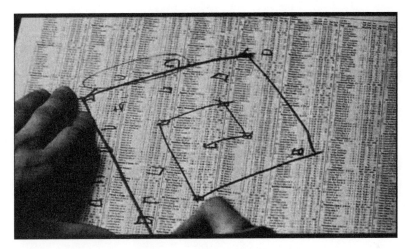

Figure 9.2 Max looks at the Fibonacci patterns of the stock market.
From the film *Pi* (1998) by Darren Aronofsky.

In *Pi*, these mathematical figures are not just the theme of the film. They are also repeated in the style of the film: circles, spirals, and Fibonacci sequences are frequently expressed in the mise-en-scène and in the camera movements. The title sequence is a graphic design of circles, spirals, and other figures such as neurons displayed on a sequence of the number pi. Elements of these graphic figures reappear later in the film. Spirals appear in the mise-en-scène, for example, in the milk in a cup of coffee, in the smoke of a cigarette, and in the arrangement of a game of Go that Max's friend Saul leaves behind when he has committed suicide. Fibonacci sequences are drawn on the financial paper and by the Jewish numerologists, and circles are featured, for example, in zeros on the computer screen, while at the same time the camera encircles Max in a 360-degree pan. In a consistent style, form and content repeat each other or are enfolded in one another.

Pi has several endings; at least there is an ambiguity about what actually happens, which is again a characteristic of the neuro-image that it shares with the time-image. We don't know whether Max actually sees the divine light, whether he actually drills his own brain, whether from a hyperactive state of positive symptoms of schizophrenia he falls into a catatonic state, or whether he actually has freed himself from his "brain power" and can accept life on a phenomenological scale of enjoying

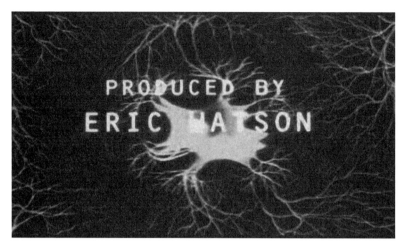

Figure 9.3 The title sequence includes images of neurons displayed on a sequence of the number pi.
From the film *Pi* (1998) by Darren Aronofsky.

nature as it appears and the company of the neighbor girl. A final observation to make about *Pi* is the presence of ants. As already indicated Max
finds an ant in his computer, and in fact his apartment is swarming with
ants—even the brain he attacks at the end is crawling with them. In the
DVD commentary, Aronofsky tells his motivation for including this
theme: when on a holiday in Mexico he visited a small unknown Mayan
temple and discovered that it was literally covered with ants. He suddenly saw that humans (the important civilization of the Mayans) and
ants were all the same; he saw the groundlessness of the "I," which is also
a groundlessness Max discovers the closer he comes to the mysteries of
the universe. Aronofsky's story and the way the ants are present in *Pi* also
resonates with Deleuze's conclusions of *Difference and Repetition*, where
he develops a nonrepresentative, preindividual way of thinking about
difference: "The ultimate, external illusion of representation is this illusion that results from its internal illusions—namely, that groundlessness
should lack differences, when in fact it swarms with them. What, after
all, are Ideas, with their constitutive multiplicity, if not these ants which
enter and leave through the fractured I."[32] I will return to *Difference and
Repetition* in the last part. For now, it is important to mark a connection
between *Pi*'s way of dissolving Max's identity by introducing ants in the

Figure 9.4 Max finds an ant in his computer.
From the film *Pi* (1998) by Darren Aronofsky.

image. Now, the ant is no longer a metaphor for a "bug" in the system but a rhizomatic connection between different forms of life without a determined "I" (subjectivity).

The Fountain: The Belief of a Brain Surgeon

Just like *Pi*, Aronofsky's more recent film is a particular kind of science fiction film. Moving between three layers of time (sixteenth-century Spain, twenty-first-century North America, twenty-fifth century somewhere in space), it is basically the story of the same couple, played by Hugh Jackman and Rachel Weisz. (Information for the film refers to *The Fountain* [dir. Darren Aronofsky, 2005], DVD edition, Warner Bros. Home Video [2006]; Special Features: "Inside *The Fountain*: Death and Rebirth" [production story].) In the twenty-first century, Tommy is a brain surgeon who tries to find a cure for his wife, Izzy, who has a brain tumor. This story unfolds into the past where conquistador Thomas wants to save Spain and her Queen Isabelle by finding a holy tree in the New Spain, and into the future, where the astronaut Tom travels through space in a biospheric "bubble-ship" (he forms an organic unity with the tree in his spaceship) and tries to deal with the previous stories. *The Fountain* is inspired by many elements, such as the myth of the fountain of youth, Spanish conquistador stories, and ancient

Figure 9.5 *Tom travels through space in a biospheric "bubble-ship."*
From the film *The Fountain* (2006), by Darren Aronofsky.

Mayan culture, as well as David Bowie's "Space Oddity's Major Tom" and cowriter Ari Handel's PhD in neurosciences. "I'll take different threads from different ideas and weave a carpet of cool ideas together," Aronofsky says.³³ This rhizomatic way of thinking and creating has led to the story of Tommy and Izzy, but the film exceeds its narrative on all sides and levels.

The Fountain is a neuro-image not simply because of its references to neurosciences and the biology of the brain. Even though the film is partly science fiction, partly happening in outer space, it is actually taking place in inner space. As with *Pi*, the film gives us a mental landscape of its main male character, only this time we are not in a mad brain, but in a metaphysical brain that reaches out into the past and the future. As a new type of image, the film shows how the organization of space has become very different as an omnidirectional space, most obviously in the futuristic parts where bulbs, spheres, and lights float in and out the frame from all directions, which are repeated in the mise-en-scène of the lights in the other parts as well. As I will show in the next section, many elements in the composition of the image return. The elements that create spatial omnidirectionality, therefore, are part of the other layers of time as well. *The Fountain* no longer gives us a window to the world, but it has become a "brain-information" table. And again, as in *Pi*, this brain has to be seen as a very sensuous one:

the senses of touch, smell, and taste are frequently emphasized in close-up shots. But all these combined elements do not give us a window onto the world or reality, but they form a new tapestry of thoughts and affects. That this is not so much dependent on the numerical possibilities of contemporary cinema will soon become clear. Finally, it can be argued that *The Fountain*'s characters are like Resnais's zombies (only more sensuous). This is especially the case with the character of Izzy, who in fact is actually dead most of the time we see her and who returns in feedback loops. I will return to this point as well.

Perhaps the most important general characteristic of the film that makes it a new type of film is that, despite its dealing with dead and seemingly outer space issues, it makes us believe in the world, in love, and in life. This is the true quest of the brain surgeon, who wants to find a cure to the "disease of death" but will discover beauty in *believing* love and life will continue.

The Problem and Fractal Style

Whereas *Pi* is a theorematic film, *The Fountain* is "problematic." Its central problems are the Big questions of life and death. What does it mean to live? What would it mean to live forever? What does it mean to die? Obviously the film cannot give any answers, but these universal questions are enfolded in the singular love story of the two main characters. The film proceeds to unpack the thoughts and affects connected to the problem of life, love, and death by repeating the same story in different layers of time. In *Difference and Repetition,* Deleuze has argued that there are several forms of repetition:

> Beyond the grounded and grounding repetitions, a repetition of *ungrounding* on which depend both that which enchains and that which liberates, that which dies and that which lives within repetition. Beyond physical repetition and psychic or metaphysical repetition, an *ontological* repetition? . . . Perhaps the highest object of art is to bring into play simultaneously all these repetitions, with their differences in kind and rhythm, their respective displacements and disguises, their divergences and decenterings; to embed them in one another and to envelop one or the other in illusions the "effect" of which varies in each case. Art does not imitate, above

all it repeats; it repeats all the repetitions, by virtue of an internal power.[34]

Deleuze develops these points by arguing that all repetitions are ordered in the pure form of time (that create different forms of differences in the repetition). He argues that the "Before" and "During" depend on the third time, the Future, which is the proper place of decision ("a decision on which everything depends, deeper than all the explanations that can be given for it," quoted earlier in connection to the Outside of the problem). This takes time "out of joint" as a repetition within the eternal return:

> There is only eternal return in the third time: it is here that the freeze-frame begins to move once more, or that the straight line of time, as though drawn by its own length, re-forms a strange loop which in no ways resembles the earlier cycle, but leads into the formless, and operates only for the third time and for that which belongs to it.[35]

Deleuze has already developed these ideas in *Difference and Repetition* as the "three syntheses of time": habitual time of the During or Present, time of recollection of the Before or Past, pure time of the eternal return of the Future.[36] They can also be recognized in the movement-image (first and second synthesis of time) and the time-image (third synthesis of time).[37] The time-image is already concerned with the third time, but it seems that the neuro-image, as third image type ("Cinema 3") brings together the other synthesis of time as well.

The neuro-image involves a form of time in which the three syntheses are playing in "strange loops," repeating and differentiating them in a sort of culminating or vortical movement of all times. In any case, Deleuze's ontological ideas on repetition, art, and time shed a light on the ontological questions and problems that *The Fountain* poses. The film clearly brings into play the different kinds of repetitions, physical, metaphysical, and ontological. Here, too, it is only in the third time, the literally ungrounded future (where everything is floating), that "all times" come together, as Deleuze suggests. Only in the future do we see the "Historical Isabelle" from the Before and "Present Izzy" from the During appear in Tom's hallucinations and feedback loops that are repeated several times, leading up to Tom's final decision to end the other two times by choosing the eternal return. Paradoxically, the eternal return happens by accepting death and returning to the Unicity of being.

Many scenes are repeated several times throughout the film. Most striking, perhaps, is the scene where Izzy suddenly appears, dressed in a white winter coat and a white knitted cap, and says, "Take a walk with me." This exact scene is repeated three times. The first time Tom replies from the future (Tom looks different in every layer of time), saying "Please, Izzy," it is as if he wants her to leave him alone. The second time we see this scene, it is Tommy (in the present) who replies "Please, Izzy" and explains that his colleagues are waiting for him for an operation. We move more deeply into that layer of time, discovering "the problem" and how Tommy is obsessed with changing that fate. The third time we see this scene, Tommy changes his mind and does follow Izzy into the snow. This will lead to Tom(my)'s final decision to finish the story of the conquistador in the past (a story that Izzy was writing and that she repeatedly asks him to finish), to finally die in the future (the climax of the film, where Tom dies in the nebula of a dying star and becomes a celestial particle), and to accept her death by planting a seed on her grave in the present. The final image of the film is another repeated scene from the present, an extreme close-up where Tommy whispers in Izzy's neck "Everything is fine": the eternal return has selected the affirmative powers of love, life, and belief.

As in *Pi*, a mathematical order seems to underlie all these scenes. In *The Fountain*, the formal mathematical principle that gives the film its particular style is another recurrent geometrical figure of the neuro-image, the fractal. Deleuze refers to Mandelbrot's fractals in *The Fold* in relation to Leibniz's philosophy and his Baroque mathematics.[38] As is well known, fractal formulas produce complex geometric shapes (very different from the Euclidian geometric lines and points of Renaissance perspective). Fractals can be subdivided into parts, each of which is a differentiated reduced-size copy of the whole. Again, we see here a logic of "difference and repetition" translated into mathematical language. We can also understand how the screen as a window projecting onto a plane (following the Renaissance perspective) of the movement-image and the time-image now, in the neuro-image, has turned into a table of data, when Deleuze explains the Baroque mathematics: "Transformation of inflection can no longer allow for either symmetry or the favored plane of projection. It becomes vortical and is produced later; deferred, rather than prolonged or proliferating: the line effectively folds into a spiral in order to defer inflection in a movement suspended between sky and earth."[39] This is one of the many occasions that show that Deleuze's own philosophy is in a sense fractal, where similar patterns and principles are repeated in endlessly complex variations throughout his entire work.

The Fountain's style is fractal on several levels. I've already mentioned the striking repetition of entire scenes. Also cinematographically and in the composition of the image, patterns recur throughout the whole film. The different layers of time are connected through formal shapes and figures by stylistic enfoldings. And each layer of time also has its own particular predominant figure. Throughout the whole film, low and high camera angles (characters looking up into the celestial starfield, the camera looking down on the scene below) are repeated frequently, emphasizing the infinity of the cosmos, the abstract beauty of the composition of the scenes on earth, and the connection between the two. Microcosmos and macrocosmos are also repeatedly connected purely visually, for instance in the image of a brain cell under a microscope that is very similar to the movements and lights in the sky. I already mentioned the particular arrangements of lightning in the first and second layers of time that match the omnidirectional cosmic light bulbs in the third. Other elements in the mise-en-scène are also very subtly repeated, such as the pattern on Isabelle's royal dress that is like the roots of a tree, which connect her to the tree in the space bubble and to the tree of life. And the whole design of the film is the shape of a crucifix (or "cruciform," also including up and down movements) that returns in all layers.[40] But, as indicated by Aronofsky, each layer for itself also has a different predominant figure. In sixteenth-century Spain and the Mayan civilizations, the triangle (the three-point star in Mayan cosmology, arches in the Queen's palace) is recurrent and sometimes enfolded in a picture on the wall in Tommy and Izzy's apartment in the twenty-first century. In this layer of time, the most repeated forms are the rectangle and square (computer screens, windows, pictures, doorways, etc.), emphasizing our screen culture. And in the third layer of the future it is the circle, the bulb, and the sphere that are presented in many variations.

Finally the film stock itself is used in a fractal way. Although the idea of fractal logics goes back to Leibniz, fractals can only actually be produced by means of computer technology with huge calculating powers. So, in a sense, it seems logical that the neuro-image, which has access to the endless possibilities of CGI, would be fractal. And yet, the power of The Fountain is certainly also due to the fact that Aronofsky has made only very limited

Figure 9.6 The triangle, a recurrent figure in sixteenth-century Spain and the Mayan civilizations, is here seen in a picture on the wall in Tommy and Izzy's apartment. From the film *The Fountain* (2006), by Darren Aronofsky.

use of digital effects. This is an indication that technology is not the cause of aesthetic change, even though it can be profoundly related to it. Most strikingly the third layer of time, the cosmic images, are not computer generated, even though that would be the current way of showing outer space. Instead, Aronofsky and his team hired Peter Parks, a specialist in macro photography, who brewed chemicals and bacteria to create a fluid dynamics on the film stock, which affected the substances photographed. Parks explains: "When these images are projected on a big screen, you feel like you are looking at infinity. That's because the same forces at work in the water—gravitational effects, settlement, and refractive indices—are happening in outer space."[41] Without any computer image, even the ontological status of the film material of *The Fountain* itself is in this way deeply fractal.

The Baroque House and the Monad

The Leibnizian Baroque enfoldments are characteristic for the neuro-image and also can be analyzed in the way matter and soul fold into one another. In *The Fold*, Deleuze describes an allegory of the Baroque house. The lower floor has windows, several small openings that stand for the five

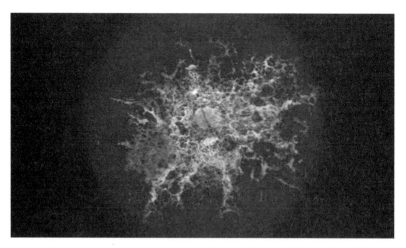

Figure 9.7 Peter Parks, a specialist in macro photography, brewed chemicals and bacteria to create a fluid dynamics on the film stock.
From the film *The Fountain* (2006), by Darren Aronofsky.

senses and are connected to the "pleats of matter." The upper floor has no windows, is decorated with a "drapery diversified by folds": "Placed on the opaque canvas, these folds, cords or springs represent an innate form of knowledge, but when solicited by matter they move into action."[42] Again, the repetition with the cinema books is striking, especially when Deleuze continues that Leibniz constructs "a great Baroque montage" that "moves between the lower floor, pierced with windows, and the upper floor, blind and closed, but on the other hand resonating as if it were a musical salon translating the visible movements below into sounds up above."[43] In this sense *The Fountain* can be seen as a "Baroque house" where the different layers of time are like the floors in a Baroque house. The images of the Future literally also are "monadic," including all series and states of the world, but whose organizing principle lies outside the monad itself and outside the world.

Deleuze suggests that the virtual resides in the soul, but it also needs matter in order to be actualized and incarnated in the subject, repeating the folds of the soul in matter. One could now argue that the subject is formed in the folds of matter and soul, physically and metaphysically, but that its formative principle lies outside both these points, in the mathematical principle. If the movement-image gives us material aspects of subjectivity (physical) and the time-image gives us immaterial aspects of subjectivity (metaphysical), the neuro-image goes beyond subjectivity, opening up to the infinite possibilities of universal series (mathematical).[44] As indicated earlier, the arrival of the neuro-image does not imply the extinction of the other two images. They remain possible variations of the image, but they will also increasingly be implicated in the third image, the image of the Third time, the future.

In the final analysis, the search for the principles of infinite possibilities is the fundamental theorem and problem of *Pi* and *The Fountain*. As Claire Colebrook argues, Deleuze has a double commitment: everything begins from the sensible but the task of thinking is to go beyond the sensible into the potentials that make the sensible possible, into the extension of any possible series outside actual experience.[45] This search for the "beginning of the universe" is also the reason both *Pi* and *The Fountain* refer to the book of Genesis. In *Pi*, the first page of Genesis in Hebrew and numerical translations appear on Max's computer screen when he is close to breaking through. *The Fountain* refers to the tree of life (as opposed to the tree of knowledge) that is described in the book of Genesis. In both cases the implication of a universal mathematical pattern of infinite possibilities is the force of the virtual that is immanent within the power of the image. *Pi's* theorematic nature brings the neuro-image to its most dangerous pole where a breakthrough turns into a breakdown of madness. *The Fountain's* presentation of the problem of death resurrects life and love in a repetition of the eternal return and a truly becoming-imperceptible, becoming-world, or becoming-cosmos in a metaphysics that reaches into a cosmic ontology. As such these films can be considered as the two most extreme poles of the contemporary neuro-image, with infinite possible variations in between. Most strikingly the neuro-image seems to refer to an increasing consciousness in the three domains of thinking (art, science, and philosophy) that we are only temporary subjects, formed by the encounters and experiences we have in the world. But beyond the groundedness of our being we can experience in the first and second synthesis of time, we are connected in a universal and ungrounded eternal return of a fractured I, "swarming with difference," into the infinite virtual potentialities of mathematical calculations that are at the basis of our madness and metaphysics.

· IV ·

Science and the Brain

CHAPTER 10

The Image of Thought and the Sciences of the Brain after *What Is Philosophy?*

Arkady Plotnitsky

The parts of the walls that were covered by paintings of his, all homogenous with one another, were like the luminous images of a magic lantern which in this instance was the brain of the artist.

—Marcel Proust, *The Remembrance of Things Past*

"Casting Planes over Chaos": Philosophy, Science, Art, and the Nature of Thought

In their *What Is Philosophy?* Deleuze and Guattari define thought as a confrontation with chaos. It is their concept and part of their *image of thought*— "the image of thought that thought gives itself of what it means to think."[1] The architecture of this concept and the lineaments of this image are multifaceted and complex. My aim here is to explore those facets of this concept and image that relate to the sciences of the brain—neurology, physiology, psychology, and others. I shall focus primarily on neuroscience, following Deleuze and Guattari's argument in their conclusion to *What Is Philosophy?* ("From Chaos to the Brain"), which extends their concept of thought to a correlative *philosophical concept* of the brain, in their sense of "philosophical concept." A philosophical concept in this sense is not an entity established by a generalization from particulars or "any general or abstract idea" (11–12, 24) but instead a complex phenomenal configuration: "there are no simple concepts. Every concept has components and is defined by them. It therefore has a combination [*chiffre*]. It is a multiplicity. . . . There is no concept with only one component" (16). Each concept is a multicomponent conglomerate of concepts (in their conventional senses), figures, metaphors, particular elements, and so forth, which form a unity or have a more heterogeneous, if interactive, architecture that is not unifiable. Philosophy, as a particular form of thought's confrontation with chaos, is defined by the invention of new concepts, which is in turn a complex process involving such components as the plane of immanence or consistency, conceptual personae, and so forth. Each such concept is also seen in terms of naming and posing a problem, one of the hallmarks of Deleuze's philosophy, which, from *Difference and Repetition* to *What Is Philosophy?* defines philosophical thinking as thinking by posing problems. While this aspect of thinking equally pertains to science (including mathematics), the confrontation between the *scientific,* rather than philosophical, thought and chaos, is, according to Deleuze and Guattari, defined by the invention of functions and frames of reference, rather than the invention of concepts (117–18). How we pursue science, however, often depends on concepts and images (philosophical, artistic, or other) related to scientific ones that we possess or form—in the case of neuroscience, those of thought itself. The reverse is, of course, equally true because scientific ideas enter and shape the architecture of our philosophical concepts. This essay takes advantage of this reciprocity in the case of Deleuze and Guattari's philosophical thinking and neuroscience. Their concept and image of thought and the corresponding concept of the brain are shaped by neuroscience or biology in general (and other fields of mathematics and science); on the other hand, this concept and image of thought, or the brain, suggest new trajectories, "lines of flight," for scientific thinking about the brain.

The concept of thought as a confrontation with chaos also requires a concept of chaos, even if only as that of the impossibility of forming such an image or concept, as found, for example, in Martin Heidegger or Jacques Derrida. Although potentially relevant to Deleuze and Guattari's argument, this is not their primary understanding of chaos, which they approach by means of a particular and, especially in philosophy, unusual concept and image.[2] According to them, "Chaos is defined not so much by its disorder as by the infinite speed with which every form taking shape in it vanishes. It is a void that is not a nothingness but a *virtual,* containing all possible *particles* and drawing out all possible forms, which spring up only to disappear immediately, without consistency or reference, without consequence" (118). This concept of chaos, *chaos as the virtual,* indeed appears to be essential to our understanding of thought and, hence, philosophy, art, and science. I would argue, however, that thought also confronts other forms of chaos, which require corresponding concepts of chaos, two such concepts in particular. Both are at least implicit in Deleuze and Guattari's argument, and in any event, adding them allows one to retain and would amplify their argument concerning the nature of thought. The first concept is that of *chaos as chance or disorder,* disorder defined by chance, or chance by disorder. Deleuze and Guattari's definition of chaos, just cited, invokes this concept and suggests or implies that, while chaos is "defined *not so much* by its disorder" (vis-à-vis a void that is a virtual), it might still be defined, at least to some degree, by disorder and, accordingly, chance. The second concept is that of *chaos as the incomprehensible,* mentioned above, which can be traced to the ancient Greek idea of chaos as *areton* or *alogon*—that which is beyond all comprehension. This "concept" disables any possibility of forming a concept or image of chaos, ultimately even as something unimaginable or unconceptualizable, apart from certain traces left by chaos in this sense upon the (phenomenal) world we experience.[3]

Deleuze and Guattari see chaos as a grand enemy but also as the greatest friend of thought and its best ally in its yet greater struggle, that against opinion, *doxa,* always an enemy only, "like a sort of 'umbrella' that protects us from chaos" (202). As they say, "the *struggle against chaos* does not take place without an affinity with the enemy, because another struggle develops and takes on more importance—the struggle *against* opinion, which claims to protect us from chaos itself. . . . [T]he struggle with chaos is only the instrument in a more profound struggle against opinion, for the misfortune of people comes from opinion. . . . And what would *thinking* be if it did not confront chaos?" (203, 208). Philosophy, art, and science are the primary means of this confrontation between thought and chaos, the confrontation through which thought keeps an affinity with chaos and, together with chaos, wages a war *against* opinion: "But art, science, and philosophy require more [than opinion]: they cast planes over chaos. These three disciplines are not like religions that invoke dynasties of gods, or the epiphany of a single god, in order to paint the firmament on the umbrella, like the figures of an *Urdoxa* from which opinions stem. Philosophy, science, and art want us to tear open the firmament and plunge into chaos" (202).[4]

As indicated above, this confrontation between thought and chaos takes a specific form in each of these endeavors: a creation of concepts and planes of consistency or/as immanence in philosophy; a creation of sensations or affects and planes of composition in art; and a creation of functions and planes of reference or coordination in science. The specificity of the workings of thought in philosophy, art, and science makes them different from each other, and part of the book's project is to develop an understanding of this specificity, most especially in the case of philosophy. And yet, the interrelationships among philosophy, art, and science appear to be equally significant for Deleuze and Guattari, and this significance compels them to develop a more complex—heterogeneous, yet interactive—landscape and history of thought in relation to which each of these fields is positioned.

As it introduces a new concept of the brain, the book's conclusion, "From Chaos to the Brain," also invokes "interferences" between philosophy, art, and science. These interferences make it no longer possible for a given field to maintain its identity as defined by its particular mode of confronting chaos, and make thought enter more complex forms of this confrontation, for example, by shifting it to other modes of thought: philosophy to science, art to philosophy, science to art, and so forth. There are, moreover, interferences that "cannot be localized" or absorbed by a given field "because [at a point of such an interference] each distinct discipline is, in its own way, in relation with a negative"; it says "a No" to itself, but without switching to another field and its alternative mode of confronting chaos (217–18). As a result, even though thought at such points still confronts chaos, this confrontation is no longer enacted by means of art, science, or philosophy, or interactions between or among them. Thought loses the cohesion governed by the "laws" of art, science, or philosophy, or of their (localizable) interferences: the law of affects and planes of composition in art; the law of functions and planes of reference or coordination in science; and the law of concepts and planes of consistency in philosophy. At the same time, however, the emergence of nonlocalizable interferences and their relationships to the brain ("cerebral plane") lead Deleuze and Guattari to an intimation of a different future of thought, still as a confrontation with chaos, but perhaps no longer linked to art, science, and philosophy, and of a different type of people. As they say, "In this submersion [of thought into chaos at an interference point] . . . there is extracted from the chaos the shadow of the 'people to come' . . . mass-people, world-people, brain-people, chaos-people" (218). Indeed, as will be seen, this type of thought may be seen as more primordial (in a logical rather than ontological sense), and it has been Deleuze and Guattari's concern all along, reflected in their concepts of schizophrenic, nomadic, minor, and related forms of thought.

"The Magic Lantern": Thought-Brain

An unexpected and intriguing part of the book's conclusion is a new conception of the *brain,* which mediates the appearance of the "shadow of people to come" and which appears in an almost sudden shift in Deleuze and Guattari's argument from thought to the brain. "*The brain,*" they say, "*is the junction*—and not the unity—*of the three planes*" through which art, science, and philosophy, each in its own way, cut through chaos and thus help us to counteract the forces of opinion (208). This is an extraordinary conjecture, most especially because it relates art, science, and philosophy to certain specific (although, as yet, not biologically specified) forms of neural functioning of the brain itself. In other words, art, science, and philosophy or, at least, something that *neurologically* defines each as a particular form of the confrontation between thought and chaos, are now seen as more primordial forms of thinking (as more immediately linked to the brain's neural functioning) rather than more mediated products of thought. Opinion, too, or what defines it neurologically, appears to be an equally primordial form of this confrontation, which, however, "blocks" chaos altogether through relatively simple forms of order, rather than continuing to maintain a relation to chaos in the way (real) thought does. The cave paintings may be seen as already manifesting these workings of thought in its confrontation with chaos (at least, artistic thought, but quite possibly also proto-philosophical and even proto-scientific thought) and its struggle, together with chaos, against then prevailing "cave" opinion, as was indeed demonstrated by George Bataille's analysis of the cave paintings as "the birth of art."[5] Could they be seen as *more primordially* manifesting these (primordial) workings of thought or the brain than our art, science, and philosophy do? It would be more accurate to see them as manifesting the same capacities of thought and the brain, given that the corresponding functioning of thought and the brain have to be socially mediated to become art, science, or philosophy qua art, science, and philosophy. They, or opinion, can only be defined as such as cultural practices, even if they arise from particular neurological capacities, linked to affects and planes of composition in art, functions and planes of reference or coordination in science, and concepts and planes of consistency in philosophy. Some cultural mediation was at work at the time of cave painting as well, as Bataille's analysis also demonstrates by relating these paintings and other "prehistoric" (hardly the right word here) art works to eroticism, death, prohibition, ritual, and so forth.

It is a complex question how far beyond the human brain Deleuze and Guattari want to push their conjecture, and it is an equally complex question how far, given the neurological specificity of the human brain, one can push it. For example, to what degree could one see the behavior of a bird like the *Scenopoetes dentirostris* (the brown stagemaker) as that of "a complete artist" in Deleuze and Guattari's sense of art, as a manifestation of thought's or the brain's confrontation with chaos by actually creating a plane of composition in the way human art does (184)?[6] "Every morning the *Scenopoetes dentirostris,* a bird in the Australian rain forests, cuts leaves, makes them fall to the ground, and turns them over so that the paler, internal side contrasts with the earth. In this way it constructs a stage for itself like a ready-made; and directly above, on a creeper or a branch, while fluffing out the feathers beneath its beak to reveal their yellow roots, it sings a complex song made up from its own notes and, at intervals, those of other birds that it imitates: it is a complete artist" (184). It would, I think, be difficult to doubt that, to the extent one can speak of the thought of animals other than humans, this thought would have to be seen as a confrontation between the brain and some form of chaos. It is, however, more difficult to ascertain that the nature of this confrontation is thought in Deleuze and Guattari's sense of casting planes over chaos, thought that maintains active or productive relationships with chaos as part of the working of thought and the brain. "Perhaps art begins with the animal," write Deleuze and Guattari (183). This is a "great perhaps," and the meaning of this "beginning" is part of this question and its great complexity. While it is not possible to pursue this question here, I will make some suggestions justifying this animal beginning of art later.

Deleuze and Guattari's conjecture requires a different conceptualization of the brain's functioning and new sciences of the brain, as against current neurological sciences or, conversely, phenomenological approaches to the mind, at least as these stood in the time of the book's publication around 1990 (a point to which I shall return later). For, they argue, both science and phenomenology appear only to be able to theorize the brain as an organ of the formation or communication of opinions. Before stating their case, Deleuze and Guattari offer a brief foray into the sciences of the brain, from experimental and cognitive psychology to the biology of the brain, whose contribution they acknowledge and credit. They agree, in particular, that "when the brain is considered as a determinate function it appears as a complex set both of horizontal connections and vertical integrations reacting on one another, as is shown by cerebral maps [as developed by the current neurological approaches to the brain]" (208). Given this basic architecture, "The question, then, is a double one: are the connections preestablished, as if guided rails, or are they produced and broken up in fields of forces. And are the processes of integration locali-zed hierarchical centers, or are they rather forms (*Gestalten*) that achieve their conditions of stability in a field of which the position of center itself depends?" (208). Deleuze and Guattari contend that, regardless of how these questions are answered, most current (in the early 1990s) neurologi-cal approaches to the brain can at most offer a theory of the brain as "an organ of the formation and communication of opinions," not "of philos-ophy, art, and science (that is to say, vital ideas)" (209). Nor, in spite of its important insights, for example, against scientific objectivism, would (most of?) phenomenology do much better because it "hardly gets us out of the sphere of opinions." At most it leads us to "an *Urdoxa*, posited as an original opinion, or meaning of meaning" (208–9). Deleuze and Guattari suggest a possible alternative:

> Will the turning point not be elsewhere, in the place where the brain is the "subject," where it becomes the subject? It is the brain that thinks and not man—the latter being only a cerebral crystallization. We will speak of the brain as Cézanne spoke of the landscape: man absent from, but completely within the brain. Philosophy, art, and science are not the mental objects of an objectified brain but the three aspects under which the brain becomes the subject, Thought-brain. They are the three planes, the rafts on which the brain plunges into and confronts the chaos. (210)

In other words, philosophy, art, and science are mental-material, thought-brain processes that make the brain the subject by virtue of their confron-tation with chaos and by extracting thought from this confrontation. The invocation of Cézanne at this juncture also brings to mind Marcel Proust's extraordinary passage, used as my epigraph: "The parts of the walls that were covered by paintings of his, all homogenous with one another, were like the luminous images of a magic lantern which in this instance was the *brain* of the artist."[7] Now, "what are the characteristics of this brain, which is no longer defined by connections and secondary integrations [of the current theories of the brain]?" According to Deleuze and Guattari:

> It is not a brain behind the brain but, first of all, a state of survey without distance, as ground level, a self-survey that no chasm, fold, or hiatus escapes. It is a primary, "true form," as Ruyer defined it: neither a Gestalt nor a perceived form but *form in itself* that does not refer to any external point of view, anymore than retina or striated area of the cortex refers to another retina or cortical area: it is an absolute consistent form that surveys *itself* independently of any supplementary dimension, which does not appeal therefore to any transcendence, which has only a single side whatever the number of its dimensions, which remains copresent to all its determination without proximity or distance, traverses them at infinite speed, without limit speed, and which makes of them so many *inseparable variations* on which it confers an equipotentiality without confusion. . . . The brain is the *mind* itself. (210–11)

The brain is the only "mind," insofar as we understand by "mind" that which is responsible for thought and the very possibility of thought as a confrontation with chaos. It is difficult to assess this extraordinary program (only sketched in very broad strokes by Deleuze and Guattati) as concerns its significance or feasibility for the future of the sciences of the brain. It is also hard to predict what role this type of image of thought as confrontation of chaos, as manifest in art, philosophy, and science, or in their interferences taking us beyond them, and the corresponding view of the brain's functioning will play in the development of these sciences. Deleuze and Guattari offer a *philosophical* concept and a philosophical image of thought and the brain, or a set of such concepts and images, which, a brilliant intuitive guess as it may be, may or may not play such a role. In addition, *What Is Philosophy?* published in 1991, reflects the state of neuroscience two decades ago. These decades have redefined the future of neuroscience, and continue to do so, virtually daily, as I am witnessing while writing this essay, for example, in a recent article by Tianming Yang and Mihael N. Shadlen, "Probabilistic

Reasoning by Neurons," which reports the discovery of this reasoning in primates.[8]

I would argue, however, that Deleuze and Guattari's argument remains relevant. Indeed these new developments suggest that it has become more inviting to consider, for example, when thinking of "*probabilistic* reasoning by neurons," also a remarkable phrase and concept by virtue of locating *reasoning* in neurons. This view is close to Deleuze and Guattari's concept of the brain by virtue of the manifest presence of a confrontation with chaos (here chaos as chance) in this reasoning. I shall return to this particular aspect of the brain's functioning below. As I said, how we pursue sciences of the brain depends on the concept and image of thought we form, which can come either from science itself or from elsewhere. Deleuze and Guattari's image of thought is clearly shaped by modern biology, even if mostly by earlier ideas, such as von Uexkühl's[9] conception of nature, also significant for those current neurological theories that I relate here to Deleuze and Guattari's work.[10] In particular, Deleuze and Guattari use this conception to support their insight that art may be something that begins with the animal, by carving out a territory and constructing the house, creating a habitat (*Umwelt*). These processes are always more than spatiotemporal. They are also qualitative or compositional, enabling the construction of a plane of composition, which is art's way of confronting chaos. "The spider web contains 'a very subtle portrait of the fly,' which serves as its counterpoint" (185). This argument connects the argument of *What Is Philosophy?* to the key aspects and concepts of Deleuze and Guattari's thought developed in their previous work. Indeed, the argument itself is already suggested in *A Thousand Plateaus*, where von Uexkühl's ideas are invoked and the same examples, such as that of the "art" of the *Scenopoetes dentirostris,* are used.[11] Associating the nonhuman animal behavior or experiences of the world just described with *art* is, as I said, a complex question. On the other hand, the relationships between the animal's (human or not) neurological activity and territorializations, reterritorializations, and deterritorializations, in short with a certain ontology that also involves a confrontation with chaos, may prove to be important for neurological sciences. I shall now discuss certain contemporary neurological theories, those advanced by Rodolfo Llinás and by Alain Berthoz, which appear to me to exhibit significant affinities with Deleuze and Guattari's argument concerning thought and the brain in *What Is Philosophy?* and with their related ideas, in particular, the ontology of thought developed in *Anti-Oedipus* and *A Thousand Plateaus*.

"Lines of Flight": The Brain, Movement, and the Ontology of Thought

Llinás and Berthoz have developed powerful arguments for the significance of physical movement in our understanding of the brain's functioning, which they indeed see as grounded in movement and, hence, action.[12] According to Llinás: "a nervous system is only necessary for multicellular creatures (not cell colonies) that can orchestrate and express active movement—a biological property known as 'motricity.'"[13] I would argue that Llinás's and Berthoz's theories can be linked to Deleuze and Guattari's ontology of thought and, correlatively, to their concept of thought-brain, grounded in the confrontation between it and chaos, as considered here. I shall mostly follow Berthoz's work, which appears to me closer to Deleuze and Guattaris, even though some of Llinás's ideas from the late 1970s on spatio-temporal representation created in the brain are especially relevant to Deleuze and Guattari's ontology of thought that I shall discuss later. Also, Llinás builds his theory in terms of thinking (as "internalized movement" [ix]), which links this theory more immediately to Deleuze and Guattari's argument, while most of Berthoz's theory deals primarily with perception. As will be immediately apparent, however, thinking is equally at stake in his theory, and, according to Berthoz, perception is already thought and indeed thought as a confrontation with chaos.

Berthoz argues that, by focusing primarily on the connectivities within the brain, current neurobiological and neurophysiological theories by and large fail to take into account the motion- and environment-oriented workings of the brain, which he sees as primary and as fundamental to the brain's development, functioning, and evolutionary emergence. Accordingly, Berthoz aims, first, at developing a more rigorous understanding of how nervous systems and the brain explore the world, and how animals move. He grounds his project in the argument that perception is not only and not so much an interpretation of sensory messages but is also an internal *simulation* of action and specifically motion. "*Perception is simulated action*," he says.[14] This view not only implies that perception and motion are irreducibly interconnected, but also that the former defines the latter because of the definitional "motricity" of the creatures endowed with nervous systems. One might say that our perception and thought (perception is, again, always thought) are defined by "lines of light," traversing and interacting with chaos. It may well be that the "line" (and specifically straight line) as movement precedes our perception and (this comes first) simulation and, it follows, visual construction of all spatiality, including that of, or related to, (the image of) line itself.[15] Created, traced, and imaged by our movement, the line becomes a "movement-image" that, although perhaps not quite a priori in Kant's sense (only movement itself may be), is prior to and defines our image of space. This precedence is *imaged* by the early history of cinema, as movement-image (*kinematic* image), as considered by Deleuze in *Cinema 1: The Movement-Image*, via Bergson, to whom, as will be seen, Berthoz appeals as well because Bergson, too, thinks that "perception springs from . . . a sort of question addressed to motor activity."[16] A similar argument may be made about motion and temporality, as considered (in the context of post–World War II cinema) by Deleuze in *Cinema 2: The Time-Image*, which takes Kant on time as its philosophical point of departure. Berthoz also invokes Husserl, inevitably (given the "protention" aspect of Husserl's phenomenology of time) although sporadically, and Merleau-Ponty throughout the book, beginning with the introduction of his main thesis, just cited, "*perception is simulated action.*" Berthoz links this view to Merleau-Ponty's insight that "vision is the brain's way of touching ['vision is palpation by gaze']."[17] We construct space and spatiality, or time and temporality, or everything that we construct or dream (including literally), beginning with our (brains') image of movement, *because we move.*

In expressing his preference for "simulation" vs. "representation," Berthoz qualifies that for him "the brain is a simulator in the sense of *flight simulator* and not in the sense of computer simulation" (which Berthoz appears to see as more representational, although one might contest this point).[18] "Flight simulator" is not merely a useful metaphor, for, especially in the context of Berthoz's argument as a whole, it also names a concept, rooted in the idea of motion, "lines of flight." It is also a philosophical concept in Deleuze and Guattari's sense, albeit also transformed by Berthoz into a scientific framework of functions, frames of references, and so forth. As he says: "Simulation means the whole of an action being orchestrated in the brain by internal models of physical reality that are not mathematical operators but real neurons whose properties of form, resistance, oscillation, and amplification are part of the physical world, in tune with the external world" (18).

Berthoz uses this idea to argue that "internalization of the properties of the physical world constrains perception, that we perceive the external world through the laws of the physical world [the laws of classical mechanics] integrated within the functioning of neural networks" (175). Even our most basic perceptions are products of complex strategies and calculations that our nervous system must perform to negotiate these constraints and to take its chances in the world. It is worth considering Berthoz's elaboration leading to the fundamental conclusion just cited, which proceeds via von Uexkühl and his "concept of *Umwelt,* or environment," and Merleau-Ponty's commentary on this concept: "for von Uexkühl, *Umwelt* signals the difference between the world as such (*Welt*) and the world as the realm in which this or that living thing exists. It is a transitional reality between the world as it exists for an absolute observer and a purely subjective domain."[19] It would be better to say "the world as it *would* exist for an absolute observer," if such an observer existed, and, as I explain later, still further qualifications are necessary. Berthoz comments on von Uexkühl as follows:

The sea urchin is an example of an animal that is no more than reflexes and a repertoire of independent synergies; the sea urchin, says von Uexkühl, is a "republic of reflexes" that exists in an *Umwelt* that represents things that are often dangerous but to which it is so well adapted that it truly lives as if there were only one world. Its nervous system contains no image of openness. On the other hand, for higher animals, *Umwelt* is not closure but openness. The world is possessed by the animal. The external world is distilled by the animal which, differentiating sensory data, is able to respond to them with fine motor action; and these differentiated actions are possible only because the nervous system is equipped like a replica of the external world (*Gegenwelt*), like a copy. It is a mirror of the world (*Weltspiegel*). The *Gegenwelt* is itself divided into *Merkwelt,* which depends on how the sensory organs are arranged, the world of perception, and the *Wirkwelt,* which is the world of action. Today it is clear the sea urchin is more than a republic of reflexes, but von Uexkühl's ideas have the benefit of setting the debate.[20]

As indicated previously, the epistemology of the situation appears to be more complex and less mirror-like or realist as concerns the world.

In particular, one can see the *Welt* as the material world, or more accurately, some of its parts, that could be experienced by living creatures as an *Umwelt* in each case (this *Umwelt* can, of course, include other living creatures, each of which has an *Umwelt*). The *Umwelt*, thus, may be seen as *related* to the part of the world with which a particular animal interacts and as the image of this part such a creature constructs in order to act successfully in the world (the *Umwelt* or the *Welt* because at least humans can construct the image or unimage of both). It may not be possible to see this image as a mirror of the world (*Weltspiegel*), invoked by von Uexküll, or even an image *of* the world (*Welt* or *Umwelt*). By analogy with Deleuze's "movement-image" and "time-image," one might better speak of a world-image or image-world. It may indeed not be possible to use any concept of "world" that we can form, for example as a (single) whole of which only a part can be perceived or engaged with. The "world" may be and appears to be ultimately quite different from whatever image of it we as living creatures, humans or other, can form, including even as the Kantian alterity of things in themselves, the point stressed by Nietzsche from his early work on, including on biological and even neurological grounds. On the other hand, this view is Kantian, as opposed to a Lockean view, insofar as it implies that, even in perception, our *brain mediates and constructs* the (phenomenal) world rather than *immediately perceives it in a mirror-like fashion*. Kant was also aware of the role of the biological machinery involved, especially, given the contemporary developments in biological sciences, by the time of *The Critique of Judgment*. Our interaction with the world also involves a confrontation with the unknowable and even ultimately unthinkable, with chaos as the incomprehensible. Our world-images or image-worlds work and help our actions even under these epistemological conditions, which, however, also complicate our thinking by making us face chance and use probabilistic predictions.

Berthoz indeed links the brain's functioning to prediction. So does, even more centrally but, again, from the same starting point of "thinking as internalized movement," Llinás, according to whom "prediction is the ultimate function of the brain" and the self itself is *"the centralization of prediction."*[21] Berthoz speaks of "a memory for predicting," again taking one of Merleau-Ponty's insights: "the movement as it is perceived . . . proceeds from its point of arrival to its point of departure."[22] In other words, our movements, as they are perceived, proceed from our estimation of what will happen, which is inevitably probabilistic (although sometimes close to being certain) to the beginning of our movement or other action. Berthoz then writes:

Perception is essentially multisensory: it uses multiple, labile frames of reference adapted to the task at hand. It is predictive; receptors detect derivatives, and the brain contains a library of prototype shapes of faces, objects, and perhaps movements and synergies. Nature has devised simplifying laws among the geometric, kinematic, and dynamic properties of natural movements. But the predictive nature of perception is also—perhaps especially—due to memory. For memory is used primarily to predict the consequences of future action by recalling those of past action. (115)

Berthoz discussed, first, how this thesis may be justified by the workings of various types of spatial memory, beginning, via Darwin's remarkable insights, with "navigational memory" (117–20). He then considers the anticipatory and predictive nature of perception, thinking, and knowledge, or action in other key neurological phenomena, such as natural movement, or strategies of capture or other goal-oriented action (such as "the art of braking"). Although these are not Berthoz's primary terms in the book (chance is only invoked briefly as a disruption of the "game of regularity" [255]), or those of Llinás, the argument for the essentially predictive nature of perception or thinking gives chance and probability central roles in our thinking and makes it a confrontation with chaos as chance.[23] "Ours" here may need to be extended well beyond the human. To cite Yang and Shadlen: "Our brains allow us to reason about alternatives and to make choices that are likely to pay off. Often there is no correct answer, but instead one that is favored simply because it is *more likely* to lead to rewards. . . . [We] show that rhesus monkeys can also achieve such reasoning. We have trained two monkeys to choose between a pair of coloured targets after viewing four shapes, shown sequentially, that governed the probability that one of the targets would furnish a reward."[24] Of particular significance here are the neural correlates of this reasoning, in humans and primates alike: the "probabilistic reasoning by neurons" in question was discovered by tracking sixty-four *individual* neurons of the monkeys.

This evidence further supports the argument that taking chances and betting on outcomes, and hence interacting with chaos as chance, is something that our brain does all the time but that our culture is reluctant to accept at least at the ultimate level.[25] We continue to seek causality behind chance, be it in our brain or mind, nature (dead or alive) or spirit, or God, who does not play dice, as Einstein (thinking of nature) believed or hoped to be the case, although, as one of the founders of quantum theory, he should have known better. In the present view, however, our interaction with and success (individual or collective, including evolutionary) in the world is defined by taking chances, one after another, from one point of a rhizome-like network of events to another. By the same token, our movement in the world is defined by and defines these rhizomes of possibilities, sometimes by using illusions or even hallucinations, to create lines of flight, even though the world, including through movements and actions of our fellow animals (human or other), may channel our rhizomes in particular ways. Indeed, illusions and hallucinations may be all that we may have about the world. Some of these illusions and hallucinations may, however, also lead to material or mental obstacles to others, and thus (along with material obstacles) constrain our motions and actions.

Berthoz regards illusions as solutions and not as errors, taking a cue from Bergson's view, mentioned above, advanced against "the mistake of those who maintain that perception springs from what is properly called the sensory vibration, and not from a sort of question addressed to motor activity" (242).[26] Berthoz contends that "perceptual illusions are solutions devised by the brain to deal with the sensory messages that are ambiguous, or that contradict either each other or the internal assumptions that the brain makes about the external world" and suggests that "illusion is not an error or a bad solution, but rather the best possible hypothesis" (242). Much of the information we have about the world is in fact illusory or, as I shall explain presently, even hallucinatory, rather than real or mirror-like, thus reflecting the presence of chaos behind it (as a virtual field of quickly arising and disappearing forms, and as the incomprehensible), and thus making illusions into thoughts confronting chaos. Berthoz argues, following I. Krechevsky's view of the rats' perception (thinking?), that "'hypothesis' is a situation in which the animal exhibits multifaceted behavior: it is systematic, goal-oriented, to a degree abstract, and not entirely dependent on the immediate environment, either to be undertaken or to be carried out. Finally, [according to Krechevsky]: 'A hypothesis is a person's interpretation of data and not a phenomenon that is derived from the data themselves.' Hypothesis is thus inference, to use Helmholtz's terminology" (243).

Now, Berthoz does differentiate illusions from hallucinations: "Illusion is a solution to an incongruity, to the loss of perceptual coherence [induced primarily, but not only, from the outside]. Hallucination, which is *a creation of the brain [itself]*, is a different story entirely. Hallucination is not the result of sensations that the brain fails to integrate into a coherent perception but of the sudden combination of endogenous memories of perception. In some sense, hallucination is a waking dream, the autonomous functioning of internal circuits that normally work to simulate the consequences of action" (253). As, however, the last sentence suggests, while *different* from illusion, hallucination is actually *not an entirely different story*. Both illusions and hallucinations are part of the *same* story of perception or thought and movement or action told by Berthoz. Indeed hallucinations or, analogously, dreams may be more crucial to it than illusions and perhaps make the latter possible, thus providing a fitting finale to Berthoz's main argument in the book. He adds a brief invective against "architects," who "have forgotten the pleasure of movement" (255–60) and the conclusion, summarizing the book. First of all, both illusions and hallucinations are *hypotheses* because in the case of hallucinations, too, "somewhere in the brain a purely internal activation initiates perceptual hypotheses." Berthoz appeals to an interesting parallel with the painter of Lascaux, which comes, again, courtesy of Merleau-Ponty, invoked on both subjects, Lascaux paintings and perception as a simulation of action as against representation (136). Berthoz writes:

> Like the painter of Lascaux for whom shadows suggested the
> images of the animals he hunted and that he projected onto
> the cave walls, here a pathological internal activity triggers the
> hallucination. Hallucinations are such that they are projected onto
> the world, *exemplifying a fundamental property of the brain.* Whereas
> illusion provides the solution to a conflict, *hallucination constitutes
> clear evidence for the central theme of this book: the anticipatory,
> projective function of the brain imposes its assumptions and memories
> on the world, and it reconstructs movement based on the slightest hint
> of change.* (253; emphasis added)

This is a crucial point. Hallucinations manifest the most fundamental capacity of the brain and its sense of movement, including in its anticipatory (and hence also chance-taking, probabilistic) capacity, and illusion would itself appear derivative of this capacity, when it is deployed in response to an incongruity. This view may be seen as parallel or even correlative to Deleuze and Guattari's "*schizophrenic*" ontology of thought, productive of hallucinations or illusions and their lines of flights—in the cognitive or thought-ontological sense, rather in a psycho-pathological sense, which, however, is also a manifestation of this ontology, even if a pathological one (not always an easy or unequivocal determination). As I shall suggest below, this more primordial hallucinatory capacity may also be responsible for illusions in the world and our *illusion of* the world. These illusions are our response to the overall incongruence, chaos (as the virtual, as chance and disorder, and as the incomprehensible) of the world that we had to confront in our motion and action already long before we became humans.

From this perspective, the parallel with the painter of Lascaux acquires a greater significance, along the lines of Deleuze and Guattari's view of art in *What Is Philosophy?* Each plane of composition in art is defined "as an *image* of a Universe (phenomenon)"—a cosmos or, to use Joyce's famous coinage, chaosmos, or a constellational assemblage of "affects and percepts"—appearing in the field of thought, in which thought of art intersects with the chaos of forms that are born and disappear with an infinite speed.[27] Thus, there are the universes, "vast planes of compositions" (188), and "composition is the sole definition of art" (191): "Rembrandt-universe [and] Debussy-universe" (177); Monet's "cosmos of roses" (180); the Universes of Klee (184–85), and so forth, from the chaosmoses of cave painters to the art and the beyond-art of "the people to come." It is crucial, however, especially in the present context, that such an image is *not a representation* of any real world, "since no art and no sensation have ever been representational" (193). Instead, it is a hallucination and a simulated action, still in part responding, just as those of cave paintings did, to the incongruous chaos of the world by creating a cosmos or chaosmos, an image of the world—first hallucinatory and then illusionary, although this sequence can be reversed. In this sense, especially if everything begins with movement, art may indeed begin with an animal, and the moving animals of Lascaux may also be allegories of perception as simulated movement. According to Berthoz, building, again, upon Merleau-Ponty: "Perception is not representation: it is [hallucinatory?] simulated action, projected onto the world. Painting is not a set of visual stimuli, but a perceptual action of the painter who has translated, through his gesture in a limited medium, a code that evokes the scene he perceives, not the scene represented" (136). In Deleuze and Guattari's view of art, this *action* in turn leads to a hallucinatory creation, a "sublime error," of a composition, cosmos or chaosmos, which embodies not representation but what they call percepts and affects, which enable subsequent perceptions or feelings (by which we respond to the work) but are not reducible to them. This "hallucinatory" projection may also be seen in terms of a cinema-like projection of the movement-image on a cave wall, thus also allowing one to speak of our brain as a cinema-brain.

"Dreaming," Berthoz adds, also provides the "essential evidence" of this primordial hallucinatory capacity and activity of the brain (253). This capacity also enables our waking life, although, as it constrains our movement and action, this life of course shapes our hallucinations, forcing us to have illusions to avoid life's incongruities. This argumentation bears crucially on the *relationships* between the Freudian or Lacanian "interpretation of dreams" and theory of desire (as reflecting an Oedipal illusion) and Deleuze and Guattari's schizophrenic, hallucinatory ontology of desire. I stress "relationships" because it is far from clear how avoidable our Oedipal illusion(s) are, even if it may be desirable to live with schizophrenic hallucinations (again, in the thought-ontological sense of Deleuze and Guattari) and alternative illusions they help us to create. In any event, it is remarkable to realize that, while we may move best, most freely and with greatest (infinite?) speed in our dreams, we also dream because we move.

The essentially illusory and hallucinatory nature of our thinking (again, in this general sense rather than referring to medically pathological phenomena) gives a greater, and more Deleuzean, complexity to Berthoz's argument concerning the brain and motion. Berthoz, we recall, takes as his point of departure a view that every moving body must follow the laws of classical mechanics, which compels the brain to invent strategies to make calculations, and, hence, to internalize the basic laws of Euclidean geometry and kinematics. It is true that the animals' movement in the world and the worlds themselves that they, including us, confront and construct are all largely defined and constrained by geometry and classical physics. Such is the case, even though the macro-world in which we exist and of which, having emerged during the long history of dead and then living matter, our bodies (including our brains) are part, is made of atoms. Hence, the ultimate constitution of this world is defined by the laws of quantum physics. Quantum physics was, of course, also developed by us, along with the understanding of this ultimate constitution as quantum or, as it appears, the ultimate impossibility of understanding of this constitution (chaos as the incomprehensible, as well as chaos as chance because quantum phenomena and correspondingly quantum theory are irreducibly statistical). It is, however, this constitution that ultimately enables our perception and interaction with the world, described by the law of classical physics and simulated as such by our neural machinery. Indeed, one might argue, as Niels Bohr and Werner Heisenberg did, that the whole conceptual structure of, first, Euclidean geometry and then of classical physics, and our corresponding physical-mathematical image of the world, may be seen as arising, via suitable mathematical refinement, from the phenomenal image of the world. This image is created by the brain, as an organ developed in organisms capable of motion, and by the brain's capacities of remembering or, importantly, forgetting the past and predicting the future. One might note that motion is a defining concept of classical physics (in quantum physics, the application of this concept involves significant complexities and may not be possible); and in this sense, classical physics naturally represents our sense of the world, as defined by our sense of movement.[28] In other words, very late in this history our thinking, as a confrontation with the world and its chaosmos, also led to the creation of geometry and classical physics, in this case by slowing the chaos down by means of frames of reference, freeze-frames, and a set of functions to go with it, as considered in *What Is Philosophy?* (117–34). This process itself, however, and other forms of confrontation with chaos, such as by way of proto-art (creating planes of composition and affects or percepts) or proto-philosophy (creating planes of immanence and concepts), are found much earlier.

One might, accordingly, suggest that what, at least, the human brain, as the product of the evolution of moving animals and their brains, creates more primordially is not a *phenomenal world* and ontology *described* by Euclidean geometry and classical physics. Instead, through its multi-faceted—art-like, science-like, philosophy-like—interaction with chaos, our brain creates, first, a Deleuzean phenomenal world and (phenomenal) ontology, the world of schizophrenic or rhizomatic possible movements, lines of flight, territorializations and deterritorializations, and so forth.[29] These movements may or may not be actualized or even virtualized (in Deleuze and Guattari's sense), but even when they are, they retain their hallucinatory character. This phenomenal world is, as I have stressed, not a mirror or even an image *of* the "world" as it actually exists, but rather an image-world or world-image, with the world itself and its existence beyond our image or concept (even beyond that of Kant's "thing in itself"). This ontology is largely unconscious and largely illusory, and it is generally richer or more complex or stranger than the world we consciously perceive as the physical world, which does not prevent its partial coming into consciousness or its interaction with the material world or "reality," or the construction of its (conscious) images. It is the ontology of desiring machines in motion and in continuous confrontations with chaos in the world as what Deleuze and Guattari see as "a body without organs." The latter, in turn, provides both energy resources for our life and forces of resistance to our desires and actions, from basic forms of the reality principle, beginning with gravity as a force governing our motion and shaping the memory of neurons (responsible for the pleasure and fear of acceleration, vertigo) to the emergence of Oedipal desiring machines.[30] In sum, our biological constitution appears to be especially suited for creating this world-image or image-world and succeeds in the world by working with this image and negotiating with the world through it, again, primarily and inevitably by taking our chances. There is nothing else we can do, even though we may believe otherwise, in part because sometimes our chances prove to be close to certainty, but *only* close, for nothing is ever absolutely certain. The "people to come," invoked by Deleuze and Guattari in closing *What Is Philosophy?*—mass-people, world-people, brain-people, chaos-people (hence also chance-people)—may prove to be defined by different *forms* of this interaction between the brain and the world, but it will be defined by them, nevertheless, and thus by the capacity of the brain to make our thought cast planes across chaos, against the never-abating forces of opinion.

The argument just sketched is neurological or cognitive-psychological, even if only hypothetically so, insofar as it aims to suggest a possible scientific investigation of the brain following upon—this is the *after, après* of my title—Deleuze and Guattari's philosophical ideas here discussed. I follow not only Llinás and Berthoz, but also Freud, both the early Freud of his *Project for Scientific Psychology* (the ideas of which have acquired new currency in contemporary neuroscience) and the Freud of *Beyond the Pleasure Principle*. Freud started his scientific career by his studies of nerve cells in relatively simple organisms and published articles on this subject before switching to psychotherapy, and he continued to follow the key developments in neurobiology during his lifetime, some of which were revolutionary. Freud knew that the psychoanalytic situation that he spent a lifetime trying to understand is ultimately defined by our neurological machinery and its interaction with the world, the machinery that creates the mental "world" (hallucinatory in character, in the earlier sense) that both enables and threatens our mental life. He thought that we are better protected against the outside world than against the damage it is capable of inflicting upon itself from within. As he said in *Beyond the Pleasure Principle*, in part devoted to this last vulnerability: "Biology is truly a land of unlimited possibilities. We may expect it to give us the most surprising information and we cannot guess what answers it will return in a few dozen years to the question we have put to it. They may be of a kind which will blow away the whole of our artificial structure of hypotheses."[31] This assessment still applies to our speculations, including the one offered here. Given, however, the intervening developments in biology and neuroscience since this statement, especially those of recent decades and the most surprising information they gave us, Deleuze and Guattari's concept and image of thought may put forward questions that the science of the brain might have to confront. It is even possible that the answers will not blow away the *philosophical* (a more appropriate term than "artificial") structure of this hypothesis, although one can hardly doubt that some of these answers will be surprising.

CHAPTER 11

Deleuze, Guattari, and Neuroscience

Andrew Murphie

THE CELEBRATION OF THE BRAIN forms a surprising conclusion to Deleuze and Guattari's writing together at the end of *What Is Philosophy?* Published at the beginning of the so-called "decade of the brain" of the 1990s, *What Is Philosophy?* is prescient concerning a series of contemporary questions regarding neuroscience in culture. In *What Is Philosophy?* the role of the brain sciences is clear. "It is up to science to make evident the chaos into which the brain itself, as subject of knowledge, plunges."[1] For Deleuze and Guattari, this plunging into chaos is the brain's strength. It allows the brain an intense engagement with the virtual.

In opening up the brain to us, the brain sciences make us rethink our relation to chaos, to the virtual. More generally, developments in the brain sciences increasingly allow for a culture that targets the nervous system more directly and precisely than before. In response, there is a critical literature emerging concerning the brain sciences and culture (by authors such as Jonathan Moreno, Joseph Dumit, Brian Massumi, Elizabeth Wilson, and Steven Rose).[2] At the same time, contemporary artists such as VJ Mark Amerika have begun to draw on neuroscience to develop a more positive, "postcognitivist"[3] concept of the aesthetic. Amerika's advice is to "Think of . . . locating the breakout potential of your neuroaesthetic self."[4]

This chapter draws on the history of Deleuze and Guattari's discussion of the brain to explain why they think science should make evident the chaos into which the brain plunges, why the brain itself is a "subject," what this might mean for increasingly "neural" cultures, and why much of this might involve aesthetics.

Mouth, Brain, and Images of Thought

There are many points of convergence between Deleuze and Guattari and those working within the brain sciences.[5] Deleuze's interest in questions regarding the brain was long-standing. Much of his philosophy is an attempt to work out the relations between the event, the brain (and thinking processes), and perhaps more directly and politically, subjectivity. Eventually, Deleuze concluded that "subjectification, events and brains are more or less the same thing."[6] So the brain was fundamental to Deleuze's materialist philosophy. Nevertheless, he wanted to allow for a metaphysics within this materialism,[7] and here the brain was to become crucial.

The problems involved are outlined in *The Logic of Sense,* precisely when discussing the relations between the corporeal (bodies, actions) and the incorporeal (events, sense, meaning "as an encounter of force fields").[8] Deleuze writes of the confusion (in the mouth) of speech, language, and eating, or: "The struggle between the mouth and brain . . . eating, on the one hand, and thinking, on the other, where the second always risks disappearing into the first, and the first, on the contrary, risks being projected onto the second."[9]

Understanding the materiality of the brain eventually arises as a possible explanation for such confusion. In 1985, while discussing the *Cinema* books, Deleuze commented,

> It's not to psychoanalysis or linguistics but to the biology of the brain that we should look for principles, because it doesn't have the drawback, like the other two disciplines, of drawing on ready-made concepts. We can consider the brain as a relatively undifferentiated mass and ask what circuits . . . the movement-image or the time-image trace out, or invent, because the circuits aren't there to begin with.[10]

The themes of the importance of the brain, an emphasis on microbiology, and the brain's openness as a form of organization are repeated many times. In an interview in 1988, Deleuze commented that

> our current inspiration doesn't come from computers but from the microbiology of the brain . . . It's not that our thinking starts with what we know about the brain but that any new thought traces uncharted channels directly through its matter, twisting, folding, fissuring it.[11]

Here we find a crucial difference between Deleuze's approach to the brain and at least some of both the cognitive and neurosciences. For Deleuze, that which is referred to throughout *What Is Philosophy?* as "opinion" or "doxa"—a mode of thinking via what we might call pregiven "charted channels"—is not really thinking. Real thought, for Deleuze, involves "uncharted channels." More than this, in *Difference and Repetition*, it involves the destruction of the received images we have of *what it is to think:*

> Generalised thought process . . . can no longer be covered by the reassuring dogmatic image [of a neat assemblage of thought processes, of "common sense"] . . . henceforth, thought is also forced to think its central collapse, its fracture . . . Artaud pursues in this the terrible revelation of a thought without image.[12]

Even the attempt to define the faculties—since Kant, arguably the philosophical basis for much of the cognitive and neurosciences to come—is suspect, at least in its finality.[13] Deleuze sees the possibility of new faculties, or of older faculties surprising us, or dissolving. Again it will come down to the crucial question of "new cerebral pathways," with science's responsibility to "discover what might have happened in the brain for one to start thinking this way or that."[14]

The fixed images of thought to which Deleuze objects should not necessarily be confused with the new images of thinking processes—if that is what they are—produced by PET [positron emission tomography] and fMRI [functional magnetic resonance imaging]. Such images can indeed bolster older images of thought. They can appear, as Joseph Dumit puts it, to "picture personhood";[15] however, the very fact of seeing these images disrupts many previous images of thought, calls for new forms of adaptation both by and to the brain, ruptures opinion and doxa, indeed multiplies some images of thought as it dissolves others. In comments on "pop videos" in 1988, which we can see as applicable to brain imaging, Deleuze said:

> What was interesting about pop videos at the outset was the sense you got that some were using connections and breaks that didn't belong to the waking world, but not to dream either, or even nightmare. For a moment they bordered on something connected with thought. This is all I'm saying; *there's a hidden image of thought that, as it unfolds, branches out, and mutates, inspires a need to keep*

on creating new concepts, not through any external determinism but through a becoming that carries the problems themselves along with it.[16]

This hidden, mutating "image of thought" is both stimulated and never quite captured in new forms of image production. This suggests that the real activity of the brain might not be totally amenable to overdetermined "models" or "pictures" of thinking processes, or to many cultural actions and politics modeled on these. The problem is the brain's creativity. "The problem is that [creative] activity isn't very compatible with circuits of information and communication, ready-made circuits that are compromised from the outset.... [T]he brain's the hidden side of all circuits, and these can allow the most basic conditioned reflexes to prevail, as well as leaving room for more creative tracings, less 'probable' links."[17]

Sean Watson rightly points to the challenge posed to common cognitive/communicational models by Deleuze's thinking: "communication in this context cannot mean information exchange or systems of representation. Instead, *communication is a matter of structural modulation of the body and nervous system. Communication is a mutual adjustment of bodies.*"[18] This adjustment can be banal. Deleuze notes that things can too quickly become a kind of "organized mindlessness . . . what happened with pop videos is pathetic" and "most cinematic production, with its arbitrary violence and feeble eroticism, reflects mental deficiency rather than any invention of new cerebral circuits."[19] Deleuze's fascination with neuroscience dwells on the possibility of something more than this banal deficiency. In *L'Abécédaire de Gilles Deleuze* (and here I am relying, gratefully, on Charles Stivale's summary):

Deleuze says that neurology is very difficult for him, but has always fascinated him. [Ideas] don't proceed along pre-formed paths and by ready-made associations, so something happens, if only we knew. That interests Deleuze greatly since he feels that *if we understood this, we might understand everything, and the solutions must be extremely varied.* He clarifies this: *two extremities in the brain can well establish contact,* i.e. through electric processes of the synapses. And then *there are other cases that are much more complex perhaps, through discontinuity in which there is a gap that must be jumped. Deleuze says that the brain is full of fissures, that jumping happens constantly in a probabilistic regime. He believes there are relations of*

probability between two linkages, and that these communications inside
a brain are fundamentally uncertain, relying on laws of probability.[20]

As for many today in the brain sciences, for Deleuze the brain is not a series of smooth, mechanical circuits. As announced in *L'Abécédaire*, and also as influenced by Steven Rose's popular book on the brain, *The Conscious Brain* (1976), Deleuze's "brain" performs two kinds of actions, one "basic conditioned reflexes," the other "creative tracings."[21] However, *both* of these are uncertain and probabilistic. Both of these are situated in a shifting topology of relations. Both involve the specificity of actual connections and breakdown within synaptic activity, and the importance of global *probabilities* of connection. This is the beginning of a "thread which rises towards the virtual."[22]

These two kinds of action also have their differences, however. The conditioned reflex does not maintain, strictly speaking, a smooth connection with a circuit, but there is still a *habit* of *probable* contact between elements over time in a *shifting* patterned response. Creative tracings, however, are more radical. They face up to discontinuities and making something new of them (entirely new syntheses)—"in which there is a gap that must be jumped." Of course, both of these will always be found in combination, contributing to the complexity of the probabilistic nature of synaptic activity. I will deal with habit in more detail.

Habit

It is true that, for Deleuze, activities in the brain "don't proceed along pre-formed paths and by ready-made associations." There is, however, a kind of continuity, *only* found in habit, as he explains in *Difference and Repetition* (75). Habit is a crucial concept for Deleuze. As *continuity,* however, habit is the dynamic and shifting *contraction* or *synthesis* of experience over time, in fact *as* time. It "contracts," synthesizes, "contemplates," or to put it simply, *unifies,* the duration *between,* and extending *through,* repetitions. It is not simple repetition. This is partly because there is no simple repetition. There is always an intensive differentiation—a change—within a duration (71ff). Simply put, "difference inhabits repetition" (76). Indeed, in that it is never simple or exact, even repetition is *part of the continuity* "contracted" in synthesis (if a kink or fold in it). In sum, habit is an ongoing adaptation to changing circumstances, even itself a force within those circumstances. It is not a way of doing things exactly the same every time.

Deleuze's discussion of habit and synthesis in *Difference and Repetition* suggests that *we are a complex mix of habits,* passive syntheses, at a number of levels. These habits include both "sensory-motor habits that we have (psychologically)" and, more basic "primary habits that we are: the thousands of passive syntheses of which we are organically composed" (74). The latter involve the ongoing composition and functioning of the body. Even before this, "we are made of contracted water, earth, light and air" (73). Something like a series of microcontemplations is necessary to contract this tumult of passive synthesis: "A soul must be attributed to the heart, to the muscles, nerves [we assume brain] and cells, but a contemplative soul whose entire function is to contract a habit" (74).

Our entire "primary vital sensibility" (73) is based on this arrangement of habits of contemplation/contraction/synthesis in the organic and the sensory-motor (perception/action). It is within this vital sensibility that "the lived present constitutes a past and future in time." Contractions of previous events into a habit are directed toward the future in an expectancy. Often this is not conscious (passive synthesis). Sometimes, building on the passive, it is conscious ("active syntheses of a psycho-organic memory and intelligence [intuition and learning]") (73). In that all of this is always at least somewhat probabilistic—and at that a shifting mix of many probabilistic events—it is hardly a clockwork model of ecological events involving brain, body, and world.

It is in these complex relations between organic and perceptual, passive and active syntheses that we find Deleuze's "rich domain of signs." Just as there is no simple repetition, here the sign is not a matter of the simple recognition or representation (which often rely on a simple notion of repetition) necessary to cognitivism. The sign is a dynamic envelope (synthesis) of events, as in some connectionist or dynamicist theories of cognition, or the theory of neural networks, both in its neurological and sensory-cultural form.[23] The material domain and action of signs is not that of the symbolic processing of representations. It is an ongoing synthesis of syntheses, of complex durations and mixtures. It is, again, "probabilistic" and uncertain. More than this, it involves both active and passive synthesis, a work on habit. It is therefore often "discontinuous" and calls for "jumping in a probabilistic regime." Signs, considered as syntheses, "always envelop heterogeneous elements and animate behaviour." This is because

each contraction, each passive synthesis, constitutes a sign which
is interpreted or deployed in active syntheses. The sign by which
an animal "senses" the presence of water does not resemble the elements which its thirsty organism lacks.[24]

Rather, there is a contraction of a complex series of primary series involving the water, the animal, and thirst, contracted as a probabilistic field of
microevents that are drawn together by this field.

In short, habit, *including habit at the level of synaptic activity,* is an ongoing and changing adaptation of brain to world, and vice versa. It must, in
the process, negotiate a complex set of real continuities and discontinuities, fissures and jumps through time, that are also part of thinking processes—what Deleuze calls "creative tracings." As he writes in *Difference
and Repetition,* "We do not contemplate ourselves, but we exist only in
contemplating—that is to say, *in contracting that from which we come*" (74,
emphasis added). Indeed, and here we find an early version of what will,
in *What Is Philosophy?* become "micro-brains." For Deleuze, "everything
is contemplation, even rocks and woods, animals and men . . . even our
actions and needs" (75).

We need to emphasize the crucial complication to this. Synthesis, or
"contraction . . . takes place not in the action itself, but in a contemplative self which doubles the agent." Here perhaps the difference is between
the direct physical action, such as the electro-chemical actions involved
in the firing of the synapse for example, and the shift in probabilistic pattern of connection—a contemplative self—or overall field of connections
involved in the brain. This contemplation contracts within, across and yet,
in a sense, separate from actions. It thus, like thought, *as thought,* "never
appears at any moment during the action" (and cannot be seen in brain
imaging). It thus not only unifies a field of probability, but can be seen *to
create a virtual structure to actions that is subjectivity.* This virtual structure
does not belong to us. "Subjectivity is never ours, it is time, the soul, or the
spirit, the virtual" (82–83).

In the light of this, consider V. S. Ramachandran's famous use of mir-
rors to break the habits of phantom limb pain, in which the mirror allowed
the sufferer to "see" a limb, a reflection of the remaining limb, where it is
in fact missing.[25] This is not so much a re-recognition as a synthesis of continuity and discontinuity.[26] We can also consider, in this light, the famous
"mirror neurons"—neurons that fire either when an animal acts or when

it observes another performing a similar act.[27] These neurons should not be seen, first, as operators of resemblance—a recognition of a picture—between pregiven agents. They are first dynamic points of synthesis of an interrelation, of a movement in an abstract field including both animals.[28] This movement reworks the probabilistic "diagrams" of future affectivity and movement (thus a stimulation to/like action). Mirror neurons involve a "contemplation" doubling and synthesizing action. This does not occur in a simple reflection of the action in the "observer," but in the new synthesis formed between the action and its observation. In other words, the mirror neuron's key element is not resemblance but the *arousal* of circuits of action and, differently, *virtual restructuration of futurity.* Two sensibilities become contiguous via a perceptual synthesis, so that "two extremities" *between* different bodies/brains "can well establish contact" in the contraction of a duration. After this, perhaps, something like "recognition" occurs, although this might be better thought of as an integrative meeting of something like that which Brian Massumi calls "biograms."[29] These "retain a privileged connection" (186) to the strangest but most fundamental set of synthetic habits we know. These are found in proprioception. Biograms are "less cartographic," more "lived diagrams based on already lived experience [though as a kind of abstract diagram of experience], revived to orient further lived experience. Lived and relived."

All these processes are a constitutive, material part of the work of the brain in the world, and *of the work of the world in the brain.*

Self-Organization

To the previous we must add the influence of Raymond Ruyer on Deleuze. Paul Bains gives an important account of the way that Deleuze and Guattari take up Ruyer's concept of an immanent auto-formation, in a kind of self-organizing *self survey* of an entire set of relations at *infinite speed,* the speed of thought perhaps.[30] It is via this self-formation without an external point of reference that the brain becomes subject, although this is, in a sense, a "subjectless subjectivity,"[31] one that diminishes the role of what is normally taken to be the subject in favor of the material brain itself. This is synthesis writ large. Here, there is the emergence, at infinite speed, of a "kind of global transconsistency or existential grasping whereby a fragmentary whole emerges . . . a unity in multiplicity, an absolute survey that involves no supplementary dimension."[32]

This could be seen as a more cybernetic or informational model of thinking processes (here the influence of Gregory Bateson); however it would be more correct, perhaps, to see cybernetics as a *restricted* attempt at modeling and deploying the dynamics of topological folds and syntheses that Deleuze will see as fundamental to the brain-world relations. Cybernetics does seem closer to the reality of feedback or dynamic folding in the networks between nervous system and world than the cognitivism that was to overwhelm it.[33] Yet cybernetics was also more restricted by its desire to enhance forms of control, and that in a largely military context, if not that of duck shooting.[34] These restrictions have proved to be highly effective in the application to performance culture (increasingly including cognitive performance),[35] including the use of brain sciences in performance cultures, and to what Samuel Weber has discussed as the reduction of cognitive processes to "targeting."[36]

Superfolds

Perhaps surprisingly, for Deleuze, science is sometimes the solution here as much as the problem. For example, in response to the, for Deleuze, reductive categorization of forms of life in the nineteenth and early twentieth centuries, "biology had to take a leap into molecular biology, or dispersed life regroup in the genetic code."[37] Genetic code, of course, opens everything up to variation again, rejuvenates the power of life itself. This changes both the nature and power of the "forces in play." Biological forces are liberated from their schematization (and from opinion and doxa) and subsequently from the forms of control involved, although this, of course, also raises questions of the manner in which these new relations of forces allow for new forms of control.

For Deleuze, that which has come into play is a force allowing the possibility of an ongoing series of auto-formations across an entire field. The potential is for what he calls an "unlimited finity." John Marks explains this: "Quite simply, an infinity of beings can apparently be constituted from the finite number of four bases from which DNA is constructed."[38] The unlimited finity evokes "every situation of force in which a finite number of components yields a practically unlimited diversity of combinations. It would be neither the fold nor the unfold but the *Superfold*, as borne out by the foldings proper to the chains of the genetic code, and the potential of silicon in third-generation machines."[39]

Transported to the brain sciences, this explains what Deleuze means when he says that in understanding the molecular biology of the brain, "we might understand everything" *and* that "the solutions must be extremely varied."[40] Neuroscience gives increased access to a new series of potentials for an ongoing, "unlimited finity," the product of a Superfold, in the brain and the nervous system, considered as onto-genetic, generative of worlds. The ethical and political issues involved in "neural cultures" will be located, for Deleuze, in the potentials of the neuroscientific superfold(s): the probabilistic nature of the synapse, the topology of a brain that can self-survey at what is, to all intents and purposes, infinite speed, along with the "unlimited finity" that neuroscience can provide access to in relation to the brain. In all these respects, the brain sciences' opening up of the processes of thinking could be a more radical event than that of the discovery of the "superfold" of the genome.

Let us consider the nature of the folds and probabilistic activity of the brain more carefully.

Dark Precursors and Graspings

We recall that the syntheses and related "contemplations," actual and virtual, micro and macro, as processes of auto-formation, occur at all levels, throughout a dynamic ecology. *They are all there is in a world considered as ongoing dynamic becoming, a world in which relations and contractions form dynamic "objects" and not vice versa.* These contractions, or syntheses, lie behind many of Deleuze's concepts of the obscure from which events arise—the "dark precursor,"[41] for example, or the depths of the fissures of the brain, the depths of the monad, and so on.

Let us take the "dark precursor." This is what precedes a bolt of lightning. We could say it is what precedes, or is even the first stage of, a *notable* contraction of synthesis. The dark precursor is "invisible, imperceptible," but it determines the thunderbolts' "path in advance but in reverse, as though intagliated," that is, in relief—in effect momentarily creating a *channel* in the chaos for the thunderbolt.[42] Of course, lightning bolts and electrical firings in the synapses can be thought of in similar ways.

For Deleuze, every "system contains its dark precursor which ensures *the communication of peripheral series*"—in synthesis as self-survey. It is this that might allow, for example, "two extremities in the brain [to] well establish contact," as Stivale writes.[43] The dark precursor is not an opening up of a "communication" in the sense of a message. It is rather, as Watson puts it "a matter of structural modulation of the body and nervous system."⁴⁴ The dark precursor modulates body and nervous system down to the level of the ongoing genesis of channels between synapses—in dark precursors of the elements of microperception.

Here we might combine the dark precursor with the subsequent firing of the synapses, in what we could also see as a *grasping* toward the coming together of an event. This is what Whitehead called an "actual occasion" or "concrescence."⁴⁵ "Grasping" is, via Whitehead, the term adopted by Guattari, precisely in the context of what he calls the "*synaptic.*"⁴⁶ For Guattari, a kind of "grasping," like the dark precursor, establishes some coherence in chaos. It finds unity or connection within it, in Deleuze's terms in "contemplation," without actually taming the chaos per se.⁴⁷ Guattari here quotes Whitehead directly: "Each new phase in the conscrescence means the . . . growing grasp of real unity of feeling."⁴⁸ Whitehead called the "dark precursor" the "lure to the creation of feelings." The final result is the *literal* creation of feeling, the sudden adjustment of the nervous system, a *unity of feeling,* not a "logical outcome." For Deleuze, like lightning, this unity of feeling is spectacular. It is partly spectacular because it undoes the temporary constraints of the channels formed temporarily by the dark precursor or the lure to feeling.⁴⁹ "Once communication between heterogeneous series is established, all sorts of consequences follow within the system. Something 'passes' between borders, events explode, phenomena flash, like thunder and lightning."⁵⁰ Synapses firing, but more than that. A new auto-formation, self-survey at infinite speed, contacts between extremities. Thought.

Of course, very little of this is conscious. Here, Deleuze's description of the "depths" of the Leibnizian monad is strikingly resonant with a Deleuzean concept of the depths of the brain. We see that syntheses are a kind of fold, contemplations perhaps a kind of inflection in this fold (the fold in self-survey).

> It is as if the depths of every monad were made from an infinity
> of tiny folds (inflections) endlessly furling and unfurling in every
> direction, so that the monad's spontaneity resembles that of agi-
> tated sleepers who twist and turn on their mattresses.⁵¹

This agitation involves much more than the "organic connections and integrations" assumed sometimes of the brain as "a constituted object of science . . . an organ only of the formation and communication of opinion."[52] Things are perhaps somewhat more chaotic. The thunder and lighting of the brain is "nonobjectifiable" in any entirely systematic way. Even thinking science is a form of creation.

> If the mental objects of philosophy, art, and science (that is to say, vital ideas) have a place, *it will be in the deepest of the synaptic fissures, in the hiatuses, intervals, and meantimes of a nonobjectifiable brain,* in a place where to go in search of them will be to create . . . [53]

Here again the brain as a kind of "superfold" (and not necessarily only in consciousness), dynamized by a rolling and chaotic series of "microperceptions."[54] These are "little folds [or, we might say, syntheses] that unravel in every direction, folds in folds, following folds, like one of Hantaï's paintings or Clérambault's toxic hallucinations." As Deleuze puts it, "the task of perception entails pulverizing the world, but also spiritualizing its dust" (87). These "tiny perceptions are as much *the passage from one perception to another* as they are components of each perception" (87, emphasis added). In addition, this ongoing chaos of microperceptions is not contained within what we take for the present moment (what William James called the "specious present").[55] Nor is this chaos necessarily smoothly integrated into a linear sequence of macroperceptions to which it can be reduced, although the macroperceptions are "nourished" by the chaos of microperceptions.

If the brain has any ongoing unity then, even as self-survey, it is not as a clear organic series of connections and integrations functioning seamlessly. It is rather as a *chaosmos,* a chaos that also nourishes auto-formation and de-formation. This is a chaosmos—which we could also call *the virtual*—in which what counts are the *notable* or *remarkable* events of difference that emerge (some of which becomes unities of feeling). This should be something scientists should like. As Deleuze puts it, "We have to understand literally—that is, mathematically—that a conscious perception is produced when at least two heterogeneous parts enter into a differential relation that determines a singularity."[56]

If microperceptions give rise to conscious perception, this is not in the manner of mechanical parts in an overarching machine. Microperceptions arise from difference creatively, and are "requisites or genetic elements" for further creation.[57] Again, like a synthesis or contraction, or a fold or inflection, a perception is a *produced differential intensity*. This is why every macroperception is not a recognition, but is something startling (or, we could say, even recognition is startling). If not startling, a macroperception would not arise. Startling differentials, then, make up "the object as a perception, and the determinability of space-time as a condition," and "differential calculus is the psychic mechanism of perception."[58]

The key notion here is that "clear perception as such is never distinct,"[59] as in a very clear and accurate picture of an object. It is rather a "distinguished . . . remarkable or notable" result of a genesis of differential series. *What is noted, in fact produced, is difference differing*. What is more, these remarkable differential events of perception can occur at different levels depending on the "perspective" involved. Deleuze suggests they occur within a protein.[60] In this context the same applies to a neuron, a mirror neuron most obviously (if indeed mirror neurons act as speculated), but in fact any neuron, considered as a sensing cell in its own right. "We" are built from this, with larger animals possessing the ability to assemble "increasingly numerous differential relations of a deepening order . . . determining a zone of clear expression that is both more extensive and increasingly hermetic."[61] This is suggestive of a brain not just larger than other animals, but allowing, at "infinite speed," a more extensive differential survey of its "unlimited finity," or ongoing "Superfold."

In sum, Deleuze presents a shifting topological model of the relations within the brain—and between brain, body, and world—of variable and shifting relations of "continuous and discontinuous communications." This is also a concept of thinking in which physical mechanisms are mutually immersed in, but not entirely equal to, shifting differential patterns or intensities. Deleuze here articulates a complex theory of perception, and indeed of the generation of "ideas." It allows for both the mutual independence and dependence of the psychic and the more obviously material (although in another sense, this qualifies the nature of matter, to which we have to add a virtual aspect).

The Brain and Electronic Images

It is with this in mind that we can briefly examine Deleuze's comments on the brain in *Cinema 2*. John Mullarkey notes that much of the cinema books is "inspired by the microbiology of the brain" and that "Deleuze sees cinema as operating at the molar level to shock the brain into forming new synapses, connections, and pathways,"[62] which, as we have just seen, will also allow for new differential series, new "dark precursors" and new forms of "thunderbolts," in short new forms of thought, new ideas. For Deleuze, real thought is impossible without this shock. The cinema is remarkable in the way it approaches the conditions of thought as shock—technically. For example, Deleuze thinks that "Montage is in thought "the intellectual process" itself, or that which, under the shock, thinks the shock."[63] The shock is a "totally physiological sensation," with a "set of harmonics acting on the cortex which gives rise to thought."

At the same time, there "is as much thought in the body as there is shock and violence in the brain. There is an equal amount of feeling in both of them."[64] Although the brain has its own specificity, for Deleuze, "brain-ness"—and thought—are not restricted to brains per se. On the other hand, sensation is not restricted to the body (the neurons are full of sensation). More than this, shock and violence, "thunderbolts" and "lightning" are found everywhere. Or, thought/sensation is extended throughout the entire ecology of events.

This means shock, violence, and thought/sensation are configured differently within different relations to different biological, social, or technical forms.[65] Deleuze is particularly concerned here with the new electronic imaging. It poses a challenge to the brain to adapt, and seems, on the other hand, to come closer to the brain's own forms of dynamic organization. The new imaging therefore, like many aspects of neuroscience (remembering that neuroscientific images themselves are exactly these new electronic images), poses the possibility of new, extended forms of control and possible freedoms in a more complex engagement with the brain—in what I have begun to call "neural cultures." In this context, the brain is both accelerated and possibly exceeded by the technics that are to some extent based on concepts, in the development of "thinking machines," of the brain. These technics augment the brain, analyze it, work it differently. The question becomes: "cerebral creation or deficiency of the cerebellum,"[66] or both at the same time?

For Deleuze, "neural cultures" are not, and will never be, those cultures that finally know how thinking processes work in an absolute sense. They will be rather those that work the powers of the ongoing, and differential, genesis of experience and action, of thought-sensation, in which experience and action, inside and outside, begin to fold over each other at dizzying speeds. They are accelerating cultures, in a "perpetual reorganization." As he writes in *Cinema 2,*

> The new images no longer have any outside (out-of field), any
> more than they are internalized in a whole. . . . . They are the object
> of a perpetual reorganization, in which a new image can arise from
> any point whatever of the preceding image.[67]

Deleuze seems attracted to this perpetual reorganization, for reasons that should by now be obvious. At the same time he has doubts about a rationalist, informational co-opting of this, in which "the shot itself is less like an eye than an overloaded brain endlessly absorbing information" (267). He wonders if the brain is adequate to the new image cultures *in their extreme informational, rationalist modes.*

As we have seen, Deleuze's concept of the brain is very different to that of a rationalist, cognitivist informatics. Yet if not an "informationalist" organ, it can be very much at odds with informationalist demands. Writing earlier in *Cinema 2* about Hal the computer's famous breakdown in Kubrick's *2001: A Space Odyssey,* Deleuze comments, "if the calculation fails, if the computer breaks down, it is because the brain is no more reasonable a system than the world is a rational one. The identity of world and brain, the automaton, does not form a whole but rather a limit, a membrane which puts an outside and an inside in contact, makes them present to each other, confronts them or makes them clash . . ." (206). John Rajchman points out that this leaves us not with a (hyper) rational brain, but a "neuroaesthetic brain,"[68] a flexible, sensate brain working with "unities of feeling," but one always with a set of limits, if flexible limits.

This gives art—and media—a special role in the reorganization of the nervous system in response to new technical and social demands. "Creating new circuits in art means creating them in the brain too."[69] As Rajchman puts it, and against some of the more naturalist accounts of art in neuroscience,

> art is less the incarnation of a lifeworld than a strange construct
> we inhabit only through transmutation or self-experimentation, or
> from which we emerge as if refreshed with a new optic or nervous
> system.[70]

This makes all art and media a kind of (proto-cognitive) architecture even, we might say especially, with regard to the folding of contacts in the brain, "if askew or non-Euclidean." With its digital imaging, mappings of brain dynamics, and restructuring of brain, body, world relations, neuroscience could also be seen as necessarily aesthetic, or, more appropriately, a kind of ongoing architectonics of the nervous system in tune with this nervous system, not imposing a cognitivist pseudoscientific paradigm upon it.

Ethical questions become highly complex questions of the design of the new as much as a loss of old certainties. If the brain's sciences were to focus only on "consequence of discovery" issues, without addressing architectonic issues, they would face the obvious problems of being unprepared for what they make available bioculturally.

All this does not mean, as in many misunderstandings of Deleuze and Guattari's concept of schizoanalysis, an ongoing complete reassembling of the brain in some wacky, impossible way. Rather, it is a question of using the plasticity and specificity that is there, alongside the foundations, so to speak. This should not mistake these foundations for a structure more rigid than it actually is, but neither should it overdetermine the capacities of the brain to "perform" complex rationalist systems that often exclude the experimental, the flexible, that are the peculiar talent of the brain as subject. For Deleuze, "Neuroaesthetics becomes possible . . . just when sensation is freed from representation and even from phenomenological conditions to become experimental and diagnostic."[71] Thus, as Rajchman points out, the perils of neuroscience reintroducing "cognitivist" schemes for recognizing objects, or else phenomenological ones for "embodying" life-worlds. It is here that Guattari's formulation of a "machinic unconscious" is crucial.

Guattari and the Synaptic

Of Deleuze's three major direct discussions of the brain (in *Cinema 2, A Thousand Plateaus,* and *What Is Philosophy?*), two were written with Guattari. Even the *Cinema* books emerged in part from Deleuze's work with Guattari. Here I shall very briefly give an account of four concepts that seem to feed into both Deleuze's view of the brain, and Deleuze and Guattari's final statement on the brain in *What Is Philosophy?* These concepts are "metamodelization," the "machinic unconscious," the "synaptic," and the "fractalizing." All of these contain a desire to take into account the immanence of self- or auto-modeling of biological/cultural events. At the same time, they contain a deep (political and ontological) suspicion of imposed models, at the least of the less-than-dynamic limits of these models. On the contrary, there is a valorization of the opening to virtuality of the gap, or "synaptic." In addition, as with Deleuze's notion that "even rocks and woods" are a contemplation, concepts such as the machinic unconscious and the synaptic are broadly applied beyond the brain.

Let us start with metamodelization. This also involves what is essentially an adaptive metamethodology.[72] Metamodelization is

something which does not establish itself as an overcoding of existing models [as is so often the case with cognitivist models of thinking processes], but as a procedure of "automodelization" which takes over all or parts of the existing models in order to construct its own cartographies, its own references and therefore its own analytical approach and methodology.[73]

It is therefore not only an immanent automodelization, but one that intervenes and reorganizes the operations of models imposed upon/within a given event.[74] If we take metamodelization seriously, to paraphrase Guattari, we have the brain we deserve[75] (for Guattari, the unconscious we deserve). With regard to thinking processes, this implies that Deleuze and Guattari were not seeking a fixed alternative model of the brain or nervous system, although they were willing to take up parts of existing models from elsewhere (new media as well as the history of philosophy and the molecular biology of the brain).

This brings us to the machinic unconscious. This is not limited to the brain, or even to the subject. It is rather the unconscious taken "transversally"—that is, as operating across fields, bringing them together in new ways. It is directed toward the future rather than the past. It is

an unconscious whose screen would be none other than the possible itself, the possible as a flower of language, but also the possible as a flower of skin, as a flower of socius, as a flower of cosmos.[76]

This radically undefined unconscious is also open to all to understand and work with, without "distress" or "preparation."[77] This is, if you like, a very democratic view of work with the unconscious. It is unrestricted by disciplinary models. It lives in the world at large. It is informed as much by media manipulation and "cohorts of technicians" as by any kind of interior life. It is "'machinic' not because it is mechanical, but because it is productive, and also because it is not necessarily centred around human subjectivity, but involves the most diverse material fluxes and social systems." It is not "the exclusive seat of representative contents"[78] (there goes symbolic processing again), but rather "the site of interaction between semiotic components and extremely diverse systems of intensity" (a kind of "extended mind" at warp drive).[79] It does not "depend on a universal syntax" or "universal structures" but rather "processes of singularization," so that it "evolves with history."[80] It can be individual or collective, and tends to underwrite the ongoing creation of territories of existence.

Guattari complicated this creation of territories with a concept he calls the "synaptic." This arises *between* existential territories and other dynamic structures. The synaptic does not open, initially at least, onto individuation and structure, connection or temporal systematization (of territories, of forms of existence, of things that are "placed" in a system or syntax of signs). The synaptic opens instead onto the "nondifferentiable."[81] It involves a potential for movement from formed systems to chaosmos, the virtual. The synapse is "*dis*-positional."[82] Synapses are *fields* of potential, that, via these openings away from structures, are able to break off aspects of events or previous territories and set them free. An example Guattari gives from perception is when an aspect of an "organized" face, say a moustache or a "grimace," breaks free in perception, and is able to connect with everything, everywhere, as in "dreams" or "phantasms," "delirium," or "religion." Technically,

> the synapses integrate, to a deterritorialized level, the existential
> breaks resulting from the refrains . . . *the synaptic gap . . . marks . . .*
> *an explicit break . . . rich in content,* even if in a content that is muti-
> lated, made "arbitrary," rendered a-signifying,[83] that is to say cut off
> from its syntagmatic bases, and from its paradigmatic bonds.[84]

In a sense, the synapse is the open shifting field of potential that comes before Deleuze's "dark precursor" creates a channel in the space between territories for a flash of lightning (or a firing of a synapse—a micro-individuation—to occur).

If Guattari's machinic unconscious has a foundational syntax or order of events it might look like this: *synaptic—dark precursor—the flash of individuation—unity of feeling*. However, the elements of events in Guattari's terms would not have a regular syntax. In fact, the synaptic for him is precisely the possibility of "points of reversal" away from determination. (This again discounts a cognitivism or computationalism of the brain.) It is true, as with the machinic unconscious, that Guattari uses the term "synaptic" in a very broad context, far beyond its usual reference to the brain. It does eventually inform Deleuze and Guattari's view of the brain, but this is a two-way street. On the one hand, as previously mentioned, Deleuze and Guattari generalize "brain-function," beyond the brain as normally conceived, to a world full of micro-brains. "[N]ot every organism has a brain, and not all life is organic, but everywhere there are forces that constitute microbrains, or an inorganic life of things."[85] On the other hand, Guattari's synaptic does seem very resonant with their discussion of actual synapses.

In *What Is Philosophy?* Deleuze and Guattari are discussing the structure of the brain, the task of science in relation to the brain, and the relation of the functions of science to the brain. They focus particularly on the chaos, or virtuality, into which the "brain itself, as subject of knowledge, plunges." They write:

> It is up to science to make evident the chaos into which the
> brain itself, as subject of knowledge, plunges. The brain does not
> cease to constitute limits that determine functions of variables
> in particularly extended areas; relations between these variables
> (connections) manifest all the more an uncertain and hazardous
> characteristic, not only in electrical synapses, which show a statisti-
> cal chaos, but in chemical synapses, which refer to a deterministic
> chaos. . . . Even in a linear model like that of the conditioned reflex,
> Erwin Strauss has shown that it was essential to understand the
> intermediaries, the hiatuses and gaps.[86]

We have seen that, for Deleuze and Guattari, the structure of the brain is not so much "syntactic" as topological, a probabilistic series of chaoses in interaction, which are folded by ongoing connections between "extreme elements." There is also a range of synaptic "dis-positional" aspects to the brain—gaps and hiatuses, electrical and chemical synapses themselves.

This is made more complex by what Guattari refers to as a "fractalisation"[87] related to the synaptic. Massumi writes of a "fracture at the base of meaning,"[88] based on Guattari's use of the term in *Cartographies Schizoanalytiques*. As well as indicating a fracturing of meaning, however, Guattari uses the term "fractalisation" to describe the "texture" of "intermediate temporalities," the fractal effect of mixed temporalities— durations and syntheses—in "becoming."[89] Related to the synaptic, fractalization indicates a further conception of the potentialization that occurs within the brain, not only in space, but in terms of the recursive creation of time(s), within the gaps and hiatuses in the nervous system of which Deleuze and Guattari are so fond.

Man Absent from, but Completely within the Brain

We can now understand why, in *A Thousand Plateaus*, Deleuze and Guattari consider the "dendrites" (the parts of the neuron receptive to stimulus) badly labeled. The term "dendrites" is derived from the Greek for "tree," and they are committed to a less hierarchically structured (rhizomatic or "grass-like") brain. In this brain, the "discontinuity of the cells, the role of the axons, the functioning of the synapses, the existence of synaptic microfissures, the leap each message makes across these fissures, make the brain a multiplicity."[90] We understand why they favor the dynamism and fragmented nature of short-term memory over long-term memory, and radicalize this further by suggesting that short-term memory "is in no way subject to a law of contiguity or immediacy to its object; it can act at a distance, come or return a long time after, but always under conditions of discontinuity, rupture and multiplicity."[91] Short-term memory is, simply put, more dynamic, more subject to folds, strange continuities over distances as well as gaps and discontinuities, more open to the "fractalisation" of meaning and temporalities. We also understand Deleuze and Guattari's concept of the brain as involving embodiment in a very direct manner— "Sensation is no less brain than the concept."[92] However, they also add an extended materialism to the mix, one that takes the incorporeal into account. In this, they give a *qualified* materialist account of metaphysics, as literally arising within the brain.

An extended note in Deleuze's *The Logic of Sense* is crucial for understanding Deleuze's concept of the brain in relation to metaphysics. Here, he accounts for the "conversion of the physical surface of the brain into a metaphysical surface."

> Modern studies insist on the relations between areas of cortical projection and topological space. "The projection in fact converts a Euclidean space into a topological space, so that the cortex cannot be adequately represented in a Euclidean manner" . . . It is in this sense that we speak of the conversion of the physical surface into a metaphysical surface, or of an induction of the latter by the former. We can thus identify the cerebral and metaphysical surfaces: it is less a question of bringing about the materialization of the meta-physical surface than of following out the projection, conversion, and induction of the brain itself.[93]

This interaction between physical surfaces and inducted metaphysical surfaces,[94] or between Euclidean and topological spaces, is the foundation of a much fuller account of thinking processes that seeks to account for thinking processes via physical structures alone. For Alliez, it allows an account of consciousness as *"nothing other than form or, rather, active formation in its absolute existence*: consciousness and morphogenesis are one and the same thing."[95] Sometimes, "seeing the brain in action" in brain scans seems to distract from this less visible process of active formation. It does so in favor of the identification of physical locations and visible structures, often as the location of aspects of subjectivity, in which we find "our" subjective states such as "happiness" or "meditative peace."

Of course, these may not be "our" states. That which neuroscience discovers but which is hard to reconcile with humanism (the key and as yet unresolved problem of neuroethics), is that "It is the brain that thinks and not man, the latter being only a cerebral crystallization."[96] It is in the brain that new worlds—new, singular universes of reference—form, including those universes of reference that become, often briefly, if again and again, "man" or "woman." Indeed, it may surprise many how central the brain is to Deleuze and Guattari's thought, and how much it is the brain that proves to be the key substitute for "man." Deleuze and Guattari's "posthumanism" is molecular biological.

> We will speak of the brain as Cézanne spoke of the landscape: man absent from, but completely within the brain. Philosophy,

art, and science are not the mental objects of an objectified brain but the three aspects under which the brain becomes subject. Thought-brain.[97]

Or rather, for all the above, not *in* the brain, or not *only* in the brain, because "The brain is the junction—not the unity—of the three planes" (the three active planes of the functions of science, the percepts and affects of art, the concepts of philosophy).[98] For Deleuze and Guattari, the brain as junction is *a folded exteriority* (echoing the history of the development of the nervous system).[99] It is a folded exteriority still profoundly subject to forces from the outside, in part because, as with mirror neurons, or perception in general, which can occur at a distance, the brain is such a powerful organ of synthesis across ecologies. At the same time, while radically open to the exterior, the brain possesses *a very deep interiority* (as "synaptic" and "fractal") in which the coming together in the brain is reworked profoundly—literally metaphysically (topologically)—as syntheses encounter the forces of the virtual. The folding of exterior and interior together (Superfold) is what enables Deleuze and Guattari to claim, in *A Thousand Plateaus,* that "there is indeed one exterior milieu for the entire stratum." The "stratum" they refer to here is not quite that of the human, but of "the prehuman soup immersing us," and this exterior milieu, "permeating the entire stratum," is "the cerebral-nervous milieu."[100]

Deleuze and Guattari's concepts of radical exteriority and radical interiority allow for a brain that "naturally" metamodels, mutilates, or "*dis*-positions" rigid schematic ontologies, including those concerning thinking processes, toward a pragmatic open to ontogenesis. In this respect, Alliez gives perhaps the most accurate definition of the brain according to Deleuze and Guattari. The brain is

that operation of being that composes the meta-stable system of phases belonging to a pre-individual world in formation—in the midst of individuation—qua ontogenesis of itself. Following the principle of a conversion of the cerebral surface into a metaphysical surface, we could almost say that the brain is ontology delivered over to the pragmatics of being.[101]

In many ways, such a definition of the brain lies precisely at the intersection of the rigors of a neuroscience concerned with the constitutive elements of the pre- (and post-) personal on the one hand, and social/

political diagnosis of the way thinking is materially constituted within biological and social practices or modes of living.

What Is Philosophy? concludes that the brain even poses a challenge to
preconstituted areas such as science, art, and philosophy. If the brain is the
"junction," then this is a junction in which neither science in general, nor
neuroscience in particular, should expect to remain themselves. As Arkady
Plotnisky puts it, Deleuze and Guattari see the "possibility of a different
future of thought, in which the boundary between philosophy, art and
science and even all three themselves disappears back into the chaosmic
field of thought."[102] The most important questions to consider (especially
at the junction of the biological/social, neuroscience, and what we have
called here "neuroaesthetics"), are "the problems of interference between
the planes that join up in the brain."[103] This interference seems particularly
acute in neuroscience's case.

Beaks Against the Window

We like to play with models of mind. Yet sometimes it seems that we
return too often to different models of recognition as the objects of our
play. We might even multiply these models of recognition as we play. Yet,
as Deleuze writes of Kant, this does not go far enough. "Far from overturn-
ing the form of common sense, Kant merely multiplied it."[104] At the heart
of this problem is the changing of a politics of being with regard to models
of thought. At the heart of this is perhaps the creativity with which we create transversal movements between fields, disciplines, and practices, when
this is necessary. And we can take the following, in which Deleuze quotes
Guattari, as perhaps a basis for an alternative future politics of being to
that of neuro-cognitive controls. Here,

> symptoms are like birds that strike their beaks against the window.
> It is not a question of interpreting them. It is a question instead of
> identifying their trajectory to see if they can serve as indicators of
> new universes of reference capable of acquiring a consistency suf-
> ficient for turning a situation upside down.[105]

Such symptoms as these birds indicate the general social and biological
urgency of the call of the virtual, no matter how faint that call might be.
This is a call which is always as close to home as it is far away.

CHAPTER 12

Mammalian Mathematicians

Clark Bailey

DELEUZE SUGGESTS that in order to understand a philosophy, in order to grasp the concepts it creates, we must return to the problem these concepts confront. In the case of Deleuze himself, Manuel DeLanda has proposed that the problem he confronts throughout his philosophy is one of avoiding essentialism; he tries to conceive our world without recourse to transcendent essences. There are a variety of ways of restating this—we could say that it is a question of pure immanence or the univocity of Being, or of creating a philosophy of self-organization and evolution—but I like to see it as a question of how mammals can do mathematics. All of these cases illustrate the same recurring problem: how can we explain why we find stable forms in the world, whether these are material things, animal behaviors, or mathematical truths, without invoking something outside it?

In the case of animals, of course, we have been antiessentialists for more than a century now. Unfortunately, despite the (occasional) success of evolutionary theory in breaking down the essentialism that once clung to the concept of species, we continue to have essentialist leanings in many other respects. Democracy and capitalism are celebrated as the unique logical end of history as much as string theory and the Big Bang lay claim to being its incontrovertible beginning. It is precisely this widespread interpretation of science as specifying transcendent and essential truths that motivates DeLanda's reconstruction of Deleuze's antiessentialist ontology in *Intensive Science and Virtual Philosophy*. I would like to begin with a simplified sketch of this reconstruction that we can then apply to our own variation on the antiessentialist problem. This involves making an analogy between the antiessentialist ontology DeLanda describes and the objective structure of a conscious brain as hypothesized in the neuroscientific research of Gerald Edelman and Christof Koch, among others. We might think of this as trying to naturalize our epistemology in such a way that we can see how our process of thinking could become isomorphic with Being.[1] Perhaps then we can make some progress toward explaining what seems a striking problem for any philosophy of immanence: how one particular symbol system of one particular hairless ape could come to predict something as remote as the movement of the planets, without needing to invoke the ape's unlikely access to a set of transcendent and essential truths. By putting the problem of antiessentialism in this evolutionary and scientific light, I hope to suggest that the neomaterialist Deleuze we find in DeLanda can provide the concepts we need to understand precisely this point where the ape touches the cosmos—the brain.

Following the author's own terminology, I have said that DeLanda outlines a neomaterialist reading of Deleuze. This idea might appear paradoxical, given that, despite the abundant analogies drawn from mathematics, biology, and physics, the first hundred pages of *Intensive Science and Virtual Philosophy* are almost entirely spent asserting a metaphysical distinction, a distinction of *kind,* between the actual states of a material world and the never-actualizable, or virtual, singularities that *structure* the causal processes that shape that world. While typical of our conception of science, drawing a fundamental distinction between the laws and the matter obeying them seems a strange basis for a truly materialist philosophy. How can a materialist insist on the equivalent reality of the virtual, an expressly immaterial category?

Before we try to answer this question, we need a better idea of exactly what we mean by the virtual—what is its structure and how does it relate to the actual? DeLanda illustrates the distinction between actual and virtual with a detailed discussion of *phase space,* a concept drawn from the branch of physics known as dynamic systems theory. Briefly, this theory seeks to represent the complex state of an actual material system as a single point in a high dimensional space. As aspects of the system change over time, this point traces out a trajectory through the space, thus representing the history of the system as a succession of points, that is, as a curve. The phase space is meant to capture every *possible* state of the system, and the lines drawn in it every *possible* history, each one corresponding to a particular starting point.

For our purposes, the important thing to understand is that the phase space of many systems has a characteristic shape (more generally a topology). If, for example, we think of a system that inevitably ends up at a unique equilibrium point no matter what its initial conditions are, the corresponding phase space landscape will be shaped like a funnel. The point at the center of the funnel is known as an attractor, or a singularity, in this case corresponding to the stable equilibrium point of the system. Naturally, there are many more complicated phase space topologies, each defined by the existence and distribution of its characteristic set of singularities and each relating to the possibility space of a particular system.

Despite this impoverished description, it gives us a clear idea of how DeLanda conceives the distinction between the actual and the virtual.[2] What we normally refer to as the "real" system, that is, some physical system in a particular state, with its parts disposed in a certain manner, as well as other possible states this system *could* occupy at some other time, *all* line up on the same side of our metaphysical divide—the actual. On the other hand, the virtual refers to something qualitatively different. It is not an actual or even a possible state of the system, but a *structure* of these possibilities; it refers to the structure of the processes that shape a system's possible behaviors, to a force behind or beneath actual possibilities that provides a consistent way of thinking about an entire set of them at once. Deleuze, of course, repeatedly warns against confusing the virtual with the possible, and insists on its direct structural reality. Thus, in our mathematical analogy, the virtual is a set of singularities that structure the evolution of a system and associate its phase space with a particular topological form. In this sense, it plays a role analogous to the one we often attribute to the immaterial laws of physics that nevertheless somehow govern the interaction of actual material bodies.

From our perspective, the most important reason for introducing these mathematical concepts into a discussion of the virtual is that they provide a very concrete way of understanding the slippery structure of the virtual itself. The main idea is that the topological forms that characterize phase spaces can themselves be transformed into one another as the systems they describe undergo a series of phase transitions. While at first it may seem extraneous to its definition and its distinction from the actual, this internal structure of the virtual helps distinguish Deleuze's theory from one that simply takes the topology of a system's phase space as its eternal Platonic essence, while reducing the actual system to a mere instance of these preexistent fundamental laws.

To understand this structure of the virtual, we need to know more about the way the topological singularities that structure a phase space can be used to describe complex systems. In fact, the analogy DeLanda makes between the mathematics of phase space and the structure of the virtual only gathers real force when we begin to consider nonlinear and nonequilibrium systems that can typically show semistable patterns of behavior beyond simple static equilibria. These dynamic equilibria still, however, correspond to singular attractors in the phase space of a system.[3] If we imagine a system that is gradually taken further away from equilibrium, these new attractors emerge as stable patterns at certain critical thresholds. In addition, they emerge together in sets. That is, in passing a given threshold, or bifurcation point, multiple distinct possible patterns appear simultaneously, with the actual system falling into one or another based on chance. A simple example would be the vortex patterns that can emerge if a liquid is heated past a certain point; both clockwise and counterclockwise rotations emerge as stable patterns at the same threshold.[4]

DeLanda proposes that we interpret the Deleuzian idea of a multiplicity, the basic "unit" of the virtual, in light of these complex systems. He defines a multiplicity as this entire unfolding sequence of stable patterns, "A multiplicity is a nested set of vector fields related to each other by symmetry-breaking bifurcations, together with the distribution of attractors which define each of its embedded levels."[5] This definition contains some implications about the structure of the virtual that I want to explicitly emphasize.

First, by referring to bifurcations and sets of singularities, it immediately puts the virtual in the realm of the nonequilibrium and the nonlinear, the realm of process. As DeLanda points out, a simple equilibrium situation, or a predictable linear system, covers over the virtual aspect of a system. By contrast, even in a very simple nonequilibrium case such as a vortex, the dynamic regularity we observe immediately requires us to speak of a *process* of production. Now that we are required to investigate this process as part of our explanation, we quickly discover a defining feature of the virtual, the existence of other well-defined but unactualized patterns alongside the one we are observing: "A system with multiple attractors, in short, has a greater capacity to express or reveal the virtual."[6]

Second, each embedded level of a multiplicity constitutes a separate topological space characterized by the actualized and unactualized attractors structuring the system's phase space at that level. Of course, Deleuze's choice of the term "multiplicity" is partly motivated by its connection with the mathematical spaces Bernhard Riemann called "manifolds," the technical name for a phase space. These are surfaces that form spaces in their own right, constructed entirely from local relations, without the need to embed them in a global space of higher dimension as part of their definition.[7] In another context, Deleuze refers to these as "absolute surfaces or volumes" defined strictly by internal relations that cohere enough to create a consistent topological form.[8] They are self-consistent immanent forms that arise like a vortex, tying together heterogeneous variations into relatively stable patterns defined from within by the way these variations form inseparable parts of an emergent whole. We will return repeatedly to this image of the vortex; it is one of the simplest of these self-defining forms, yet it elegantly illustrates how a global form can emerge from and react back upon local interactions. DeLanda suggests that these absolute surfaces and the unactualized attractors required for their definition are the "spatial" dimension of the virtual.

The corollary of this internal definition of a multiplicity as "filling all of its dimensions," however, is that it can only be defined up to the limit at which it transforms into another. Each of these transformations corresponds to a phase transition between the levels of a multiplicity, a line marking the emergence of one irreducible topological form from another as some threshold of intensity is crossed. True to form, Deleuze uses numerous terms to refer to these transformations by which the various levels of a multiplicity unfold; but whether we prefer to call these the operations of a quasi-cause, becomings, or lines of flight, the important point is that they are always conceived as the *necessary inverse* of the self-defining aspect of a multiplicity. The *absolute*-ness of a multiplicity, that is, its not being defined *relative* to another or to a library of preexisting forms, means that adding or subtracting anything, modifying its internal relations, causes it to transform into something else.[9] These transitions between levels would correspond to a characteristic virtual "temporality," and in this essay I will refer to them as becomings.

Finally, this means that the definition of a multiplicity is necessarily *recursive*; a more complex multiplicity is defined in terms of the simpler structures it can differentiate into, which it thus contains internally in the same way a greater intensity contains a lesser one. Each level of a multiplicity is a multiplicity, with a similar structure in its own right. Moving in the opposite direction, the capacity of a multiplicity to transform means that its definition is open-ended. It can only be defined up to the point at which it would again transform, and leads us inevitably toward more and more complex multiplicities, toward the limit of the cosmos, even if we cannot think of this limit as a telos, but only in the form in which the limit always appears in Deleuze—as a line marking a transformation, as a phase transition. Thus, DeLanda's definition of a multiplicity actually refers to two types of singularity, singular attractors within a level and singular phase transitions between levels. Here we see a fundamental ambiguity we will explore in more depth later—a becoming that from one perspective appears to be an interaction between two individual multiplicities can be seen from another perspective as the crossing of a strictly internal threshold, a "voyage in intensity."

We should take a moment to observe how different are the "units" that populate this virtual world from those of our everyday world. In place of individual point-like particles that enter into combinations with one another (mixtures), we find a world characterized by lines of transformation marking how one structure flows into others by crossing thresholds of intensity. It is not that two multiplicities are indistinguishable, but simply that, under determinate conditions, they can both be seen as embedded parts of a particular emergent *fusional* multiplicity, a more intense multiplicity. The virtual is marked by the fluidity of its units, ordered precisely according to their capacity for transformation or differentiation. It is a realm of intensities that do not combine linearly and that do not divide without changing in nature, making it by the same turn a place where truly novel structures can unfold. By contrast, in the actual world nothing emerges or flows, and instead of imagining that "more is different," we imagine that there are merely different collections of colored balls, varying mixtures of predefined forms.[10]

To summarize, then, the virtual has three parts: multiplicities, becomings, and the plane of consistency. Though we will return to the role of this third component in more detail later, we can already see how the plane of consistency is implied by the recursive definition of a multiplicity. It serves as a sort of space-time fabric in which two multiplicities can interact in a becoming that links them together to unfold a consistent new form. Of course, as DeLanda's appendix bears witness, this same triadic structure occurs in so many guises in Deleuze's philosophy that it is hard to know what its canonical name should be.[11] To further complicate matters, we have to take Deleuze and Guatarri seriously when they say that a philosophical concept is created by establishing such intimate connections between its components that they become indiscernible, each term flowing into and presupposing the others in a sort of dynamic conceptual equilibrium, making them impossible to separate.[12] The virtual is a concept insofar as it posits this internal consistency and inseparability of its three components, while at the same time throwing off connections to other concepts—in this case, the concept of the actual.

Hopefully we are now in a position to apply DeLanda's interpretation of the virtual to a particular concrete problem. *Intensive Science and Virtual Philosophy* focuses on *actualization,* or the manner in which one virtual multiplicity can progressively differentiate into others, and finally even into actual entities, through a cascade of phase transitions, inadvertently leaving us with a sort of Big Bang theory of the virtual. I would like to explore the opposite path. Considering *virtualization* may give us a more concrete idea not only of what exactly a self-consistent absolute form might *be,* but also how it could arise spontaneously in a material system, and what this immanent organization might *do.* In fact, a virtualizer is a particularly interesting and useful device.

Rather than beginning with the more traditional philosophical puzzle of how matter could give rise to mind, the philosopher Robert Nozick begins his discussion of "the human world as part of the objective world" in *Invariances* by asking what consciousness would be good for once it had arisen. In other words, what would its function be for an evolved organism?[13] The hypothesis he settles on is that consciousness functions as a sort of "zoom lens," enabling an organism to pick out subtle features in, and more finely attune its behavior to, an environment. This ability becomes useful in the context of an environment sufficiently complex or unstable that the organism's best responses cannot be evolutionarily hard-wired.[14] In these cases, the organism needs to integrate disparate and ever-novel data about its environment in order to base its actions on the correlations it uncovers. Nozick suggests, "Consciousness involves the meshing of different streams of data into a unified synthesis or representation. Objectivity involves the approachability of the same fact or object from different angles. Whether different streams of data can be given a unified representation is a test of whether there is some objective fact present that is approachable from different angles. . . . To test for objectiveness may be (part of) the function of subjective consciousness."[15] He goes on to make the case that this integration of data is inherently aimed at action, and that however it is formed, it needs to be registered *as* an integration by the parts of the brain that need to act on the particular combination of circumstances involved.

Of course, it is precisely this bottleneck that tempts many to introduce some sort of homunculus into the brain that can simultaneously form a picture of what is going on and function as a central dispatch to communicate this to the appropriate centers of action. If we are to avoid this internal subject, we have to imagine some way for data to be integrated without an integrator, and registered and acted upon without a commander. Nozick proposes a solution to this problem that is relevant to our understanding of a multiplicity:

> There is an important notion in the literature on game theory, the notion of "common knowledge," wherein each person in a group knows that p, each knows that each of the others knows that p, each knows that each of the others knows that each of the others knows that p, and so on up the line. . . . Conscious awareness, I hypothesize, involves registering all the way up the line. This involves p registering everywhere (in the relevant parts of the brain), its registering everywhere that it registers everywhere, its registering everywhere that it registers everywhere that it registers everywhere, etc. It involves common registration. We might say that consciousness is the brain's common knowledge.[16]

The idea that consciousness is analogous to common knowledge neatly ties the problem of integrating to the problem of acting upon that integration, and in the process eliminates the homuncular middleman. As a result, we see that consciousness *is* exactly what it *does*; the idea of common knowledge not only hypothesizes the form of consciousness as information integration, but simultaneously explains why it is useful. As Nozick puts it,

> In a disaggregated system in which unified simultaneous action produces good results while staggered action or action of only some parts might lead to disaster, common knowledge is needed to coordinate the simultaneous joint action. Consciousness is what coordinates a distributed system in simultaneous joint action—that is the function of consciousness. That is accomplished by common knowledge, common registering within the system, and it is this common knowledge which . . . constitutes consciousness.[17]

From this description, we can quickly see how consciousness could be a multiplicity (or more precisely, one level of a multiplicity) in the sense we defined earlier. In unifying the various specialized processes going on in a brain, consciousness draws an "absolute" form, defined from within by an infinite regress of reciprocal relations, without reference to a supplementary dimension or a prior plan that would lend its unity to the construction. Instead, it arises in the manner of a vortex, whose structural essence is inseparable from its effect on the world around it.

In its evolutionary context, this neural integration can be seen as an "objective model" of (part of) an animal's environment, a neurobiological realization of a singularity in an environment's space of possibilities. Consciousness brings together previously separate information about the environment, gathered by specialized sense organs, in a way that immediately illuminates and acts on the structure of this information. Obviously, if this integration of information occurring within the brain did not lead to adaptive actions in the objective world, our organism would be rather short-lived. Nozick's description of this state of integration as a "representation," however, can lead to confusion, insofar as the term is often linked to the correspondence theory of truth, and other similar theories involving little men in our heads who watch movies—theories Deleuze's antihumanism and anti-Oedipalism are squarely aimed at dismantling. We can isolate some key differences between what I would like to call (for want of a more elegant term) an "objective model" of a system (in this case an animal's environment), and its representation.

First, a model *forms itself* from the consistent interaction among some neural components causally connected to the world. It is articulated from within without needing a subject, instead emerging as a singular form from some material, thus breaking with our usual understanding of matter as a homogeneous inert substance that simply receives whatever form we impress on it. In place of a preexistent representational mold, we have a spontaneous capacity for the self-organization of forms, in place of a knowing subject, a working brain.

Second, this type of model bears no relation of resemblance to the system it models. Our common understanding of a representation is as a sort of *picture* of the world, with one term imitating the other in the way a copy imitates an original. To this picture we attach a name or a symbol that somehow allows us to accurately "call up" and refer to all of the original features of a situation. Representation is thus tied to a specifically *symbolic* form of memory, along with the correlated assumption of some decoder that understands how to interpret the symbols. An objective model, instead of bearing a resemblance to the system modeled or symbolizing it for some assumed subject, is an actual process *isomorphic* with what it models. It is a topologically equivalent alternate realization of some aspect of a system, a simulation that shares, up to a certain point, the same phase space singularities as the original.[18] This of course is the evolutionary purpose of consciousness—to assemble isolated facts about the objective world into a functionally integrated *neural* process that lets you successfully cope with it. An isomorphic model can contain information about the world without directly referring to or symbolizing it because it has its own internal informational structure.

Consciousness as common knowledge allows a brain to contain information about the world by containing information about itself. As Nozick suggests in what sounds almost like a reductio ad absurdum of the idea of symbolic reference, "Evolution selects for such isomorphism in the structure of representational experience because this enables aspects of the world to be read off aspects of experience."[19]

The objective model then, does not stand outside the actual world as in some supplementary ideological dimension, but arises immanently, as a continuation of the world, capturing in one coherent and unified form certain aspects of the world, slicing through it to find and act upon a consistent set of objective correlations useful to the animal. This operation is, nevertheless, *still part* of the world; it is always realized in some material, and while it reveals a virtual structure of the environment on one hand, it is at the same time an actual physical system. Being part of the actual world, having an actual effect on it, a model cannot help but be taken up as the object of further models in a potentially endless sequence of models modeling models modeling . . . Obviously, this possible recursive expansion is the second part of what relates the objective model to the idea of a multiplicity we have already discussed.

Despite the terminological differences, I find Nozick's idea of common knowledge an apt illustration of how a system could exhibit a virtual structure and what use this might be put to. In addition, Nozick goes on to speculate about a neurobiological mechanism that would support something like the infinite regress of consciousness as common knowledge, suggesting, "There has been recent investigation of the hypothesis that consciousness involves synchronized firing at 40 megahertz. If the rate of firing of each affects the rate of firing of the others, then synchronized firing, aided by the reentrant networks emphasized by Gerald Edelman, might constitute the brain's being in a state of common knowledge. This common firing is a global property of the brain, dependent upon the action of the individual parts and upon their synchrony."[20]

Exploring this type of synchrony, and the way in which it can spontaneously give rise to consistent global forms based on merely local interactions, is an active topic of research in a number of branches of science.[21] While here we focus on the science of consciousness, any of these areas could potentially benefit from a more careful look at the philosophical problems surrounding the idea of self-organization, and hence from the philosophy of Deleuze; after all, if "immanence is immanent only to itself," all organization is self-organization.

Recent neuroscientific work has naturally gone far beyond the tentative suggestion Nozick makes about synchronized firing; however, insofar as one can speak of a scientific *theory* of consciousness, the hypothesis that a conscious state corresponds to a spontaneous and transient synchronization of distinct groups of neurons captures the basic idea. Though a number of authors have put forward versions of this hypothesis, one of the most fully developed theories is summarized by the neuroscientists Gerald Edelman and Giulio Tononi in their recent book, *A Universe of Consciousness*.[22] These authors set out to explain what they call "primary consciousness," that is, the type of conscious experience other mammals such as monkeys and dogs have, rather than trying to directly tackle the linguistic self-consciousness of humans.

A Universe of Consciousness centers around the concept of the "dynamic core," a hypothesized neural mechanism meant to explain what Edelman and Tononi see as the basic phenomenological properties of all conscious experience, "such properties include unity—the fact that each conscious state is experienced as a whole that cannot be subdivided into independent components—and informativeness—the fact that within a fraction of a second, each conscious state is selected from a repertoire of billions and billions of possible conscious states, each with different behavioral consequences."[23] Their theory thus takes an interesting philosophical form—Edelman and Tononi insist that consciousness is a *process*, and that they seek not merely to uncover what neural structures are *correlated* with that process but to find a *neural* mechanism that explains *why* this *mental* process has the structure it does. Though seemingly straightforward, this point is important enough and overlooked often enough that it bears repeating. What we want to explain is not simply how consciousness, defined abstractly as a certain type of information processing system or a certain sort of functional flow-chart, happens to be instantiated in a physical system like the brain. Instead, we seek a physical process that is isomorphic to a conscious process, so that we can see how they could share the same features. Though this idea may at first sight appear to be a form of dualism, the intent is rather to give the two sides a common ground in one and the same process; consciousness is a mental process as much as a physical process, and these two are structurally identical—they are the same process, virtually, the same "thing."

The dynamic core hypothesis makes two basic claims about the material side of this equation: first, that the neural activity supporting consciousness is widely distributed but tightly integrated, and second, that this activity is also highly differentiated, meaning that there are many different possible integrated activity patterns available to the brain. Integration is established by a type of rapid feedback between specialized brain regions that causes these regions to temporarily synchronize their firing patterns.[24] Edelman and Tononi suggest that the density of reciprocal connections in the cerebral cortex exists to provide the basis for this integration, and that the topological complexity of this network of connections is precisely what makes the brain a special piece of gelatinous matter. At the same time as these connections allow for integration of distributed regions, they also create a functional separation between the regions involved and the rest of the brain. A subset of elements within a system will constitute an integrated process if, on a given time scale, these elements interact much more strongly among themselves than with the rest of the system.[25] A conscious thought, then, is this process of coherent integration between widely distributed specialized brain areas that occurs on the time scale of a few hundred milliseconds, creating what Edelman and Tononi call a "functional cluster."

Clearly, these integrated clusters are the same spontaneous immanent forms structured by and structuring a mutually consistent set of interactions that we spoke of before as levels of a multiplicity or as common knowledge. They reflect the phase space singularities of the dynamic system we call the brain, and they constitute momentarily stabilized modes of global function self-organizing out of local interactions, as well as very concrete examples of the same absolute self-defining forms created by a population of variations become inseparable.

The second aspect of the dynamic core hypothesis reveals the necessary correlate of this immanent integration; it exists alongside other semistable integrated states that it is distinct from but capable of transforming into. In this context, Edelman and Tononi discuss consciousness as informative because of its diversity, and they relate this to a measure of the complexity of the brain: "The ability to differentiate among a large repertoire of possibilities constitutes information, in the precise sense of 'reduction of uncertainty.' Furthermore, conscious discrimination represents information that makes a difference, in the sense that the occurrence of a given conscious state can lead to consequences that are different, in terms of thought and action, from those that might ensue from other conscious states."[26] The information of consciousness is thus not primarily information about the world, but information about the brain, about which of its potential states the brain is in. It arises from the nature of the feedback connections that create a consistent state; the consistency, as with common knowledge, lies precisely in the fact that if anything were different about the activity of one of the components, the global state arrived at would have to be different. The variations involved have become inseparable, so that the very existence of one conscious state simultaneously distinguishes it from any other possible state, thus creating, "information from the point of view of the system itself—the number of states that make a difference to the system itself."[27] This amounts to saying that consciousness is necessarily a highly dynamic system with multiple attractors.

Fortunately, even though we cannot count or even classify these attractors, Edelman and Tononi go on to propose a mathematical definition of neural complexity based on evaluating the number of different states of each part of the system that are relevant to the behavior of the whole. "The fact that subsets of the system can take many different states means that individual elements are functionally segregated or specialized (if they were not specialized, they would all do the same thing, which would correspond to a small number of states). On the other hand, the fact that different states of a subset of the system make a difference to the rest of the system implies that the system is integrated (if the system were not integrated, the states of different subsets of the system would be independent)."[28]

Consciousness, then, must be a complex phenomenon, in the technical sense of having many independent parts that can still be integrated—a system balanced between order and chaos, a "fragile and delicate mechanism" where as many differences as possible, in the brain and ultimately in the world, can make a difference. Here, we begin to see the value of applying

Deleuze's ideas about the virtual to the brain; Edelman and Tononi have backed into a specifically *virtual* definition of consciousness, a definition depending not just on the actual activity pattern of some neuron or set of neurons, but on the simultaneous existence of objectively distinct *unactualized* patterns. The virtual has precisely this reality appropriate to what isn't, to "extra-Being."[29] It is the reality of these virtual patterns that distinguishes consciousness from mere feature detection or form recognition, and that gives it the ability to find structure in a novel and uncertain world.[30] A conscious brain is a virtual structure, or more precisely, an actual material structure that extracts and preserves the virtual structure of the processes implicit in the world around it, a feat of perception made possible by the brain's capacity to preserve its *own* virtual structure. Edelman and Tononi are at their most Deleuzian when they say

> extrinsic signals convey information not so much in themselves, but by virtue of how they modulate the intrinsic signals exchanged within a previously experienced neural system . . . At every instant, the brain goes "far beyond the information given," and in conscious animals its response to an incoming stimulus is therefore a "remembered present."[31]

Though we will return in a moment to the role of memory in consciousness, we can already see how thinking of the brain as a virtualizer that goes continually beyond the actual is merely a rephrasing of the idea that consciousness is a *process*. A "virtual philosophy" is inherently a philosophy of processes and structures of processes that can never be adequately captured in a single snapshot. The reality of a conscious state as a process of information integration appeals to a virtuality structuring the distribution of other possible outcomes of this process.

Here, we can return to the idea of the objective model. Consciousness is useful to an animal insofar as it identifies a situation in the world on what must be incomplete evidence. Being forced to act in a complex world, and in the absence of solid statistical correlations, an animal is constantly guessing how its scant data fits into an overall pattern. Edelman and Tononi give us this example: "Consider an animal in a jungle, who senses a shift in the wind and a change in jungle sounds at the beginning of twilight. Such an animal may flee, even though no obvious danger exists. The changes in wind and sound have occurred independently before, but the last time they occurred together, a jaguar appeared; a conscious connection, though not provably causal, exists in the memory of that conscious individual."[32]

Instead of simply processing the world in a step-by-step fashion in order to extract a predetermined form, consciousness depends on an entire internal mechanism whose dynamics need to reveal the way the world might be in order to know how it is. The internal process is structured by attractors that are merely *triggered* by some environmental information that consciousness is continually going beyond.[33] Going beyond the information given to form a question is, of course, precisely what we use models for. We go beyond a question in the propositional form, "Is there a jaguar outside my cave now?" to arrive at understanding an entire problem, "When do these terrible beasts attack?" DeLanda has already characterized the way in which a Deleuzian epistemology would be a "problematic" rather than an "axiomatic" system, an approach dependent on setting up a relevant space of possible explanations and where "understanding a system is not knowing how it actually behaves in this or that specific situation, but knowing how it would behave in conditions which may in fact not occur."[34] The idea of the "objective model" is simply an attempt to naturalize this problematic epistemology by relating the mental processes going on in a scientific subject or an animal adapting to its environment to the physical processes going on in the scientific animal's brain, making room for the possibility of their isomorphism at the same time as the brain becomes isomorphic with the world, and hence useful, regardless of whether the animal is a mathematician or a mammal. Echoing Deleuze and Guattari, we might say that it is the brain that models, not the scientist, though the two are isomorphic.[35]

The brain, in short, is a virtualizer. It extracts a full virtual form only implied in its actual environment; on the one hand, it is casually connected to features of that environment, and on the other, it renders these diverse facets into a consistent form through its own internal feedback relations—functioning as a sort of cosmic switchboard achieving the most unlikely connections. Unfortunately, this vision of the virtual is always in danger of passing into the transcendental essence of physical law already governing the world, or of becoming the structure of an ideal mental world that somehow magically manages to represent the actual one. These are the two sides of the same many-faced humanist menace that always seeks to open a metaphysical gap between conscious thought and everything else.

To avoid conceiving the virtual as preexisting either in here or out there, we need to concretely identify the immanent mechanisms by which a real virtual model is extracted from the world at the same time as it is actualized in the brain. Or, what amounts to the same, to see how the functioning of an actual brain could become isomorphic to the virtual structure of an actual environment. This means that we need to consider a more abstract level of feedback between the actual and the virtual *itself* that would function as a selective force rendering the two consistent. Edelman and Tononi's theory of the role of memory in consciousness fits perfectly into this context, as it is a theory of selection (which they call Neural Darwinism), sketching out how the complexities of a dynamic core that necessarily goes beyond the actual can nevertheless be adapted to it.

Edelman and Tononi take the platitude that consciousness is a process, not a thing, seriously enough to look outside the 100–200 millisecond window required for conscious neural integration and explicitly consider a process extending over the entire life of an animal. Though their "theory of neuronal group selection" is both complex and controversial, their main point is that understanding the history of a brain is essential to understanding its current functioning.

Edelman and Tononi invoke the image of Darwin to hold together a three-tiered process of selection that gives rise to the repertoire of states available to the dynamic core of an adult brain: developmental selection, experiential selection, and selection of feedback connections between brain maps ("reentry").[36] Each level functions by a process of "natural selection" in which a spontaneous internal mechanism gives rise to a superabundance of possible configurations that are then pruned through interaction with the world. In each of the cases, however, the precise selective force shaping the final product is different. According to Edelman and Tononi, developmental and early experiential selection are guided by the existence of phenotypical features already selected for during evolution, so that "The mere fact of having a hand with a certain shape and a certain propensity to grasp in one way and not in another enormously enhances the selection of synapses and of neural patterns of activity that lead to appropriate action."[37]

Later stages of experiential selection that refine sensory modalities, and the selection of feedback connections between these modalities, proceed by "a series of diffusely projecting neural value systems . . . that are capable of continually signaling to neurons and synapses all over the brain.

Such signals carry information about the ongoing state of the organism (sleep, waking, exploration, grooming, and the like), as well as the sudden occurrence of events that are salient for the entire organism (including, for example, novel stimuli, painful stimuli, and rewards)."[38]

These neuromodulatory systems are the common sites of action of many pharmaceuticals associated with mental illness (recreational and otherwise) and are also closely tied to our emotional responses (hence the term "value").[39] Here, the feedback that selectively prunes the virtual picks up speed, as the force of selection no longer operates relative to a slowly changing adaptive phenotype, but to real-time adaptive actions that result in feelings of pleasure or pain for the organism. We go from a relatively fixed organism, with a few basic reactions executed without respect to their context, to a more flexible creature capable of changing its actions based on how correlations it discovers in its environment relate to a (still) relatively fixed set of value responses.

These processes of selection set the stage for an organism with memory, though a specifically nonrepresentational type of memory where, "the triggering of any set of circuits that results in a set of output responses sufficiently similar to those that were previously adaptive provides the basis for a repeated mental act or physical performance. In this view, a memory is dynamically generated from the activity of certain selected subsets of circuits."[40]

This type of memory is not based on the assumption of a latent set of codes or categories in either the environment or the mind, and it is exactly what we need to see how a virtual model is extracted from the world in step-by-step fashion, becoming isomorphic to it without representing it. Memory accumulates information about the world immanently by actively dividing it up on the basis of potential actions, creating neighborhoods or basins of attraction in the immense diversity of experience. These are dynamic reclassifications based on attractors in neural phase space that are sufficient for certain adaptive performances, imaginative connections that are continually tested against the world on the basis of the repeated actions they lead to. A memory is like a miniature objective model of the possibilities an environment affords for being around—a state of a brain-world system consistent with there being an animal, a cycle of repeated perception and action that has been selected naturally because it provided for its own *future repetition*. Thus, individual memory forms a more flexible adaptive system grafted onto the developmental and even genetic ones, meaning that each model of how to stay alive is built on top of others, modeling how things that are already models might behave, in a cascade of semistable mnemonic forms. Nevertheless, at all these levels, memories are less like the storage and recall of predefined forms than the actual creation and recreation of forms—perhaps for the jungle animal, a jaguar simply *is* a shift in wind and change in sounds at twilight.

"This should be read without a pause: the animal-stalks-at-five-o'clock."[41] Deleuze and Guattari have the same type of nonrepresentational memory in mind when they speak of a becoming as an "antimemory" that creates a "zone of indiscernibility" between previously disconnected parts of the world.[42] Becomings are the building blocks of an uncoded world, a world of assemblages that *do* things rather than forms that *are* things.[43] Deleuze makes especially clear that these are never imitations or representations, but result from composing our body with a part of the world so as to create a stable form from the interaction between the two: "Do not imitate a dog, but make your organism enter into composition with something else in such a way that the particles emitted from the aggregate thus composed will be canine as a function of the relation of movement and rest, or of molecular proximity, into which they enter."[44]

Creating a memory that is useful for life is precisely this process of extracting a consistent form, a "topological dog," a "molecular dog," that cannot be said to be either in the organism or in the world. A becoming is what happens *between* perception and action, as a sort of materialist imagination.

For Deleuze, of course, becomings do not begin with humans, any more than for Edelman and Tononi memory begins with consciousness.[45] It is only a question of composition, of finding a consistent and semistable interaction between an organism and the environment, a feedback loop of action and reaction that composes a cycle that allows for its own reperformance, a compound that "stands up on its own."[46] This creation that arises like a vortex, being a feedback loop, tends to maintain itself, though of course it is always in danger of falling apart because of slippages in its internal relations or changes in its external context. Equally, one cycle can compose with others, can augment or reduce its scope and speed, can bifurcate, or simply drift away from its original relation. These emergent stabilities based on feedback loops form the persistent furniture and organisms of our world, as well as the memories of a conscious animal.[47]

Seeing nonrepresentational memory systems as the expressive musical motifs that emerge through the selective interaction of organisms with their environment, and at the limit give birth to these very organisms, opens up a host of connections with Deleuzian concepts such as the refrain, territoriality, art, etc. As they say: "music is not the privilege of human beings: the universe, the cosmos, is made of refrains . . . The question is more what is not musical in human beings, and what already is musical in nature."[48]

However, instead of taking this more cosmic route, I would like to relate this discussion of memory back to our earlier ideas about the objective model. At the same time, we can return to DeLanda's framework and clear up what may appear to be a slippage in the terms of our analogy—a becoming as antimemory appears to correspond more closely to the stable attractors or singularities structuring each level of a multiplicity than to the transitions between levels that we previously attempted to identify it with. I want to argue that this potential confusion is part of the recursive definition of a multiplicity.

The role of memory in the objective model is neatly summed up by Edelman and Tononi's description of consciousness as a "remembered present." As we have seen, a conscious brain with a dynamic core is a mechanism for integrating an enormous amount of information in a short time—something Edelman and Tononi refer to as constructing a "scene." The information integrated in this manner is primarily information about the brain though, and only secondarily information about the world at large. Thus, for the brain to be adaptively compatible with the external world, the information it integrates about itself, about its own state, via consciousness, has to have been formed by a process of gradual accumulation of experience with the world, experience that has at every step selected for neural mechanisms that produce adaptive responses. This is to say that an animal's conscious perception of its world, in going continually beyond that world, is immediately experienced in light of past adaptive responses. Edelman and Tononi put the point thus: "The ability of an animal to connect events and signals in the world, whether they are causally related or merely contemporaneous, and, then, through reentry with its value-category memory system, to construct a scene that is related to its own learned history is the basis for the emergence of primary consciousness."[49]

It is as if all of an animal's history is superposed on its present world, its depth flattened into a meaningful surface so that, "every act of perception is, to some degree, an act of creation, and every act of memory is, to some degree, an act of imagination."[50] Reaching the "remembered present" marks another stage in the hierarchy of selective processes we mentioned earlier, thus allowing Edelman and Tononi to propose a model for the evolution of consciousness that sits naturally with their theory of neural Darwinism:

> At a point in evolutionary time corresponding roughly to the tran-
> sitions between reptiles and birds and reptiles and mammals, a
> critical new anatomical connectivity appeared. Massively reentrant
> connectivity arose between the multimodal cortical areas carrying
> out perceptual categorization and the areas responsible for value-
> category memory.[51]

Whether or not we accept this as an evolutionary explanation, we can still see it as a conceptual one. What has happened is that the feedback sta-bilizing a memory system has become internal (hence "reentrant"), and the surplus production on which selection operates now takes the form of a cloud of imaginative possibilities, or more precisely, a set of unactualized attractors. This is precisely the role of the dynamic core, which appears in this light as an intense acceleration of the very same process of selective feedback, one that pushes it over a quantitative threshold where it changes in kind—becoming-conscious. It is this fantastic increase in *speed* that makes the brain into an objective model of the world—suddenly it seems to contain everything *within* itself, almost as if it could dispense with the feedback from the rest of the universe and could rely instead on its simula-tion of this world, as a virtual machine modeling the world from above. It now allows our "hypotheses to die in our stead" because at the same moment that the present is remembered, the future can be projected, experienced immediately and virtually. One conscious moment becomes an entire superposition of past and future.[52]

I imagine that this logic is implicit in Deleuze's continual obsession with speed as an intensity, and with going "as fast as possible." At the limit, this speed would become infinite and "absolute," linking heterogeneous parts in a simultaneous process, which as we have seen is precisely the type of feedback movement necessary for a process like consciousness, both neurally and philosophically: "The absolute expresses nothing transcen-dent or undifferentiated. It does not even express a quantity that would exceed all given (relative) quantities. It expresses only a type of movement qualitatively different from relative movement. A movement is absolute when, whatever its quantity and speed, it relates 'a' body considered as multiple to a smooth space that it occupies in the manner of a vortex."[53]

These absolute movements are also the way Deleuze and Guattari suggest that one can, "become everybody/the-whole-world" which is "to world, to make a world."[54] Though one may immediately object that this idea of becoming-the-world is not at all the same thing as creating a model of it, our idea of an *objective* model, arising immanently through a selection and accumulation of singularities, is meant to answer this. This model is not a transcendental representation of the world, but more like the world recreating (part of) itself as a simulation. In this context, the metaphor of a single plane folding over on itself, knotting together spatially and temporally distinct points to create a form of interiority, seems almost inevitable.

The real virtuality of consciousness, however, is nothing without some actual material system capable of supporting it. The virtual is not an open space of preexisting *possible* ideas or preexisting *possible* worlds. It is the *structure* of a space of unactualized states lying *alongside* the actual present world, not the condition of possibility or ground of Being, but superficial extra-Being. It is built, created, and extracted step-by-step as an extension that "takes off" from the actual to go beyond it, yet whose reality is symbiotic with the actual and applies back to it. It is a strictly materialist conception of possibility.

We have already seen how this "taking off" is like the creation of an objective model, and hypothesized that it occurs in the brain of every conscious animal, extracting and consolidating the virtual piece by piece through selective forces, thus making one model already a whole cascade of nested models modeling other models. Referring back to DeLanda's abstract framework, we can see this process of extraction at work in the interaction between any two multiplicities, each of which has an internal structure in its own right. In certain circumstances, a sort of resonance can develop whereby a behavior typically unstable in one "props up" an analogous behavior in the other, in the manner of the poles of a tipi, producing a mutually consistent and relatively stable interaction. While this interaction unfolds a new collection of stable states for both systems individually, it can also be seen as the creation of a new, composite multiplicity, whose space of behaviors includes the feedback relation between the previously unstable parts as an internal singularity. Though it must proceed through the *actual* systems to which the multiplicities apply and is characterized by an *actual* threshold of interactive intensity, the emergence of truly novel singular attractors constitutes a *virtual* production. The new multiplicity contains the other two in their separated forms as sub-multiplicities, illustrating the ambiguity of a becoming that takes two distinct points as a reference only to sweep them up into a single indissoluble form that transforms both. Here, we see how the confusion between singular attractors and singular thresholds points to the deeper fact that "one" never becomes alone, and that an unfolding level of a multiplicity has inevitably been structured by a history of interaction with some actual exterior. These nonequilibrium systems always have an outside, making "becoming" and "multiplicity" go together as two sides of the same recursive concept. Similarly, the objective model is always also an actual entity built up from actual interactions, and hence a potential variation is some other model.

DeLanda only briefly describes this process of successive virtualization, a "mechanism of immanence," and suggests we identify its product with the plane of consistency, understood as a purely virtual continuum relating multiplicities to one another and providing a space for the interaction: "Deleuze's hypothesis is that such an actual system may be 'sampled' or 'sliced through' to obtain . . . the entire set of attractors defining each flow pattern and the bifurcations which mediate between patterns."[55] This would make consciousness itself into a plane of consistency—a place where the most cosmically dispersed parts of the world, past and future, are organized and laid out in the form of a virtual model.

However, I think that insofar as we identify the plane of consistency with the plane of immanence, the substance of Being, with, in short, everything, this plane cannot be purely virtual.[56] We cannot avoid some brand of transcendence if we do not explicitly make room for the *actual* on the plane of immanence. The plane needs to be interwoven, simultaneously actual and virtual, or better yet, to be the *process* of their interconversion, the rhythmic pulse of the cosmos, the very *motor* of Being: "The plane of immanence has two facets as Thought and as Nature, as Nous and as Physis. That is why there are always many infinite movements caught within each other, each folded in the others, so that the return of one instantaneously relaunches another in such a way that the plane of immanence is ceaselessly being woven, like a gigantic shuttle."[57]

No matter how absolute the speeds of consciousness, the virtual model never ceases to need some actual material brain as its support. The brain extracts the virtual on the condition of being an actual and immanent part of the world that it can never escape and can never totalize, in the same way that any multiplicity is built from the sediment of its becomings and bordered by a becoming that serves as its outside. Thus, I would like to suggest that our understanding of virtualization as recursive serves to create a zone of indiscernibility between multiplicity, becoming, and the plane of consistency, considered as components of the concept of the virtual, which naturally causes this concept to link up with the actual. The plane of immanence, a pure and all-embracing virtuality, or a preexisting virtual space-time, is an abstraction continually fleeing off into the distance, woven step by step by the alternating processes of virtualization and actualization, unfolding and enfolding, "in the fractalization of this infinitely folded up infinity."[58]

I believe there is abundant textual evidence to support this most abstract hypothesis that

> there must be at least two multiplicities, two types, from the outset. This is not because dualism is better than unity but because the multiplicity is precisely what happens between the two . . . actual states of affairs and virtual events are two types of multiplicities that are not distributed on an errant line but related to two vectors that intersect, one according to which states of affairs actualize events and the other according to which events absorb (or rather adsorb) states of affairs.[59]

However, the most important point is to see how this question is not something tacked on to the end of Deleuze's philosophy as a sort of postmodern one-upmanship, but a thought reflected in the very structure of his slippery recursive concepts. The question of the immanence of the virtual informs everything right from the beginning. "A" multiplicity contains others within it and the potential to itself be enveloped. "A" becoming turns into a whole chain of becomings—animal, plant, molecule, imperceptible. "The" plane of immanence is an infinitely interleaved fractal continually breaking up into strata. These concepts are all structured identically—the same process is at work on a whole set of levels separated only by phase transitions, and "the" concept refers to the whole *process*. This is the logic of immanence from the very beginning, the logic of a nature becoming expressive, becoming thought, which is neither anthropomorphic nor mystical.

Accordingly, our theory of consciousness has been phrased in the same fashion. The consciousness Edelman and Tononi set out to explain was the "primary consciousness" of mammals, birds, etc. Yet we have seen how this same process began long before and could continue accelerating and transforming—how the same process could be apes or humans or Google or capitalism. Perhaps this finally gives us a tiny glimpse of insight into the question that motivated the essay and enables us to see our human mathematical models in some continuity with their evolutionary ancestors, revealing how they can be both so effective and so limited at the same time. This theory of consciousness, then, covers a lot more and a lot less than we set out to explain. As a theory about the evolutionary purpose and mechanism of a material system it cannot be specific enough to explain only human consciousness, or distinguish it from that of related animals. From this perspective, we have failed to explain anything like mammalian mathematicians. Indeed, any system exhibiting these general traits of integrating information in a repeatable fashion would exhibit some form of "consciousness."[60] There would be many forms and degrees of this phenomenon, corresponding to which types of models were being made consistent in the manner described. Any idea deserving the name of a "theory" of consciousness will need to function this way. In any true theory of this type, the lines separating the human from the beast and the computer can only appear as a limit, a bridge, a phase transition. Deleuze, of course—with his nonorganic life, his territorial refrains, his dissolving simian alter-ego—already knew this.[61]

The Metaphysics of Science:
An Interview with Manuel DeLanda

Manuel DeLanda and Peter Gaffney

PETER GAFFNEY: *Deleuze once claimed that "modern science has not found its metaphysics, the metaphysics it needs."[1] I would describe your work as such a metaphysics. In any case, I am aware of no other philosopher today who has done so much to show us what such a metaphysics would look like. Do you think science needs metaphysics, and do you view your own work as addressing this need?*

MANUEL DELANDA: Yes, I do see my work as a contribution to a metaphysics of science. In a Deleuzian metaphysics, the most important thing is to get rid of the notion of "law," as in the eternal and immutable laws of science. The concept of "law" is a theological fossil from the historical period when all scientists believed in the existence of the biblical God. Naturally, a metaphysics, or to use the term I favor, an ontology, appropriate to science must eliminate that concept, as I try to do in chapter four of *Intensive Science and Virtual Philosophy.* Eliminating a word does not mean, of course, eliminating the referent of that word, at least not if you believe like I do that there is a world that exists independently of our minds. The referent of the word "law" is immanent patterns of being and becoming. The question is then how to preserve these immanent patterns without theology. In the book *A Thousand Plateaus,* Deleuze proposes to replace laws with the twin concepts of "singularities" and "affects" (408). The first concept refers to the immanent tendencies of material systems, such as the tendency of water to become ice or steam at certain critical points of temperature. These critical points are singularities, in the sense that they are rare and nonordinary; whereas nothing remarkable happens at, say, 99 degrees centigrade (or 98, 97, 96 degrees, all of which are ordinary points), something special happens at 100 degrees: water becomes steam. Affects are not tendencies but capacities: the capacities of material entities to affect and be affected by one another. Unlike properties, which are always actual, both tendencies and capacities are only virtual or potential, until they are actually exercised or manifested. Think for instance of a knife: it is either sharp or it is not. That is a property, and it is always actual. But it also has the capacity to cut, a capacity that may never become actual if the knife is never used. And when it does become actual, it is not as a state (like the property of being sharp) but as an event that is always double: to cut, to be cut. In other words, for a capacity to affect, to become actual, it must be coupled with a capacity to be affected. A metaphysics or ontology of science based on these two concepts is an entirely different thing than the official metaphysics of science based on laws.

In the epigraph for this collection, Deleuze describes a category of "inexact" notions that scientists must make rigorous "in a way that's not directly scientific" (Negotiations, 29). In books like War in the Age of Intelligent Machines *and* Intensive Science and Virtual Philosophy, *you seem to be doing precisely this kind of thinking, on behalf of science but also on the part of philosophy. Is this your aim? What kinds of scientific notions would you say remain inexact? How does your approach make them more rigorous?*

The creation of mathematical concepts that are "inexact yet rigorous" goes back to the nineteenth century and were already discussed by Bertrand Russell (who influenced Deleuze on this subject). Take the difference between numerical series, some of which are cardinal (one, two, three . . .) while others are ordinal (first, second, third . . .). In the first we can make judgments of exact equality, while in the second we can only make rigorous judgments of differences (e.g., that a series is greater or lesser than another series but not by exactly how much). A similar point applies to geometric spaces. Metric spaces, like those of Euclidean geometry, are spaces in which the notions of exact length, area, or volume are basic and that allow for exact quantitative comparisons. Nonmetric spaces (projective, differential, topological geometries) progressively depart from this exactness (and lengths, areas, volumes are no longer basic concepts) but still allow one to make rigorous judgments of differences. When these spaces are used to model physical phenomena (as in the use of "phase space" to visualize the space of possible states for a system) we also give up exact quantitative judgments but not rigorous qualitative ones about the system, such as that the system has a unique attractor that defines its long-term tendencies (with "long-term" having only a qualitative not quantitative significance). It can be argued that ordinal series are more basic that cardinal ones (the numbers from the latter can be obtained from the pure nonnumerical order of the former, as shown by the mathematician Dedekind) and that nonmetric spaces are more basic than metric ones (lengths, areas, volumes can be obtained from topological concepts following a cascade of broken symmetries, as first proved by the mathematician Felix Klein). In Deleuze, the virtual plane of immanence would be like a topological or ordinal space from which the actual world with its quantitative relations would emerge by a similar process of progressive differentiation.

In Intensive Science and Virtual Philosophy, *you suggest that there is an "intimate relationship between epistemology and ontology" (135), so that we might say a problem concerning knowledge of the world, on the one hand, is no different than the process of individuation or adaptation we might find in a living organism or species. In other words, scientific thought is "isomorphic" with real virtual (physical) processes. What do you think this tells us about the creative (emergent, constructive, generative) potential of science?*

The question of isomorphism arises in the context of explaining the success of mathematical models in science. There is no a priori reason why a model based on differential equations should exhibit the same kind of behavior as a physical system in the laboratory. Even as late as the mid–twentieth century, physicists (such as Eugene Wigner) talked of "the unreasonable effectiveness of mathematics in the natural sciences," that is, they were still mystified by the isomorphism. But if there is indeed such a thing as the plane of immanence, inhabited by virtual diagrams or multiplicities, then it becomes possible to think of any one of those multiplicities as becoming actualized in a variety of physical systems, as well as in a variety of mathematical entities. The correspondence between models and laboratory phenomena would then be explained as due to the fact that model and phenomenon would be coactualizations of the same multiplicity.

Have you found that the scientific community has begun to acknowledge this potential, and if so, what kind of reaction has it provoked?

The problem with scientists is that many of them embrace an impoverished philosophy of science known as "positivism," according to which they are not in the business of revealing the true ontology of the world but only creating compact descriptions that are useful to make predictions and to increase our degree of control over laboratory phenomena. This is good for public relations (if an entity postulated to exist, such as the nineteenth-century "ether," turns out not to exist, scientists can wash their hands) but it makes for bad philosophy. Thus, for the kind of neomaterialism that Deleuze's work makes possible to impinge on actual science, the first thing that needs to go is positivism, but that will take a long time to happen.

You have given some examples in your work that suggest that science itself is becoming more Deleuzian. I am especially interested in your recent lectures on connectionism and its application to the design of artificial intelligence, a field historically dominated by symbolic theory. Does this development indicate a more general change in scientific thought, and would you say it corroborates a Deleuzian view of science and the virtual?

Neural nets exemplify a paradigm of the mind that is very different from the one based on symbols and categories. In particular, the conception of memory in the two paradigms is quite different. In regular computers, a memory is an actual piece of data or symbol that is stored as such and retrieved by its address. In neural nets, data is not stored as a representation (what is stored is the means to reproduce that data via a pattern of the strength of synapses), and it is retrieved by its content (a small piece of the original data, fed into a neural net that has the right pattern of synaptic strengths, can reproduce the full content of the memory). Now, neural nets are in their infancy (today, they can mimic only insect intelligence), but the way they work (not storing representations but the means to recreate them via a piece of their content) is clearly more closely related to a Humean or Bergsonian conception of the mind than the Kantian conception, this being the symbolic paradigm.

In spite of Deleuze's own investment in science, there are some who criticize the "scientist" interpretation of Deleuze's thinking. For example, in "Science and Dialectics in the Philosophies of Deleuze, Bachelard and DeLanda," James Williams argues that science-based definitions of Deleuzian terms do not sufficiently take account of the "scientific revolutions," which tend to invalidate one theory (and its definitions) and replace it with another. Philosophical concepts, on the other hand, are not susceptible to these same revolutions and tend to work independently of any particular theory (whichever theory is valid at a given time). Can you respond to this criticism?

There are two deeply wrong things with that essay. The first one is that the author does not reveal his ontological commitments: is he an idealist, someone who does not believe in an objective world, or is he a realist? The reason why this matters is that if you are an idealist then you will be biased toward a relativist epistemology. Evidence that Williams is an idealist is in the opening paragraphs of his essay where he takes me to task for using the notion of "falsehood." I claim that it is simply false to say that essences exist, a claim that I would certainly defend. The problem here is that if he is indeed an idealist, then his objections are not so much that I use scientific ideas as evidence in support of Deleuze's position, it is that I claim that Deleuze is a realist. In that same paragraph he suggests that Deleuze was an empiricist (since his first published book is on Hume). Now, an empiricist (or a positivist) believes in the mind-independent existence of only those entities that we can perceive directly. Unlike me, for example, Hume treated causality not as an objective relation in which one event produces another event, but as the observed constant conjunction between two events. Empiricists, of course, get into trouble the moment they need to account for telescopes, microscopes, and any imaging device that goes beyond unaided perception, and that is one of the reason not to adopt that metaphysics. Now, whether Williams espouses an idealist or a positivist ontology is unclear, but I am willing to bet he is an idealist, in which case he should just come out and say it.

The second problem, one that unfortunately is also present in Deleuze's own work, is to believe that there is such a thing as "science in general." I do not share that belief: in my ontology, terms such as "the market," "the state," "science," and "culture" are all reified generalities, not real entities. All that exists in the actual world are individual entities, that is, unique, historically produced entities. Instead of "science," therefore, I would speak of a plurality of individual scientific fields, each with its own methods, models, and history. Moreover, the number of fields is increasing not decreasing as if they were converging on a final truth: the population of scientific fields is diverging, as they track a reality that is itself divergent and open-ended. In this interview I have used the word "science" but always in a context where, with the right paraphrasing, we can replace it with the term "scientific fields" in the plural.

When you combine these two problems (an idealist tendency toward a relativist epistemology with a reification of "science in general") you end up with nonsense. This is easier to explain if instead of Bachelard we use Popper, another philosopher whose ideas have been used by idealist Deleuzians to critique my work. Popper is famous for having said that what characterizes a scientific statement is that is always "falsifiable." This is true if taken to mean that no scientific statement can be a priori. I certainly reject all a priori claims to knowledge (hence my amazement at Williams's use of the word "rationalist" in relation to my reconstruction of Deleuze). But Popper's dictum is false if taken to imply that all scientific statements are equally falsifiable: while the existence of dark matter (to use Williams's example) is indeed merely hypothetical, and the hypothesis may indeed turn out to be false, the existence of oxygen (or carbon) is beyond question. To falsify the statement "oxygen exists," for example, we would also have to falsify "water molecules are made of two atoms of hydrogen and one of oxygen," and thousand of statements like that. I cannot envision a situation in which thousands of statements and laboratory results would be falsified all at once. That the existence of oxygen is beyond falsifiability does not mean, of course, that we know everything about oxygen: we may know about its properties and tendencies (both of which are finite) but not about its capacities to affect and be affected because these form an infinite set (think of oxygen's combinatorial capacities: the composite molecules it can form, the larger molecules these composites can form and so on). Finally, the criterion of falsifiability should be used carefully, or it can lead to misrepresent the history of science, to believe, for example, that the theory of relativity falsified Newtonian mechanics. That is nonsense. All that Einstein did was to limit the field of validity of Newton: the immanent patterns of change Newton discovered (e.g., the "laws" of classical mechanics) are still valid but only for speeds that are slow relative to the speed of light. If a future conceptual revolution "falsified" the periodic table of the elements, the situation would be similar to this: the new view of atomic species would likely preserve the old one (oxygen's existence included) as a special but still valid case.

Now, the question of how wise it is to use scientific ideas in my materialist reconstruction of Deleuzian philosophy can be rephrased like this: which ideas in *Intensive Science* are like "dark matter" and which are like "oxygen"? Several of the processes used as illustrations of Deleuze's ideas (such as the two examples in chapter 2 about embryological processes used to illustrate his concepts of "extensities" and "qualities") are like "dark matter": they come from cutting-edge science and may indeed turn out to be wrong in detail. But because they are only illustrations, they can be replaced in the future (either by me in future editions, or by readers using their own illustrations) without affecting the main argument. On the other hand, ideas from group theory and symmetry (a two-hundred-year-old discipline with many successes), those from dynamical systems theory, such as phase space and its attractors (a one-hundred-year-old equally successful discipline), and those about the relations between metric and nonmetric spaces (which are one hundred and fifty years old) have penetrated every field of hard science and are highly unlikely to be proved wrong in the future, although they may, of course, be revised and their validity limited. It is these much more robust results that do play a constructive role in the argument, in the sense that one can hardly be a realist with respect to entities like the "plane of immanence" and its "virtual multiplicities" without them. If those ideas turn out to be wrong, then the philosophical concepts, as I reconstructed them, will also lose their validity. But in that case, so much would have to be rethought in philosophy and in science that the fact that my book has lost its validity would be the least of humanity's problems.

Your interpretation of Deleuze's concept of the virtual plays a large part in your approach to science (and vice versa). In our previous discussions, you have also mentioned art as a discipline that engages the virtual in a way that is almost scientific. Can you explain? Would you say that there are certain forms of art that "do the work" of science, and if so, does this suggest any expansion of the role of art with respect to the exact sciences?

What I was referring to in those conversations is that scientists tend to keep things simple (to be able to prove theorems about a particular model, for example), while artists can use the same technology to create complex things without fear of losing analytic tractability. Thus, a typical scientific discussion of neural nets will stick to the simplest designs, while an artist can dare to push the envelope. The best example I can think of is Karl Sims's use of neural nets (in combination with genetic algorithms) to breed swimming and walking creatures: they have a neural net in each one of their limb and body articulations, such that the synaptic strengths can evolve over many generations to generate coherent motion. The creatures that result (they can be seen in an animation in his Web site[2]) are probably too complex to be useful as science, but as an artist he does not have to worry about this, and we are all the better off because of it: he is exploring computer actualizations of virtual diagrams that scientists would probably not tackle for decades.

Notes

Preface

1. Bruno Latour, "Why Has Critique Run Out of Steam? From Matters of Fact to Matters of Concern," *Critical Inquiry* (Winter 2004): 225–48.

2. Ibid., 242–43.

3. "The Metaphysics of Science: An Interview with Manuel DeLanda" (Afterword in the present volume).

4. "Interstitial Life: Some Remarks on Causality and Purpose in Biology" (chapter 4 in the present volume).

Introduction

1. Gilles Deleuze and Félix Guattari, *What Is Philosophy?* trans. Hugh Tomlinson and Graham Burchell (New York: Columbia University Press, 1994), 112.

2. Ibid., 112.

3. Ibid., 118.

4. Deleuze and Guattari use this expression in *A Thousand Plateaus* to describe the relationship between the human body, tools, and products: "The hand as a general form of content is extended in tools, which are themselves active forms implying substances, or formed matters; finally products are formed matters, or substances, which in turn serve as tools. Whereas manual formal traits constitute the unity of composition of the stratum, the forms and substances of tools and products are organized into parastrata and epistrata that themselves function as veritable strata and mark discontinuities, breakages, communications and diffusions, nomadisms and sedentarities, multiple thresholds and speeds of relative deterritorialization in human populations. For with the hand as a formal trait or general form of content a major threshold of deterritorialization is reached and opens, an accelerator that in itself permits a shifting interplay of comparative deterritorializations and reterritorializations—what makes this acceleration possible is, precisely, phenomena of 'retarded development' in the organic strata." Gilles Deleuze and Félix Guattari, *A Thousand Plateaus,* trans. Brian Massumi (Minneapolis: University of Minnesota Press, 1987), 61.

5. Ibid., 119.

6. See Alfred North Whitehead, *Process and Reality* (New York: The Free Press, 1978), 32–34, and Steven Shaviro's analysis of appetition in the present volume. This term is also used by Leibniz in his *Monadology* in reference to the "action of the internal principle which brings about the change or the passing from one perception to another." In *Discourse on Metaphysics, Correspondence with Arnauld, Monadology*, ed. Paul Janet, trans. George Montgomery (La Salle, Ill.: Open Court Publishing Company, 1990), 253–54.

7. "L'Intelligence organise le monde en s'organisant elle-même." Jean Piaget, *La Construction du réel chez l'enfant* (Paris: Delachaux et Niestlé, 1937), 311.

8. Manuel DeLanda, *Intensive Science and Virtual Philosophy* (New York: Continuum, 2002), 135.

9. Gilles Deleuze, *Bergsonism*, trans. Hugh Tomlinson and Barbara Habberjam (New York: Zone Books, 1988), 16.

10. Gilles Deleuze, *Difference and Repetition*, trans. Paul Patton (New York: Columbia University Press, 1994), 63.

11. Deleuze and Guattari, *What Is Philosophy?* 38.

12. Gilles Deleuze, *Foucault*, trans. Sean Hand (Minneapolis: University of Minnesota Press, 1988), 114–17.

13. Deleuze and Guattari, *A Thousand Plateaus*, 369–74.

14. "A distinction must be made between the two types of science, or scientific procedures: one consists in 'reproducing,' the other in 'following.' The first involves reproduction, iteration and reiteration; the other, involving itineration, is the sum of the itinerant, ambulant sciences. Itineration is too readily reduced to a modality of technology, or of the application and verification of science. But this is not the case: *following is not at all the same thing as reproducing*, and one never follows in order to reproduce. The ideal of reproduction, deduction, or induction is part of royal science, at all times and in all places, and treats differences of time and place as so many variables, the constant form of which is extracted precisely by the law." (*A Thousand Plateaus*, 372).

15. See Jacques Lacan, "Truth and Science," trans. Bruce Fink, *Newsletter of the Freudian Field* 3, no. 1 & 2 (Spring/Fall 1989): 4–29. Lacan argues here that modern logic is "the strictly determined consequence of an attempt to suture the subject of science, and Gödel's last theorem shows that this attempt fails, meaning that the subject in question remains the correlate of science, but an anti-nomial correlate since science turns out to be defined by the deadlock endeavor to suture the subject" (10).

16. In *The Structure of Scientific Revolutions* (Chicago: University of Chicago Press, 1996), Kuhn writes that theoretical conflicts arising between existent and emergent scientific paradigms are "terminated by a relatively sudden and unstructured event like the [visual] gestalt switch" (122). It should be noted that Kuhn is

in some ways critical of the comparison between scientific revolution and visual gestalt theory. He does not go so far, for example, as Norwood Hanson in his book *Patterns of Discovery: An Inquiry into the Conceptual Foundations of Science* (London: Cambridge University Press, 1958).

17. "Machine et structure," originally published in *Psychanalyse et transversalité: Essai d'analyse institutionnelle*, is reprinted in *Molecular Revolution: Psychiatry and Politics*, trans. Rosemary Sheed (New York: Penguin, 1984), 111–19. It is worth noting that Guattari considers the machinic nature of science not just in this article but throughout *Psychanalyse et transversalité*.

18. I am indebted here to Tom Kelso for bringing this polemic to my attention (and insisting on its importance). Todd May also discusses two ways of reading Deleuze's engagement with science in his insightful chapter on "Deleuze, Difference, and Science," in *Continental Philosophy of Science*, ed. Gary Gutting (Malden, Mass.: Blackwell Publishing, 2005), 239–57, esp. 251.

19. See Jean-Clet Martin, *Variations: la philosophie de Gilles Deleuze* (Paris: Éditions Payot et Rivages, 1993), 244–55, especially his notion of ascending backward, from crystalization to immanence: "Comme le redira Simondon à la suite de Bergson, la perception est une opération de refroidissement, de cristallisation en chaînes déterminées, laissant derrière elles toute une agitation moléculaire infiniment plus riche ... Mais au lieu de descendre cette ligne de différenciation, de refroidissement ou cristallisation en chaînes déterminées, on peut trouver des conduits de dérivation à même de contourner l'égouttoir perceptif pour remonter vers le plan lumineux d'immanence" (246). Also see Jean-Clet Martin, *L'Image virtuelle: essai sur la construction du monde* (Paris: Kime, 1996).

20. Rosi Braidotti, "Discontinuous Becomings: Deleuze on the Becoming-Woman of Philosophy," *Journal for the British Society of Phenomenology*, 24 (1993): 44–55. Cited in John Mullarkey, *Post-Continental Philosophy: An Outline* (New York: Continuum, 2006), 17.

21. Mullarkey, *Post-Continental Philosophy*, 17–18.

22. For Manuel DeLanda, see especially *Intensive Science and Virtual Philosophy* (New York: Continuum International Publishing Group, 2002); for Keith Ansell Pearson, see *Germinal Life: The Repetition and Difference of Gilles Deleuze* (London, New York: Routledge, 1999).

23. DeLanda, *Intensive Science and Virtual Philosophy*, 2.

24. James Williams, in an interview with Mark Thwaite, February 13, 2006, online at *ReadySteadyBook*, http://www.readysteadybook.com (accessed October 21, 2007).

25. James Williams, "Science and Dialectics in the Philosophies of Deleuze, Bachelard and DeLanda," *Paragraph* 29, no. 2 (2006): 113.

26. Wolfram Schommers, "Truth and Knowledge," in *What Is Life? Scientific Approaches and Philosophical Positions*, ed. Hans-Peter Dürr, Fritz-Albert

Popp, and Wolfram Schommers (London: World Scientific, 2002), 41. Bernard d'Espagnat makes a more rigorous, if exceedingly technical, argument regarding the epistemological difficulties confronted when using dispositional terms to describe a physical system (the term "magnetic," for example). Cf. *In Search of Reality* (New York: Springer-Verlag, 1983), 133–45.

27. Deleuze and Guattari, *What Is Philosophy?* 126.

28. Cf. Gilles Deleuze, *The Logic of Sense,* trans. Mark Lester and Charles Stivale (New York: Columbia University Press, 1990), 184; *Difference and Repetition,* 64, 165.

29. Deleuze and Guattari, *What Is Philosophy?* 59.

30. Deleuze and Guattari, *What Is Philosophy?* 124.

31. Cf. Deleuze, *Logic of Sense,* 76–77; *Difference and Repetition,* 224–25.

32. Deleuze and Guattari, *What Is Philosophy?* 37.

33. Jan Faye identifies three constituent positions or claims of scientific realism: ontological, semantic, and epistemological. She points out that not all realists subscribe to all positions (there are weaker and stronger versions). *Rethinking Science* (Hampshire: Ashgate, 2002), 168–79. With the term "Cartesian realism," I am referring more generally to this tendency throughout the history of philosophy, following the definition proposed by Joseph Margolis: "any realism, no matter how defended or qualified, that holds that the world has a determinate structure apart from all constraints of human inquiry and that our cognizing faculties are nevertheless able to discern those independent structures reliably. 'Cartesianism' serves as a term of art here, not confined to Descartes's doctrine. It ranges over pre-Kantian philosophy, Kant's own philosophy (quixotically), and over the views of such contemporary theorists as Putnam and Davidson." "Incommensurability Modestly Recovered," in *History, Historicity and Science,* ed. Tom Rockmore and Joseph Margolis, (Hampshire: Ashgate, 2006), 194. Of course, as Descartes was a rationalist and not an empiricist, this term may appear ill-suited in the discussion that follows. But this is only a superficial difference: Cartesian rationalism should be seen here as a form of epistemic foundationalism (the identification, in the cogito, of belief intrinsic to a foundational observation, namely the cogito), which then serves as grounds for empirical knowledge. For further reading on this problem, see Laurence Bonjour's appendix on the a priori justification of empirical knowledge, in *The Structure of Empirical Knowledge* (Cambridge, Mass.: Harvard University Press, 1985), 191–211.

34. See Faey's definitions for theories, models and paradigms in *Rethinking Science,* 144–67. For an explanation of how the physical model works in time, see Erwin Schrödinger's "The Present Situation in Quantum Mechanics," in *Quantum Theory and Measurement,* ed. J. A. Wheeler and W. H. Zurek (Princeton, N.J.: Princeton University Press, 1983), 152–53. This article was originally published in *Proceedings of the American Philosophical Society* 124 (1980): 323–38.

35. Cf. Auguste Comte, *The Positive Philosophy,* trans. Harriet Martineau (New York: AMS Press, Inc., 1974), 25–38, 399–439.

36. A. J. Ayer, *Language, Truth and Logic* (New York: Dover Publications, 1970), 100. Quoted in Richard Rorty, *Philosophy and the Mirror of Nature* (Princeton, N.J.: Princeton University Press, 1979), 337.

37. Henri Bergson, *Creative Evolution,* trans. Arthur Mitchell (Mineola, N.Y.: Dover Publications, Inc., 1998), 194.

38. Deleuze, *Difference and Repetition,* 199.

39. Eric Alliez, *The Signature of the World: Or, What Is Deleuze and Guattari's Philosophy?* trans. Eliot Ross Albert and Alberto Toscano (New York: Continuum, 2004), 35–36.

40. See note 4.

41. Mullarkey, *Post-Continental Philosophy: An Outline,* 38.

42. Gilles Deleuze, *Expressionism in Philosophy* (New York: Zone Books, 1992), 100–103.

43. Mullarkey, *Post-Continental Philosophy: An Outline,* 38.

44. Cf. Timothy Murphy, "Quantum Ontology: A Virtual Mechanics of Becoming," in *Deleuze and Guattari: New Mappings in Politics, Philosophy, and Culture,* ed. Eleanor Kaufman and Kevin Jon Heller (Minneapolis: University of Minnesota Press, 1998), 211–29.

45. Ilya Prigogine and Isabelle Stengers, *Order out of Chaos: Man's New Dialogue with Nature* (New York: Bantam Books, 1984), 9.

46. It is not difficult to find this tendency in Descartes's writings: in *Discourse on the Method and Meditations,* for example, but even more strikingly in *The World* and *Treatise on Man.* Giorgio de Santillana argues that this tendency precedes Descartes, namely in Parmenides and the Eleatics: "[Descartes's] 'extended matter' is again the One Thing, like Parmenidean Being, and it coincides with the geometrical continuum; matter, space, and motion become again analytical relations. Sense perception is for him, as it was for the Eleatics, a source of obscurity, incapable of providing a base for true knowledge." *Origins of Scientific Thought* (Chicago: University of Chicago Press, 1961), 106.

47. It is notable that this formulation is maintained by Prigogine and Stengers; see *Order out of Chaos,* 9.

48. Deleuze and Guattari, *A Thousand Plateaus,* 379.

49. Deleuze and Guattari, *What Is Philosophy?* 47. Also see Gilles Deleuze, *Negotiations,* trans. Martin Joughin (New York: Columbia University Press, 1995), 208. The discussion that follows refers primarily to 210–18 of the same text.

50. "Knowledge is neither a form nor a force but a *function*: 'I function.' The subject now appears as an 'eject,' because it extracts elements whose principal characteristic is distinction, discrimination: limits, constants, variables and functions, all those functives and prospects that form the terms of the scientific

proposition . . . Yet these elements, as states of affairs, are 'inseparable from the potentials they take from chaos itself and that they do not actualize without risk of dislocation or submergence.'" Deleuze and Guattari, *What Is Philosophy?* 214–15.

51. Cf. Guattari and Rolnik, *Molecular Revolution in Brazil,* 179.

52. Deleuze and Guattari, *A Thousand Plateaus,* 478–79.

53. Ibid., 43–45, 59.

54. Ibid., 461.

55. This interview can be found in Arnaud Villani, *La guêpe et l'orchidée: Essai sur Gilles Deleuze* (Paris: Éditions de Belin, 1999), 129–31.

56. Schrödinger, "The Present Situation in Quantum Mechanics," 152–53. This article was originally published in *Proceedings of the American Philosophical Society,* 124 (1980): 323–38. See "Superposing Images: Deleuze and the Virtual after Bergson's Critique of Science" in the present volume, chapter 2.

57. Steven Shaviro makes the point that the conception of chaos in complexity theory is a fully deterministic one, i.e., theoretically fully determined; what makes this approach different from more conventional notions of causal determinism is that it regards the development of phenomena to be "sensitive to differences in initial conditions too slight to be measured" and therefore "not actually determinable ahead of time pragmatically." "Interstitial Life: Remarks on Causality and Purpose in Biology" in the present volume, chapter 4.

58. Deleuze, *Difference and Repetition,* 206–7.

59. Recent efforts to unite the four fundamental forces (gravitation, electromagnetism, weak and strong interaction) seem to corroborate this claim. Cf. John Webb, "Are the Laws of Nature Changing with Time?" *Physics World* (April 2003): 33–34.

60. Deleuze and Guattari, *A Thousand Plateaus,* 372.

61. Ibid., 374.

62. Deleuze, *Bergsonism,* 14.

63. Deleuze, *Difference and Repetition,* 222.

64. See Richard Monastersky, "Forecasting into Chaos: Meteorologists Seek to Foresee Unpredictability," *Science News,* 137, no. 18, (May 5, 1990): 280–82.

65. Deleuze and Guattari, *What Is Philosophy?* 169.

66. Villani examines the conceptual figure of the wasp-orchid dyad at length in his book *La guêpe et l'orchidée: Essai sur Gilles Deleuze* (Paris: Éditions Belin, 1999). See especially his chapter on the Deleuzian notion of creativity, "Constitution d'un plan idéal de creation," 71–90.

67. Gilles Deleuze, "La methode de la dramatisation," in *L'Ile déserte et autres texts* (Paris: Éditions de Minuit, 2002), 134–35.

68. Wolfram Schommers, "Evolution of Quantum Theory," in *Quantum Theory and Pictures of Reality: Foundations, Interpretations, and New Aspects,* ed. W. Schommers (Berlin; New York: Springer-Verlag, 1989), 31.

69. Deleuze and Guattari, *A Thousand Plateaus,* 142.

70. Friedrich Nietzsche, *The Will to Power,* ed. Walter Kauffman, trans. Walter Kauffman and R. J. Hollingdale (New York: Vintage Books/Random House, 1968), 244. Another fragment reads: "Against positivism, which halts at phenomena—'There are only *facts*'—I would say: No, facts is precisely what there is not, only interpretations." Starting with "Richard Wagner in Bayreuth" (1875), Deleuze identifies a change in Nietzsche's attitude toward science: "He was more and more interested in the sciences: in physics, biology, medicine." *Pure Immanence: Essays on a Life,* trans. Anne Bowman (New York: Zone Books, 2001), 56. Also see Deleuze's analysis of Nietzsche, science, and the question of quantitative difference in *Nietzsche and Philosophy,* trans. Hugh Tomlinson (New York: Columbia University Press, 1983), 44–47.

71. Cf. Deleuze, *Nietzsche and Philosophy,* 95–96; *Logic of Sense,* 75; *Proust and Signs,* 16–17.

72. Deleuze, *Logic of Sense,* 278; cf. also *Empiricism and Subjectivity,* 35; *Bergsonism,* 98; *Difference and Repetition,* 6.

73. Deleuze, *Difference and Repetition,* 206–7.

74. "Desiring-production forms a binary-linear system. The full body is introduced as a third term in the series, without destroying, however, the essential binary-linear nature of this series: 2, 1, 2, 1 . . ." Gilles Deleuze and Félix Guattari, *Anti-Oedipus: Capitalism and Schizophrenia,* trans. Robert Hurley, Mark Seem, and Helen R. Lane (Minneapolis: University of Minnesota Press, 1983), 14.

75. Deleuze, *Bergsonism,* 107.

76. See Bergson, "Psycho-physical Parallelism and Positive Metaphysics," trans. Matthew Cob, in *Continental Philosophy of Science,* ed. Gary Gutting (Malden, Mass.: Blackwell Publishing, 2005), 59–68. Also see Jean Gayon's discussion of Bergson's spiritualism in the same edition ("Bergson's Spiritualist Metaphysics"): "Bergson was indeed a spiritualist. All of his writings are devoted to demonstrating the existence of the mind, the supremacy of the spiritual over 'matter,' by a careful reflection on various spheres of human knowledge and experience" (45–46).

77. Deleuze, *Bergsonism,* 98–99. Deleuze and Guattari return to this duality, between the internal cause of mutation and external cause of selection, in *A Thousand Plateaus:* "On the one hand, modifications of a code have an aleatory cause in the milieu of exteriority, and it is their effects on the interior milieus, their compatibility with them, that decides whether they will be popularized. Deterritorializations and reterritorializations do not bring about the modifications; they do, however, strictly determine their selection" (54).

78. Immanuel Kant, *Critique of Practical Reason,* trans. Mary J. Gregor (Cambridge, Mass.: Cambridge University Press, 1996), 178.

79. In the following quote, Deleuze and Guattari are most likely thinking of the duplication of DNA into RNA sequences (and of the latter into proteins),

but the discovery of genomic sequences called "introns," noncoding sections of mRNA that are removed during the process of splicing, gives us another example of such a surplus: "The vast interspecific differences in intron number between eukaryotic genomes constitute an important puzzle with which theories about the determinants of genome complexity—for example, selfish genetic elements, organismal complexity or population size—must come to terms. Indeed, introns have prominent roles in several ambitious theories of genome evolution. A long-standing theory of intron origin postulates that recombination within introns facilitated the construction of the first full-length genes, and a similar theory has been proposed for the origins of many multidomain metazoan genes. Introns have also been suggested to increase fitness by increasing intragenic recombination, or to have boosted transcript fidelity in early eukaryotes through nonsense-mediated decay (NMD). Differences in intron number have also served as the potential crowning example of the hypothesis that increases in genome complexity are often the results of, or are themselves, deleterious mutations." Scott William Roy and Walter Gilbert, "The Evolution of Spliceosomal Introns: Patterns, Puzzles and Progress," *Nature Reviews Genetics,* 7 (2006): 211–12. Deleuze and Guattari will also refer to the endosymbiotic theory of evolution, which claims that mito-chondria and plastids are inherited from viruses and bacterial cells. In *A Thousand Plateaus,* they write that "fragments of code may be transferred from the cells of one species to those of another, Man and Mouse, Monkey and Cat, by viruses or through other procedures. This involves not translation between codes (viruses are not translators) but a singular phenomenon we call surplus value of code, or side-communication" (53). On endosymbiotic evolution, see Lynn Margulis, *Origin of Eukaryotic Cells* (New Haven, Conn.: Yale University Press, 1970).

80. Deleuze and Guattari, *A Thousand Plateaus,* 53.

81. Cf. Peter Hallward, *Deleuze and the Philosophy of Creation:* "As Deleuze understands it, living contemplation proceeds at an immeasurable distance from what is merely lived, known or decided. Life lives and creation creates on a virtual plane that leads forever out of our actual world." (London: Verso, 2006), 164; and Mark Hansen, "Becoming as Creative Involution?: Contextualizing Deleuze and Guattari's Biophilosophy," *Postmodern Culture* 11, n. 1 (September 2000), online at *Project Muse,* http://proxy.library.upenn.edu:2298/journals/postmodern_culture/vo11/11.1hansen.html (accessed June 25, 2008).

82. Deleuze and Guattari, *A Thousand Plateaus,* 411.

83. Ibid., 409–10.

84. Cf. the creation of time in Plato's *Timaeus,* trans. Donald J. Zeyl (Indianapolis: Hackett Publishing Company, 2000), 23–24 (Stephanus page and section numbers 37c6–38b5).

85. Deleuze and Guattari, *A Thousand Plateaus,* 64.

86. Connectionism, in its current application within the field of artificial intelligence, rejects the symbolic paradigm for cognitive processes in favor of one based on the layered, nonlinear distribution of excitations and inhibitions: "Connectionism can be distinguished from the traditional symbolic paradigm by the fact that it does not construe cognition as involving symbol manipulation. It offers a radically different conception of the basic processing system of the mind-brain, one inspired by our knowledge of its nervous system. The basic idea is that there is a network of elementary *units* or nodes, each of which has some degree of activation. These units are *connected* with each other so that active units excite or inhibit other units. The network is a *dynamical system* which, once supplied with initial input, spreads excitations and inhibitions among its units." William Bechtel and Adele Abrahamsen, *Connectionism and the Mind: Parallel Processing, Dynamics, and Evolution in Networks,* 2nd ed. (Oxford: Blackwell Publishers, Inc., 2002), 1–2. The various layers of "nerves" in a Perceptron (artificial neural network) are, in this sense, analogous to the strata described by Deleuze and Guattari in *A Thousand Plateaus.*

87. Hans Moravec, *Robot: Mere Machine to Transcendent Mind* (New York: Oxford University Press, 1999), 76.

88. "In linguistics and psychoanalysis, its object is an unconscious that is itself representative, crystallized into codified complexes, laid out along a genetic axis and distributed within a syntagmatic structure. Its goal is to describe a de facto state, to maintain balance in intersubjective relations, or to explore an unconscious that is already there from the start, lurking in the dark recesses of memory and language. It consists of tracing, on the basis of an overcoding structure or supporting axis, something that comes ready-made." Deleuze and Guattari, *A Thousand Plateaus,* 12.

89. Johanna Drucker, "Digital Ontologies: The Ideality of Form in/and Code Storage—or—Can Graphesis Challenge Mathesis?" *Leonardo* 34, no. 2 (2001): 141–45.

90. Gilles Deleuze, *The Fold,* trans. Paul Patton (Minneapolis: University of Minnesota Press, 1993), 19.

91. Constantin V. Boundas identifies a number of different formulations in his entry on subjectivity in *The Deleuze Dictionary,* ed. Adrian Parr (New York: Columbia University Press, 2005), 268–70.

92. Steven Shaviro, July 14, 2008, online at *The Pinocchio Theory,* http://www.shaviro.com/Blog (accessed July 20, 2008).

93. Deleuze, *Logic of Sense,* 174.

94. Deleuze, *Difference and Repetition,* 70–74.

95. Ibid., 70.

96. Ibid., 96.

97. Deleuze and Guattari, *Anti-Oedipus,* 10–14.

98. Deleuze, *Logic of Sense*, 71.

99. Deleuze, *Proust and Signs*, 97–98.

100. Deleuze and Guattari, *What Is Philosophy?* 41.

101. Deleuze and Guattari, *A Thousand Plateaus*, 378.

102. Deleuze, *The Fold*, 117.

103. Deleuze and Guattari, *A Thousand Plateaus*, 26–38.

104. See Freud, *Introductory Lectures on Psycho-Analysis*, trans. James Strachey (New York: W. W. Norton, 1966), 443–44.

105. Deleuze, *Nietzsche and Philosophy*, 34.

106. These are the two instances or agencies that account for the latent and manifest content of dreams, the first presenting a wish that is censored or distorted by the second. (see Freud's chapter on "Distortion in Dreams" in *The Interpretation of Dreams*). Freud would later assimilate these two instances to the id and ego.

107. See note 4.

108. Gilles Deleuze, *Empiricism and Subjectivity*, trans. Constantin V. Boundas (New York, Columbia, 1991 [1953]), 100.

109. Ibid., 24.

110. Ibid., 88.

111. Deleuze, *Difference and Repetition*, 98.

112. Deleuze and Guattari, *Anti-Oedipus*, 16.

113. Ibid., 30.

114. Deleuze, *Logic of Sense*, 311.

115. Cf. Deleuze, *The Fold*, 19–20.

116. Deleuze, *Difference and Repetition*, 96.

117. Cf. Deleuze, *The Fold*, 97; Deleuze and Guattari, *Anti-Oedipus*, 53.

118. Cf. Deleuze, *Logic of Sense*, 172–75; *The Fold*, 60–61.

119. Deleuze and Guattari, *Anti-Oedipus*, 68–75.

120. Deleuze and Guattari, *Anti-Oedipus*, 129, emphasis original.

121. "Still we do not yet know whence this impulse to truth comes, for up to now we have heard only about the obligation which society imposes in order to exist: to be truthful, that is, to use the usual metaphors, therefore expresses morally: we have heard only about the obligation to lie according to a fixed convention, to lie gregariously in a style binding to all." Friedrich Nietzsche, "On Truth and Lying in an Extra-Moral Sense," in *Literary Theory: An Anthology*, ed. Julie Rivkin and Michael Ryan (Victoria, Australia: Blackwell Publishing Ltd., 2004), 263.

122. Deleuze and Guattari, *A Thousand Plateaus*, 76.

123. Deleuze and Guattari, *Anti-Oedipus*, 116–17.

124. Ibid., 11.

125. Ibid., 7; Deleuze, *Dialogues*, 78–79.

126. Deleuze and Guattari, *Anti-Oedipus*, 55.

127. Even Michel Foucault, who has gone furthest in showing how these virtual walls are actualized in the form of hospitals and prisons, focuses primarily on the way the body and its affects comprise the site for manifestations of power. Deleuze writes, "As the postulate of property, power would be the 'property' won by a class. Foucault shows that power does not come about in this way: it is less a property than a strategy, and its effects cannot be attributed to an appropriation 'but to dispositions, maneuvers, tactics, techniques, functionings'; 'it is exercised rather than possessed; it is not the "privilege," acquired or preserved, of the dominant class, but the overall effect of its strategic positions.'" Deleuze, *The Fold*, 25.

128. Deleuze, *Logic of Sense*, 296–97; Deleuze and Guattari, *Anti-Oedipus*, 77.

129. This manifesto, signed by eleven scientists and philosophers (including Bertrand Russell and Albert Einstein), called for the abolition of thermo-nuclear weapons. Cf. "The Russell-Einstein Manifesto" in Bertrand Russell, *Man's Peril*, 1954–55, ed. Andrew G. Bone (London: Routledge, 2003), 304–33.

130. Deleuze and Guattari, *Anti-Oedipus*, 68.

131. Mullarkey, *Post-Continental Philosophy: An Outline*, 38.

132. "[W]e can see that Deleuze treats the Spinozian system as two distinct moments, as two perspectives of thought, one speculative and another practical . . . the first moment of the *Ethics*, speculative and analytic, proceeds in the centrifugal direction from God to the thing in order to discover and express the principles that animate the system of being; the second moment of the *Ethics*, practical and synthetic, moves in the centripetal direction from the thing to God by forging an ethical method and a political line of conduct . . . Rather than a destructive moment followed by a constructive moment, Deleuze's Spinoza presents a speculative, logical investigation followed by a practical, ethical constitution: *Forschung* followed by *Darstellung*. The two moments, then, speculation and practice, are fundamentally linked, but they remain autonomous and distinct—each with its own method and animating spirit." Michael Hardt, *Gilles Deleuze: An Apprenticeship in Philosophy* (Minneapolis: University of Minnesota Press, 1993), 58.

133. Deleuze, *Foucault*, 76.

134. Gilles Deleuze, *Proust and Signs*, trans. Richard Howard (Minneapolis: University of Minnesota Press, 2000), 94.

135. Deleuze and Guattari, *What Is Philosophy?* 145.

136. Schrödinger, "The Present Situation in Quantum Mechanics." This article was originally published in *Proceedings of the American Philosophical Society*, 124 (1980): 323–38. For more analysis of Schrödinger's opposition to the Copenhagen Interpretation, see my own analysis in "Superposing Images: Deleuze

and the Virtual after Bergson's Critique of Science," in the present volume (chapter 2).

137. Schrödinger, "The Present Situation in Quantum Mechanics," 323.

138. Deleuze and Guattari, *Anti-Oedipus*, 100.

139. Deleuze, *Difference and Repetition*, 277.

140. Deleuze and Guattari, *What Is Philosophy?* 37.

141. Deleuze and Guattari, *Anti-Oedipus*, 128–29.

142. Cf. Guattari and Rolnik, *Molecular Revolution in Brazil*, 72.

143. Deleuze, *Pure Immanence*, 73.

144. Deleuze, *Logic of Sense*, 179–180.

145. Deleuze and Guattari, *Anti-Oedipus*, 16.

146. Ibid., 129, emphasis original.

147. Ibid., 68.

148. Deleuze, *Dialogues*, 77–78.

149. Deleuze, *Negotiations*, 60.

150. Cf. Gilles Deleuze, "The Brain Is the Screen: An Interview with Gilles Deleuze," trans. Marie Therese Guiris, in *The Brain Is the Screen: Deleuze and the Philosophy of Cinema*, ed. Gregory Flaxman (Minneapolis and London: University of Minnesota Press, 2000), 365–73.

151. Deleuze and Guattari, *Anti-Oedipus*, 122.

152. Deleuze and Guattari, *A Thousand Plateaus*, 49–52.

153. Ibid., 49.

154. Ibid., 50.

155. Ibid., 49.

156. Deleuze and Guattari, *Anti-Oedipus*, 67.

157. Deleuze and Guattari, *A Thousand Plateaus*, 50.

158. "Bioids" and "chemotons" are the neologisms, respectively, of Peter Decker and Tibor Gánti. Cf. Peter Decker, "Possible resolution of racemic mixtures by bistability in 'bioids,' open systems which can exist in several steady states," *Journal of Molecular Evolution*, 2, no. 2–3 (June, 1973): 137–43; and Tibor Gánti, *Chemoton Theory* (New York: Kluwer Academic/Plenum Publishers, 2003). Also see Manuel DeLanda's analysis of the machinic phylum in "Non-Organic Life," in *Zone 6: Incorporations*, ed. Jonathan Crary and Sanford Kwinter (New York: Urzone, 1992), 129–67.

159. Deleuze and Guattari, *A Thousand Plateaus*, 51.

160. Ibid., 64.

161. Cf. Keith Ansell-Pearson's excellent analysis, in *Germinal Life: The Difference and Repetition of Gilles Deleuze*, of the way Deleuze conceives of ethics and individuation "beyond the human": "It is individuation conceived as a field of intensive factors (haecceities) which informs Deleuze's rethinking of ethics 'beyond' the subject so as to move thought beyond the 'human condition' and to

open it up to the inhuman and overhuman . . . It is rather the 'I' and the self that are abstract universals that need to be conceived in relation to the individuating forces that consume them." (London: Routledge, 1999), 96.

162. Deleuze and Guattari, *What Is Philosophy?* 185.

1. The Insistence of the Virtual in Science and the History of Philosophy

1. Cf. Aristotle, *Physics* VIII, 5, 256a 29.

2. *Internel* is the neologism of Charles Péguy, as cited by Deleuze and Guattari in their book *What Is Philosophy?* (Hugh Tomlinson and Graham Burchell translate the term as *Aternal*): "In a great work of philosophy, Péguy explains that there are two ways of considering the event. One consists in going over the course of the event, in recording its effectuation in history, its conditioning and deterioration in history. But the other consists in reassembling the event, installing oneself in it as in a becoming, becoming young again and aging in it, both at the same time, going through all its components or singularities. It may be that nothing changes or seems to change in history, but everything changes, and we change, in the event: 'There was nothing. Then a problem to which we saw no end, a problem without a solution . . . suddenly no longer exists and we wonder what we were talking about'; it has gone into other problems; 'there was nothing and one is in a new people, in a new world, in a new man.' This is no longer the historical, and it is not the eternal, Péguy says: it is the *Aternal* [*Internel*]. Péguy had to create this noun to designate a new concept." *What Is Philosophy?* 111.

3. Gilles Deleuze, *Cinema 2: The Time-Image,* trans. Hugh Tomlinson (Minneapolis: University of Minnesota Press, 1989), 135.

4. Gilles Deleuze, seminar of February 17, 1981.

5. Deleuze, *Cinema 2,* 272.

6. Gilles Deleuze, seminar of January 14, 1974.

7. Ibid., 180.

8. Ibid., 254.

9. It is evident that Hölderlin, who refers to this "living whole in the thousand articulations" in a letter to his brother, is the object of a precise and constant meditation on the part of Deleuze. Further support for this interpretation can be found in *Cinema 2,* where Deleuze searches, beyond the constraints of sensorimotor perception, for "a little time in its pure state" (17). This quote, which refers to Proust's *A la recherche du temps perdu,* and which also appears in *Bergsonism* (trans. Hugh Tomlinson and Barbara Habberjam [New York: Zone Books, 1988], 85), is what Hölderlin seems to be seeking in his *Anmerkungen zur Antigonae* and *Anmerkungen zum Oedipus,* specifically with reference to his analysis of the "Father of Time."

2. Superposing Images

1. The *IUPAC Recommendations on Nomenclature and Symbols and Technical Reports from Commissions* defines *repeatability* as "The closeness of agreement between independent results obtained with the same method on identical test material, under the same conditions (same operator, same apparatus, same laboratory and after short intervals of time)," while *reproducibility* is defined as "The closeness of agreement between independent results obtained with the same method on identical test material but under different conditions (different operators, different apparatus, different laboratories and/or after different intervals of time)." L. A. Currie and G. Svehla, "Nomenclature for the presentation of results of chemical analysis (IUPAC Recommendations 1994): Commission on Analytical Nomenclature," *Pure Applied Chemistry,* 66, no. 3 (1994): 598.

2. In *The Structure of Scientific Revolutions* (Chicago: University of Chicago Press, 1996), Kuhn writes that theoretical conflicts arising between existent and emergent scientific paradigms are "terminated by a relatively sudden and unstructured event like the [visual] gestalt switch" (122). It should be noted that Kuhn is, in some ways, critical of the comparison between scientific revolution and visual gestalt theory. He does not go so far, for example, as Norwood Hanson in his book *Patterns of Discovery: An Inquiry into the Conceptual Foundations of Science* (London: Cambridge University Press, 1958).

3. For a close analysis of the "disunity" of the sciences, see Peter Galison and Peter J. Stump eds., *The Disunity of Science* (Stanford, Calif.: Stanford University Press, 1996), especially Ian Hacking's "The Disunities of the Sciences" (37–74) and John Dupré's "Metaphysical Disorder and Scientific Disunity" (101–17).

4. Gilles Deleuze and Félix Guattari, *A Thousand Plateaus,* trans. Brian Massumi (Minneapolis: University of Minnesota Press, 1987), 372.

5. Gilles Deleuze and Félix Guattari, *What Is Philosophy?* trans. Hugh Tomlinson and Graham Burchell (New York: Columbia University Press, 1994), 161.

6. Eric Alliez, *The Signature of the World: Or, What Is Deleuze and Guattari's Philosophy?* trans. Eliot Ross Albert and Alberto Toscano (New York: Continuum, 2004), 36.

7. Ibid., 36.

8. Deleuze and Guattari, *What Is Philosophy?* 118.

9. Ibid., 281.

10. Ibid., 370.

11. For a basic explanation of the science behind the Higgs boson, see Michael Riordan, P. C. Rowson, Sau Lan Wu, "The Search for the Higgs Boson," *Science,* New Series, 291, no. 5502 (January 12, 2001): 259–60; a more recent article states that further research in high-energy physics has resulted in a theory indicating

"yet-undiscovered particles that would be partners to standard-model particles, plus a family of five Higgs bosons." Peter Weiss, "Corralling the Mass Maker: Hunting Ground Shifts for Elusive Particle," *Science News*, 165, no. 24 (June 12, 2004): 371.

12. Deleuze and Guattari, *A Thousand Plateaus*, 373. This notion of problematics comes directly from *Bergsonism*, where Deleuze identifies three rules of Bergsonian intuition as a method: "The first concerns the stating and creating of problems; the second, the discovery of genuine differences in kind; the third, the apprehension of real time." Gilles Deleuze, *Bergsonism*, trans. Hugh Tomlinson and Barbara Habberjam (New York: Zone Books, 1988), 14.

13. James Williams, in an interview with Mark Thwaite, February 13, 2006, online at *ReadySteadyBook*, http://www.readysteadybook.com/Article.aspx?page =jameswilliams (accessed October 21, 2007).

14. Gilles Deleuze, *Cinema 1: The Movement-Image*, trans. Hugh Tomlinson and Barbara Habberjam (London: The Athlone Press, 1986), 78–79.

15. Deleuze, *Bergsonism*, 34.

16. "It is not enough to say that number is a collection of units; we must add that these units are identical with one another, or at least that they are assumed to be identical when they are counted. No doubt we can count the sheep in a flock and say that there are fifty, although they are all different from one another and are easily recognized by the shepherd: but the reason is that we agree in that case to neglect their individual differences and to take into account only what they have in common. . . . But now let us even set aside the fifty sheep themselves and retain only the idea of them. Either we include them all in the same image, and it follows as a necessary consequence that we place them side by side in an ideal space, or else we repeat fifty times in succession the image of a single one, and in that case it does seem, indeed, that the series lies in duration rather than in space. But we shall soon find out that it cannot be so. For if we picture to ourselves each of the sheep in the flock in succession and separately, we shall never have to do with more than a single sheep. In order that the number should go on increasing in proportion as we advance, we must retain the successive images and set them alongside each of the new units which we picture to ourselves: now, it is in space that such a juxtaposition takes place and not in pure duration. In fact, it will be easily granted that counting material objects means thinking all these objects together, thereby leaving them in space." Henri Bergson, *Time and Free Will*, trans. F. L. Pogson (London: George Allen and Unwin Ltd., 1910), 76–77.

17. Deleuze and Guattari, *What Is Philosophy?* 129.

18. "The more consciousness is intellectualized, the more matter is spatialized." Henri Bergson, *L'Évolution créatrice* (Paris: Les Presses universitaires de France, 1959), 132.

19. As in Bergson's discussion of auditory aphasia and "acoustic memory" with regard to speech and language (116–26), the main issue for Bergson, in the case of all sensori-motor perception, is that it cannot explain how sensations endure in the mind. Without the indivisible time of memory, these "molecular vibrations of the cortical substance" (24) cannot produce sensation at all.

20. See, for example, how Bergson introduces this argument in the preface of his first book, *Time and Free Will*: "We necessarily express ourselves by means of words and we usually think in terms of space. That is to say, language requires us to establish between our ideas the same sharp and precise distinctions, the same discontinuity, as between material objects. This assimilation of thought to things is useful in practical life and necessary in most of the sciences. But it may be asked whether the insurmountable difficulties presented by certain philosophical problems do not arise from our placing side by side in space phenomena which do not occupy space, and whether, by merely getting rid of the clumsy symbols round which we are fighting, we might not bring the fight to an end. When an illegitimate translation of the unextended into the extended, of quality into quantity, has introduced contradiction into the very heart of the question, contradiction must, of course, recur in the answer. The problem which I have chosen is one which is common to metaphysics and psychology, the problem of free will. What I attempt to prove is that all discussion between the determinists and their opponents implies a previous confusion of duration with extensity, of succession with simultaneity, of quality with quantity: this confusion once dispelled, we may perhaps witness the disappearance of the objections raised against free will, of the definitions given of it, and, in a certain sense, of the problem of free will itself." Bergson, *Time and Free Will*, xxiii–xxiv.

21. Bergson's discussion of joy in *Time and Free Will* is particularly instructive here: "Finally, in cases of extreme joy, our perceptions and memories become tinged with an indefinable quality, as with a kind of heat or light, so novel that now and then, as we stare at our own self, we wonder how it can really exist. Thus there are several characteristic forms of purely inward joy, all of which are successive stages corresponding to qualitative alterations in the whole of our psychic states. But the number of states which are concerned with each of these alterations is more or less considerable, and, without explicitly counting them, we know very well whether, for example, our joy pervades all the impressions which we receive in the course of the day or whether any escape from its influence. We thus set up points of division in the interval which separates two successive forms of joy, and this gradual transition from one to the other makes them appear in their turn as different intensities of one and the same feeling, which is thus supposed to change in magnitude" (10–11). This is relevant as well to a Deleuzian ethics, based on his readings of Spinoza (the formation of adequate ideas based on passive joys and desires) and Nietzsche (joy as Dionysian affirmation of active over reactive forces). See Constantin V. Boundas's

entry on "Spinoza + Ethics of Joy," in Adrian Parr ed., *The Deleuze Dictionary* (New York: Columbia University Press, 2005), 263–64.

22. Bergson, *Time and Free Will*, 2.

23. Ibid., 32–33.

24. Ibid., 71.

25. Deleuze, *Bergsonism*, 34.

26. Ibid., 34.

27. Henri Bergson, *The Meaning of the War: Life and Matter in Conflict*, ed. Wildon Carr (London: T. Fisher Unwin, Ltd., 1915), 34–35.

28. I am thinking here of the striking way in which recent "meta-technological" analyses of robotics and artificial intelligence have pointed to the necessary ascendance of technology over its merely instrumental role vis-à-vis human interest. In *Robot: Mere Machine to Transcendent Mind* (New York: Oxford University Press, 1999), robotics engineer Hans Moravec suggests that our capacity for abstract thought is increasingly rendered obsolete by machines designed to do this for us, a state of affairs that leads him to the conclusion that humanity will one day be obliged to cede its place to the machine: "by performing better and cheaper, the robots will displace humans from essential roles. Rather quickly, they could displace us from existence. I'm not as alarmed as many by the latter possibility, since I consider these future machines our progeny, 'mind children' built in our image and likeness, ourselves in more potent form . . . It behooves us to give them every advantage and to bow out when we can no longer contribute" (13). Moravec has also articulated an interesting paradox at the intersection of technology and visual culture: "it is comparatively easy to make computers exhibit adult level performance on intelligence tests or playing checkers, and difficult or impossible to give them the skills of a one-year-old when it comes to perception and mobility." *Mind Children* (Cambridge, Mass.: Harvard University Press, 1988), 15.

29. Henri Bergson, *Creative Evolution*, trans. Arthur Mitchell (New York: Dover Publications, Inc., 1998), 198–99.

30. Henri Bergson, *Matter and Memory*, trans. Nancy Margaret Paul and W. Scott Palmer (London: George Allen and Unwin Ltd., 1911), 151.

31. Ibid., 151.

32. Ibid., 185.

33. Ibid., 124.

34. Ibid., 161.

35. Deleuze, *Cinema 1: The Movement-Image*, 78–79.

36. See Henri Bergson, "Psychophysical Parallelism and Positive Metaphysics," trans. Matthew Cobb, in *Continental Philosophy of Science*, ed. Gary Gutting (London: Blackwell Publishing, Ltd., 2005), 59–68. Bergson's lecture was originally published in *Bulletin de la Société française de philosophie*, May 2, 1901.

37. Comparing scientific and philosophical forms of thought, Deleuze and Guattari claim that (in the exact sciences and mathematics): "Seeing, seeing what happens, has always had a more essential importance than demonstrations, even in pure mathematics, which can be called visual, figural, independently of its applications." *What Is Philosophy?* 128.

38. Erwin Schrödinger, "The Present Situation in Quantum Mechanics," in *Quantum Theory and Measurement*, ed. J. A. Wheeler and W. H. Zurek (Princeton, N.J.: Princeton University Press, 1983), 152–53. This article was originally published in *Proceedings of the American Philosophical Society*, 124 (1980), 323–38.

39. Bergson, *Time and Free Will*, 100.

40. Schrödinger, "The Present Situation in Quantum Mechanics," 152.

41. Ibid., 153.

42. This interpretation of Schrödinger's definition of the scientific image is not altered by his recommendation, in face of the Copenhagen interpretation, of a "deliberate about-face of the epistemological viewpoint," which consists merely in a more scrupulous attention to measurement in the absence of any certainty about a mind-independent world: "Now it is fairly clear; if reality does not determine the measured value, then at least the measured value must determine reality—it must actually be *after* the measurement in *that* sense which alone will be recognized again. That is, the desired criterion can be merely this: repetition of the measurement must give the same result. By many repetitions I can prove the accuracy of the procedure and show that I am not just playing" (158).

43. Ibid., 152.

44. Ibid., 153.

45. On the nature of this polemic, and its correspondence to the virtual in Deleuze's conception, see Timothy Murphy, "Quantum Ontology: A Virtual Mechanics of Becoming," in Eleanor Kaufman and Kevin Jon Heller eds., *Deleuze and Guattari: New Mappings in Politics, Philosophy, and Culture* (Minneapolis: University of Minnesota Press, 1998), 211–29.

46. Schrödinger, "The Present Situation in Quantum Mechanics," 157.

47. See, for example, Heisenberg's account of this problem: "one can easily see that there is no way of observing the orbit of the electron around the nucleus. The second step shows a wave pocket moving not around the nucleus but away from the atom, because the first light quantum will have knocked the electron out from the atom. The momentum of light quantum of the y-ray is much bigger than the original momentum of the electron if the wave length of the e-ray is much smaller than the size of the atom. Therefore, the first light quantum is sufficient to knock the electron out of the atom and one can never observe more than one point in the orbit of the electron." Werner Heisenberg, *Physics and Philosophy: The Revolution in Modern Science* (New York: Harper and Row, 1958), 47–48.

48. Ibid., 157.

49. This was Einstein's contention (and, implicitly, Schrödinger's). See Albert Einstein, Boris Podolsky, and Nathan Rosen, "Can Quantum-Mechanical Description of Physical Reality Be Considered Complete?" in *Quantum Theory and Measurement,* ed. John Archibald Wheeler and Wojciech Hubert Zurek (Princeton, N.J.: Princeton University Press, 1983), 138–43. This article was originally published in *Physical Review,* 47 (1935), 777–80.

50. Bergson, *Matter and Memory,* 133–34. Cited by Deleuze in *Bergsonism,* 56.

51. Wolfram Schommers, "Evolution of Quantum Theory," in *Quantum Theory and Pictures of Reality: Foundations, Interpretations, and New Aspects,* ed. W. Schommers (Berlin; New York: Springer-Verlag, 1989), 31.

52. Schommers, "Evolution of Quantum Theory," 261–62.

53. Bergson, *Matter and Memory,* 161–62.

54. Deleuze, *Bergsonism,* 40.

55. Deleuze, *Cinema 1,* 60.

56. Deleuze and Guattari, *What Is Philosophy?* 127, 159.

57. Ibid., 127.

58. Ibid., 162.

59. Ibid., 127.

60. Ibid., 129.

61. See Bernard d'Espagnat, *In Search of Reality* (New York; Berlin: Springer-Verlag, 1983), 122–29.

62. Deleuze and Guattari, *What Is Philosophy?* 169.

63. Ibid., 118.

64. Ibid., 123–24.

65. Cf. Deleuze and Guattari on the complicated process of coalescence and dissolution among the various kinds of strata: "In short, the epistrata and parastrata are continually moving, sliding, shifting, and changing on the Ecumenon or unity of composition of a stratum; some are swept away by lines of flight and movements of deterritorialization, others by processes of decoding or drift, but they all communicate at the intersection of the milieus. The strata are continually being shaken by phenomena of cracking and rupture; either at the level of the substrata that furnish the materials (a prebiotic soup, a prechemical soup . . .), at the level of the accumulating epistrata, or at the level of the abutting parastrata: everywhere there arise simultaneous accelerations and blockages, comparative speeds, differences in deterritorialization creating relative fields of reterritorialization." *A Thousand Plateaus,* 55.

66. Deleuze, *Bergsonism,* 88.

67. Deleuze and Guattari, *A Thousand Plateaus,* 40–57.

68. Ibid., 49.

69. Deleuze, *Bergsonism,* 78, 93–94.

70. For an in-depth analysis of Deleuze's break with Bergson, see Keith Ansell Pearson, *Germinal Life: The Repetition and Difference of Gilles Deleuze* (London, New York: Routledge, 1999), 74–75.

71. Deleuze, *Bergsonism*, 78.

72. Deleuze, *Negotiations*, 55. Mark Hansen outlines a similar development in Deleuze's concept of the virtual, ultimately rejecting the Bergsonian notion of differences of kind in favor of "a more primordial domain of difference—differences of intensity which constitute 'the entire nature of difference,' that is, both differences in degree and differences in kind." "Becoming as Creative Involution?: Contextualizing Deleuze and Guattari's Biophilosophy," *Postmodern Culture* 11, no. 1 (September 2000), online at *Project Muse*, http://proxy.library.upenn.edu:2298/journals/postmodern_culture/v011/11.1hansen.html (accessed June 25, 2008).

73. Deleuze and Guattari, *What Is Philosophy?* 130.

74. Deleuze and Guattari, *A Thousand Plateaus*, 141.

3. The Intense Space(s) of Gilles Deleuze

1. "The space of nomad thought is qualitatively different from State space. Air against earth. State space is 'striated,' or gridded. Movement in it is confined as by gravity to a horizontal plane, and limited by the order of that plane to preset paths between fixed and identifiable points. Nomad space is 'smooth,' or open-ended. One can rise up at any point and move to any other. Its mode of distribution is the *nomos:* arraying oneself in an open space (hold the street), as opposed to the *logos* of entrenching oneself in a closed space (hold the fort). *A Thousand Plateaus* is an effort to construct a smooth space of thought." See Brian Massumi's introduction to Gilles Deleuze and Félix Guattari, *A Thousand Plateaus,* trans. Brian Massumi (Minneapolis: University of Minnesota Press, 1987), 13.

2. Friedrich Nietzsche, *The Gay Science*, trans. Walter Kaufmann and R. J. Hollingdale (New York: Vintage, 1967), 125.

3. Gilles Deleuze, *Difference and Repetition*, trans. Paul Patton (New York: Columbia University Press, 1994), 238; Gilles Deleuze and Félix Guattari, *What Is Philosophy?* trans. Hugh Tomlinson and Graham Burchell (New York: Columbia University Press, 1994), 48.

4. See Deleuze and Guattari, *A Thousand Plateaus*, 142–45, 511.

5. Gilles Deleuze, *The Logic of Sense*, trans. Mark Lester and Charles Stivale (New York: Columbia University Press, 1990), 146.

6. Ibid., 177.

7. Jean-Clet Martin, "L'Espace sensible," *Chimères*, online at http://www.revue-chimeres.fr/drupal_chimeres/files/40chi05.pdf (accessed January 21, 2010).

8. Kant defines space as an a priori form of sensibility, and Jean-Clet Martin comments: "Affirmer cela, affirmer que l'espace est la forme de notre sensibilité

c'est la thèse inaugurale de la phénoménologie, l'expérience fameuse du cube qui montre que c'est en moi que se fait le montage des six faces lorsque le sens externe se soumet finalement au sens interne, au temps de la conscience. Dès lors il faut retrouver dans tout espace éprouvé une forme a priori subjective qui le rende possible, une condition qui sera offerte par le moi, le sujet capable de faire l'unité de la diversité spatiale au sein du temps, comme si le temps, l'ordre du temps s'imposait à la diversité de l'espace. C'est ici la notion de perspective qui domine l'image de la pensée, le perspectivisme qui pose le sujet au centre de toute chose."

9. Constantin V. Boundas, "An Ontology of Intensities," *Epoché* 7, no. 1 (Fall 2002): 18.

10. Nathan Widder, "What's Lacking in Lack: A Comment on the Virtual," *Angelaki*, 5(3) (December 2000): 128.

11. Aristotle, *Physics*, trans. Robin Waterfield (Oxford: Oxford University Press, 1996), 161 (VI: 9, 239b5).

12. Brian Massumi, *Parables for the Virtual: Movement, Affect, Sensation* (Durham, N.C.: Duke University Press, 2002), 6.

13. Ibid., 6.

14. Gilles Deleuze, *Negotiations: 1972–1990*, trans. Martin Joughin (New York: Columbia University Press, 1995), 146.

15. "To think is to experiment, but experimentation is always that which is coming about—the new, remarkable, and interesting that replace the appearance of truth and are more demanding than it is." *What Is Philosophy?* 111.

16. Deleuze, *Difference and Repetition*, 220.

17. Ibid., 216, 251.

18. Gilles Deleuze, "The Method of Dramatization," in *Desert Islands and Other Texts*, ed. David Lapoujade, trans. Michael Taormina (Semiotext[e]) (Cambridge, MA: The MIT Press, 2004), 96–7.

19. Cf. Manuel Delanda, "Nonorganic Life," in *Zone: Incorporations*, ed. Sanford Kwinter and Jonathan Crary (New York: Zone Books, 1991), 130.

20. Cf. Manuel Delanda, "Deleuze and the Genesis of Form," *Art Orbit* 1 (March 1998), online at http://artnode.se/artorbit/issue1/f_deleuze/f_deleuze _delanda.html (accessed October 21, 2007).

21. See, for instance, chapter 5 in *Difference and Repetition*.

22. Deleuze, *Difference and Repetition*, 228, 266; also see James Williams, *Gilles Deleuze's* Difference and Repetition: *A Critical Introduction and Guide* (Edinburgh: Edinburgh University Press, 2003), 169; and chapter 2 of Manuel Delanda's *Intensive Science and Virtual Philosophy* (London: Continuum, 2002).

23. Boundas, "An Ontology of Intensities," 32.

24. Deleuze, *Difference and Repetition*, 230.

25. Massumi, *Parables for the Virtual*, 157.

4. Interstitial Life

1. Cf. Michael J. Behe, *Darwin's Black Box: The Biochemical Challenge to Evolution,* 2nd ed. (New York: Free Press, 2006).

2. Richard Dawkins, *The Blind Watchmaker* (New York: Norton, 1987).

3. Michael Ruse, *Darwin and Design: Does Evolution Have a Purpose?* (Cambridge, Mass.: Harvard University Press, 2003), 274ff.

4. Edward O. Wilson, *Consilience: The Unity of Knowledge* (New York: Vintage, 1999), 291.

5. Steven Shaviro, *Connected, or, What It Means to Live in the Network Society* (Minneapolis: University of Minnesota Press, 2003), 205–12.

6. Steven Pinker, *How the Mind Works* (New York: Penguin, 1999), 43.

7. Immanuel Kant, *Critique of Judgment,* trans. Werner S. Pluhar (Indianapolis: Hackett, 1987), 269.

8. Ruse, *Darwin and Design,* 268.

9. Gilles Deleuze, *Difference and Repetition,* trans. Paul Patton (New York: Columbia University Press, 1994), 135.

10. Gilles Deleuze, *The Logic of Sense,* trans. Mark Lester (New York: Columbia University Press, 1990), 98.

11. Immanuel Kant, *Critique of Practical Reason,* trans. Werner Pluhar (Indianapolis: Hackett, 2002), 9.

12. Ludwig von Bertalanffy, *General System Theory: Foundations, Development, Applications* (New York: George Braziller, 1976); Humberto Maturana and Francisco Varela, *Autopoiesis and Cognition: The Realization of the Living* (Berlin: Springer, 1991); Stuart Kauffman, *Investigations* (New York: Oxford University Press, 2000); Susan Oyama, *The Ontogeny of Information: Developmental Systems and Evolution,* 2nd ed. (Durham, N.C.: Duke University Press, 2000); James Lovelock, *Gaia: A New Look at Life on Earth* (New York: Oxford University Press, 2000); Lynn Margulis and Dorion Sagan, *Acquiring Genomes: A Theory of the Origin of Species* (New York: Basic Books, 2002).

13. Cf. Gilles Deleuze and Félix Guattari, *A Thousand Plateaus,* trans. Brian Massumi (Minneapolis: University of Minnesota Press, 1987), 10.

14. Luciana Parisi, *Abstract Sex: Philosophy, Bio-technology, and the Mutations of Desire* (New York: Continuum, 2004), 53.

15. Alfred North Whitehead, *Process and Reality* (New York: The Free Press. 1929/1978).

16. Susannah Kate Devitt, "Bacterial Cognition," in *Philosophy of Memory* (2007), http://mnemosynosis.livejournal.com/10810.html.

17. Toshiyuki Nakagaki, Hiroyasu Yamada, and Agota Toth, "Maze-Solving by an Amoeboid Organism," in *Nature,* Volume 407(6803), September 28, 2000, page 470.

18. Anthony Trewavas, "Green Plants as Intelligent Organisms," in *Trends in Plant Science* (October 9, 2005), 414.

19. Alexander Maye, et al., "Order in Spontaneous Behavior," in *PLoS ONE* 2.5. *e443*. doi:10.1371/journal.pone.0000443 (May 2007).

20. Devitt, "Bacterial Cognition," quoting Pamela Lyon.

21. Guenter Albrecht-Buehler, "Cell Intelligence" (1998), in http://www.basic .northwestern.edu/g-buehler/cellinto.htm.

22. Eshel Ben Jacob, Yoash Shapira, and Alfred I. Tauber, "Seeking the Foundations of Cognition in Bacteria: From Schrodinger's Negative Entropy to Latent Information," in *Physica* 359 (2006), 496.

23. Whitehead, *Process and Reality*, 240.

24. Alfred North Whitehead, *Modes of Thought* (New York: The Free Press, 1938/1968) 150–51.

25. Maye, et al., "Order in Spontaneous Behavior," 6.

26. Morse Peckham, *Explanation and Power: The Control of Human Behavior* (Minneapolis: University of Minnesota Press, 1979), 165.

27. Maye, et al., "Order in Spontaneous Behavior," 8.

28. Bjoern Brembs, "Do Fruit Flies Have Free Will?" (2007), in http:// brembs.net/spontaneous.

29. Whitehead, *Process and Reality*, 105–6.

5. Digital Ontology and Example

1. Compare Johanna Drucker's priority of the *graphesis* over the *mathesis*, in "Digital Ontologies: The Ideality of Form in/and Code Storage—or—Can Graphesis Challenge Mathesis?" *Leonardo* 34, no. 2 (2001), 141–45.

2. Cf. Matt Kirschenbaum's concept of *formal materiality*, in *Mechanisms: New Media and the Forensic Imagination* (Cambridge, Mass.: MIT Press, forthcoming).

3. Cf. Mark Hansen's Bergsonian notion of the body-brain as a *frame*, in *New Philosophy for New Media* (Cambridge, Mass.: MIT Press, 2006).

4. Cf. Sandy Stone: "Even in the age of the technosocial subject, life is lived through bodies." Allucquère Rosanne Stone, "Will the Real Body Please Stand Up?" in *Cyberspace: First Steps*, ed. Michael Benedikt (Cambridge, Mass.: MIT Press, 1991), 81–118.

5. Each of these four footnoted theorists emphasizes the materiality of the digital in a context that acknowledges the complexity of this materiality, its difference from other sorts of material.

6. Gilles Deleuze, *Difference and Repetition*, trans. Paul Patton (New York: Columbia University Press, 1994).

7. Calling these moments of virtuality *stages* is potentially misleading, as they are not clearly successive and they are not immersed in the passage of time, for they are the processes by which time comes to pass. Deleuze indicates a certain ontogenetic priority for different moments of the virtual; for example, the undetermined is prior to the reciprocal determination. But these events are neither successive nor simultaneous, for they are perplicated, both singular and universal, outside of time and at the heart of temporality.

8. Brian Massumi, "Line Parable for the Virtual: (On the Superiority of the Analog)," in *The Virtual Dimension: Architecture, Representation, and Crash Culture,* ed. John Beckmann (New York: Princeton Architectural Press, 1998), 309.

9. This is a key factor in Peter Hallward's objection to Deleuze's philosophy, that it leaves no place for the intervention of a political subject. But perhaps Hallward attends too little to Deleuze's own struggle with this question. Deleuze's philosophy is an extended attempt to make a place for the subject without abandoning the Nietzschean affirmation at the heart of his philosophy. There are strategic developments in Deleuze's later works that carve out different niches for subjectivity. Witness the role of becoming in *A Thousand Plateaus,* which invites the human back into the virtual, having alienated the human from ontology in *Difference and Repetition.*

10. Similar error-tolerance measures exist for other means of digital transmission and storage. Red book (audio) CDs use multiple kinds of redundancy to ward off error and ambiguity; for instance, on a CD, fourteen "material" bits are used to encode eight bits of audio data, which inefficiency ensures that even when the audio data has many 0 bits in a row, the material representation of those data never has more than three consecutive 0s. The analog-to-digital hardware system can thus distinguish between a sample with many consecutive 0s and a malformed or blank disc.

11. Even the notion of an *object* seems to dissolve in the realm of the digital. The digital does not *object,* throw itself against or resist the interventions of the user. On the contrary, the digital submits, it does exactly what it is asked to do, what it is supposed to do.

12. Deleuze, *Difference and Repetition,* 186.

13. Greg Egan's imaginative science fiction novel, *Permutation City,* experiments with the idea of distributing the digital across time and space. In this novel, a computer-generated consciousness is unaware of the time that passes between successive simulated mental states; if it takes the computer ten seconds of "real time" to process the transition from one mental state to the next, the calculated consciousness still experiences this as an instantaneous transition. Egan ultimately divorces the digital from its material altogether, suggesting that any given calculation can be found already taking place by choosing an appropriate subset of the universe's matter to designate as the computer.

14. Claude Shannon, quoted in Wolfgang Hagen, "The Style of Sources: Remarks on the Theory and History of Programming Languages," in *New Media, Old Media: A History and Theory Reader,* ed. Wendy Chun and Thomas Keenan (New York: Routledge, 2006), 166.

15. It is tempting but probably overzealous to equate the digital as described here to Deleuze's account of negation in *Difference and Repetition.* Deleuze criticizes an image of thought in which the actual is severed from the virtual that animates it and gives it sense, and this disconnection of the actual from its genesis leaves representation to stand as the plenary speaker of ontology. What I have been calling *form* shares much with this notion of representation (and bears little relation to Deleuze's use of the term *form*), but the digital lacks the complexity of representation. Form is more like representation abstracted, representation made into a formula, so that it loses its connection with the here and now. Representation, as Deleuze relates, contains the seeds of its own undoing, for it is always excessive. The digital has crushed even this life out of representation, so there remains no challenge to the generic character of the digital.

16. Massumi, "Line Parable for the Virtual," 310.

17. This claim, that computers have no horizon of attention, no context, is the principal claim of Hubert Dreyfus's classic *What Computers Can't Do* (New York: HarperCollins, 1978). Computers are strictly tied to a predetermined set of possible inputs, they can only take in what they are expecting. The computer has no "horizon" of understanding, no milieu against which it experiences the world. Human beings make sense of their worlds hermeneutically, operating from the outside in as well as the inside out. The computer only has its inside.

18. I don't mean to elide the considerable complexity of the operation of the CPU in the real world. Though the CPU remains the central site of calculation, and continues to define the computer as a technology, a great deal of additional processing (of a similar nature) takes place in various subsystems of the computer, including, for instance, the graphics card, dedicated I/O subsystems (such as a disk subsystem), and others. The CPU itself is not monolithic: it is increasingly common to have more than one CPU in a single computer, and the latest trend in commercial personal computing is to incorporate many CPUs ("cores") on a single silicon wafer, so that they are tightly integrated. The CPU itself is subdivided into segments; it has multiple stages of operation, including a stage where it retrieves a command and another where it dispatches that command, deciding which part of the CPU should handle this particular type of command. Thus, the CPU deals with different commands differently, dedicating a particular part of its surface to the logic that governs, for instance, operations on floating point numbers as opposed to integer arithmetic.

19. Strictly speaking, 0 and 1 are rarely represented electronically as the absence and presence of the flow of electricity. Rather, they are generally

represented as two different electrical potentials, neither of which is a null or absent potential. Thus, the system will be less likely to interpret a malfunction (such as a loss of power) as the presence of a 0 bit.

20. At the time of this writing, the universal serial bus (USB) is the typical interface that connects the keyboard to the motherboard of the computer. Prior to USB there were competing standards for this connection, and there will no doubt be a new standard that overtakes USB a few years from now. The principle remains the same, however.

21. Even a fast touch-typist enters data into the computer orders of magnitude more slowly than a single cycle through a loop that checks for the presence of such data. More than 99 percent of the time there are no data waiting when the computer checks the flag.

22. Different kinds of interrupt systems work differently. Some interrupt systems include a "masking" feature, which enables software to set a flag rendering it vulnerable only to certain interrupts and not to others. This masking function does not so much eliminate the hierarchy of the interrupt mechanism as complicate it, implying a kind of self-reflection and a sense of temporality, a friction with time.

23. Some interrupt systems include means to save the state of the CPU (and the relevant data it is working on) when the interrupt occurs, so that this state can later be restored. An interrupt that saves all the relevant conditions of the operation of the computer allowing accurate and total resumption of the algorithm is called *precise*. Though perhaps still just as violent, a precise interrupt is in such a case more considerate because it allows a return to the current algorithm after the interrupting condition has been attended to.

24. Massumi, "Line Parable for the Virtual," 306.

6. Virtual Architecture

1. Gilles Deleuze and Félix Guattari, *What Is Philosophy?* trans. Hugh Tomlinson and Graham Burchell (New York: University of Columbia Press, 1994), 186.

2. Deleuze and Guattari, *What Is Philosophy?* 167–68.

3. Deleuze and Guattari, *What Is Philosophy?* 176.

4. [In using the word "pleat," we are following Tom Conley's translation of Deleuze's *The Fold* (Minneapolis: University of Minnesota Press, 1993). In an earlier translation by Jonathan Strauss, we find instead the word "coils" (Deleuze, "The Fold," in *Yale French Studies,* No. 80, Baroque Topographies: Literature/ History/Philosophy [1991], 227–47). —Trans.]

5. [Paul Patton, in his translation of *Difference and Repetition,* suggests that Deleuze's use of both *fond* and *fondement* is "connected with the philosophical

concept which is regularly rendered as 'ground' in English translations of German philosophy." (New York: Columbia University Press, 1994), xiii.—Trans.]

6. Gilles Deleuze, *The Fold,* trans. Paul Patton (Minneapolis: University of Minnesota Press, 1993), 19.

7. For a comprehensive presentation of these transformations in contemporary architecture, see the illuminating article by Marie-Ange Brayer, "Vers une architecture 'intelligente'," published in *Qu'est-ce que l'architecture aujourd'hui?* (Paris: BeauxArts/TTM Éditions, 2007), 20–27.

8. Bernard Cache, "Objectile: poursuite de la philosophie par d'autres moyens?" *Rue Descartes* no. 20, *Gilles Deleuze. Immanence et vie* (1998), 149. In the original French: "des lignes et surfaces à courbure variable comme les plis d'une sculpture baroque."

9. Deleuze, *The Fold,* 81.

10. Thom Mayne, *Morphosis: Connected Isolation* (London: Academy Editions, 1993), 7.

11. Peter Sloterdijk, *Bulles-Sphères I* (Paris, Pauvert, 2002).

12. [Antonioli cites Sloterdijk, *Bulles-Sphères I,* 64. This translation is based on the original German text in *Sphären I: Blasen* (Frankfurt: Suhrkamp, 2000), 57. Translation by Ilinca Iurascu—Ed.]

13. Félix Guattari, *The Three Ecologies,* trans. Ian Pindar and Paul Sutton (New Brunswick, N.J.: Athlone, 2000). Originally published in French as *Les trois ecologies* (Paris: Galilée, 1989).

14. "1837: Of the Refrain," in Gilles Deleuze and Félix Guattari, *A Thousand Plateaus,* trans. Brian Massumi (University of Minnesota Press, 1987), 310–50.

15. See Félix Guattari, *L'Inconscient machinique* (Paris: Recherches, 1979), 109–55.

16. See note 5.

17. "La déterritorialisation, sous toutes ses formes, 'précède' l'existence des strates et des territoires." Guattari, *L'Inconscient machinique,* 13.

18. Cf. the article "Les machines architecturales de Shin Takamatsu." *Chimères,* no. 21 (Winter 1994), 127–41.

19. Félix Guattari, *Cartographies schizoanalytiques* (Paris: Galilée, 1989), 291–301.

20. "Depuis quelques millénaires et peut-être à l'imitation des crustacés ou des termites, les êtres humains ont pris l'habitude de s'entourer de carapaces de toutes sortes." Guattari, *Cartographies schizoanalytiques,* 291.

21. "Aujourd'hui à quoi servirait-il, par exemple, dans une ville comme Mexico, qui fonce, en plein délire, vers ses 40 millions d'habitants, d'invoquer Le Corbusier! Même le baron Haussmann n'y pourrait plus rien." Ibid., 291.

22. Ibid.

23. Ibid.

24. "La forme architecturale n'est pas appelée à fonctionner comme *gestalt* clôturée sur elle-même, mais comme opérateur catalytique déclenchant des réactions en chaîne au sein des modes de sémiotisation qui nous font sortir de nous-mêmes et nous ouvrent des champs inédits du possible." Ibid., 300.

7. The Subject of Chaos

1. Gilles Deleuze and Félix Guattari, *What Is Philosophy?* trans. Hugh Tomlinson and Graham Burchell (New York: Columbia University Press, 1994), 202.

2. Gilles Deleuze, "The Brain Is the Screen. An Interview with Gilles Deleuze," trans. Marie Therese Guiris, in *The Brain Is the Screen: Deleuze and the Philosophy of Cinema,* ed. Gregory Flaxman (Minneapolis and London: University of Minnesota Press. 2000), 367.

3. Deleuze and Guattari, *What Is Philosophy?* 208.

4. Ibid., 208.

5. Ibid., 218, emphasis original.

6. Gilles Deleuze, *The Fold,* trans. Paul Patton (Minneapolis: University of Minnesota Press, 1993), 97.

7. Ibid., 97.

8. Ibid., 76.

9. Ibid., 76. Hence, in *The Fold,* Deleuze treats chaos as a cosmological approximation ("the sum of all possibles" filtered through compossibility), a physical approximation ("depthless shadows" filtered though a prism of Nature), and a psychic approximation ("the sum of possible perceptions" filtered through a differentiating perception).

10. Ibid., 86.

11. Ibid., 64.

12. See Deleuze's interview, "The Brain Is the Screen," in *The Brain Is the Screen: Deleuze and the Philosophy of Cinema.*

13. Ibid., 1920.

14. Deleuze and Guattari, *What Is Philosophy?* 210.

15. Ibid., 37.

16. Ibid., 38.

17. Ibid., 205.

18. Ibid., 36.

19. Gilles Deleuze, *Difference and Repetition,* trans. Paul Patton (New York: Columbia University Press, 1994), 56.

20. Deleuze and Guattari, *What Is Philosophy?* 203.

21. Ibid., 201.

22. Inasmuch as the engagement with chaos gives rise to an even more distressing struggle with opinion, Deleuze treats this inclination in the name of a number of different philosophers and in a number of different conceptual neighborhoods. But across these scattered discussions, and perhaps this leitmotiv, it is possible to discern the lineaments of a kind of cosmology that runs from the unfettered chaos of the universe to the evolutionary morphogenesis of *homo sapiens* to the natural history of morals and the production of a "regular and predictable animal," subject. Above all else, this cosmology concerns what we have called the "the plane of immanence," which Deleuze elaborates in virtually every one of his works but which he also elaborates, across these three siblings, along the lines of a fundamental paradox.

23. To be sure, Rorty expresses a similar complaint with philosophy insofar as it devolves into a metaphysical-moral system, but whereas this leads him to abandon the problem of philosophy and metaphysics to the panoply of literary narratives, Deleuze sees literature as the very source of health with which philosophy endeavors to nurture its transformation.

24. Gilles Deleuze, *Proust and Signs*, trans. Richard Howard (Minneapolis: University of Minnesota Press, 2000), 94.

25. Thus, even when philosophers introduce a rational method in order to get to the bottom of things, to validate reality or ensure truth, this method is always already underwritten by certain convictions. For instance, Descartes avoids defining man as a "rational animal," thereby escaping certain presuppositions (what is rationality? what is animality?), only to introduce subjective or implicit presuppositions of a different kind (what is the cogito? why does this form have a "natural" relationship with thinking?). The presuppositions of classical philosophy, which condition thought and the thinker, invariably emerge the great affirmations to which this lineage of philosophy leads: everyone knows that God exists, everyone knows that the world exists, everyone knows that the subject exists.

26. Deleuze, *Difference and Repetition*, 130.

27. Ibid., 131.

28. Ibid., 133.

29. Ibid., 133.

30. Deleuze and Guattari, *What Is Philosophy?* 145.

31. This double structure explains why, as Deleuze and Guattari say, discussion so often takes on a particular sense of expression: "'as a man, I consider all women to be unfaithful'; 'as a woman, I think men are liars.'"

32. Ibid., 146.

33. Deleuze, *Difference and Repetition*, 222.

34. Ibid., 223.

35. Ibid., 222.

36. Deleuze and Guattari, *What Is Philosophy?* 205.

37. James Williams, *Gilles Deleuze's* Difference and Repetition: *A Critical Introduction and Guide* (Edinburgh: Edinburgh University Press, 2003), 169.

38. Deleuze, *Difference and Repetition*, 224.

39. Ibid., 223.

40. Fredric Jameson, "Marxism and Dualism in Deleuze" in *The South Atlantic Quarterly* ("A Deleuzian Century"), ed. Ian Buchanon, 96, no. 3 (1997), 394–95.

41. Strict dualisms are always symptomatic of dualists themselves, who prefer the crude organization of the one or the other ("You're either with the terrorists or against them") for the play of differences and the percolation of countless micromovements that always subsist or insist beneath this simplification. As we have implied, Deleuze suggests that another name for this "dualism unto death" is the dialectic because the latter always augurs the organization of difference, as contrariety, and the ultimate resolution of difference, as the Absolute. Indeed, Deleuze responds to the dialectical impulse, from the time of *Difference and Repetition* all the way to *What Is Philosophy?* by suggesting that its appeal to dualism stages "the contradiction between rival opinions to extract from them suprasensible propositions able to move, contemplate, reflect, and communicate in themselves" (WIP 80).

In other words, the reduction of concepts to propositions, which the dialectic promises to resolve in a higher form of knowledge, that turns out to be no more than a juncture of opinions (*doxa*) according to a yet higher opinion (*urdoxa*).

42. Deleuze, *Difference and Repetition*, 37.

43. Ibid., 284.

44. Keith Devlin, *Mathematics: The Science of Patterns* (New York: Holt Paperbacks, 1996), 54.

45. Deleuze, *Logic of Sense*, 36.

46. Ibid., 59.

47. Ibid., 70.

48. Ibid., 59.

49. Deleuze refers to Prigogine and Stengers on several occasions, beginning in an interview on the subject of *A Thousand Plateaus* that he gave in 1980 (see *Negotiations 1972–90*, 29), and concluding with his invocation of chaos theory in *What Is Philosophy?* (226).

50. Deleuze and Guattari, *What Is Philosophy?* 206.

51. See Ilya Prigogine and Isabelle Stengers, *La Nouvelle Alliance* (Paris: Gallimard, 1986).

52. Deleuze, *The Fold*, 86.

53. Ibid., 86.

54. Deleuze, *Logic of Sense*, 59.

55. Ibid., 86. Also see Nietzsche's "The Dawn" (130).

8. Elemental Complexity and Relational Vitality

1. See Rosi Braidotti, *Nomadic Subjects: Embodiment and Sexual Difference in Contemporary Feminist Theory* (New York: Columbia University Press, 1994) and *Patterns of Dissonance: A Study of Women in Contemporary Philosophy* (New York: Routledge, 1991).

2. See Rosi Braidotti, *Metamorphoses: Towards a Materialist Theory of Becoming* (Cambridge, Mass.: Polity [Blackwell Publishers, Ltd.], 2002); and *Transpositions: On Nomadic Ethics* (Cambridge: Polity [Blackwell Publishers, Ltd.], 2006).

3. See Robert Bernasconi and Sybol Cook, *Race and Racism in Continental Philosophy* (Bloomington, Ind.: Indiana University Press, 2003).

4. See Ludmilla Jordanova, *Sexual Visions: Images of Gender in Science and Medicine Between the Eighteenth and Twentieth Centuries* (Madison, Wis.: The University of Wisconsin Press, 1993).

5. Keith Ansell-Pearson, *Viroid Life: Perspectives on Nietzsche and the Transhuman Condition* (New York and London: Routledge, 1997).

6. See Evelyn Fox Keller, *A Feeling for the Organism: The Life and Work of Barbara McClintock* (New York: Henry Holt and Company, LLC, 1983).

7. Hilary Rose, "Nine Decades, Nine Women, Ten Noble Prizes: Gender Politics on the Apex of Science," in *Women, Science and Technology: A Reading in Feminist Science Studies,* ed. Mary Wyer, Mary Barbercheck, Donna Geisman, Hatice Orun Otzurk, and Marta Wayne (New York and London: Routledge, 2001), 61.

8. Donna Haraway, *Modest_Witness@Second_Millennium. FemaleMan©_Meets_OncoMouse™* (London and New York: Routledge, 1997), 142.

9. Donna Haraway, *The Companion Species Manifesto: Dogs, People and Significant Otherness* (Chicago: Prickly Paradigm Press, 2003).

10. Astrid Henry, *Not My Mother's Sister: Generational Conflict and Third-Wave Feminism* (Bloomington, Ind.: University of Indiana Press, 2004).

11. Braidotti, *Nomadic Subjects.*

12. Paul Gilroy, *Against Race: Imaging Political Culture beyond the Color Line* (Cambridge, Mass.: Harvard University Press, 2000); Edward Glissant, *Poetics of Relation,* trans. Betsy Wing (Ann Arbor: University of Michigan Press, 1997); and Patricia Hill Collins, *Black Feminist Thought: Knowledge, Consciousness and the Politics of Empowerment* (New York and London: Routledge, 1991).

13. Vron Ware, *Beyond the Pale: White Women, Racism and History* (London and New York: Verso, 1992); Gabriele Griffin and Rosi Braidotti, *Thinking Differently: A Reader in European Women's Studies* (London: Zed Books, 2002).

14. Edgar Morin, *Penser l'Europe* (Paris: Gallimard, 1987); Étienne Balibar, *Politics and the Other Scene* (London: Verso, 2002); and Braidotti, *Transpositions.*

15. Deleuze and Guattari, *Anti-Oedipus: Capitalism and Schizophrenia,* trans. Robert Hurley, Mark Seem, and Helen R. Lane (Minneapolis: University of Minnesota Press, 1983).

16. Rose, "Nine Decades, Nine Women, Ten Nobel Prizes."

17. Braidotti, *Transpositions.*

18. Ansell-Pearson, *Viroid Life.*

19. Glissant, *Poetics of Relation.*

9. Numbers and Fractals

1. See Gilles Deleuze, "The Brain Is the Screen" in *The Brain is the Screen: Deleuze and the Philosophy of Cinema,* ed. Flaxman, Gregory (Minneapolis: University of Minnesota Press, 2000).

2. Gilles Deleuze, *Cinema 2. The Time-Image,* trans. Hugh Tomlinson and Robert Galeta (London: The Athlone Press, 1989), 156, emphasis original.

3. Ibid., 161.

4. Ibid., 167.

5. Ibid., 170.

6. Ibid., 172, emphasis original.

7. Ibid., 173.

8. Ibid., 174–75.

9. Ibid., 179–80.

10. See David Rodowick, *Gilles Deleuze's Time Machine* (Durham, N.C.: Duke University Press, 1997); Ronald Bogue, *Deleuze on Cinema* (New York: Routledge, 2003); Patricia Pisters, *The Matrix of Visual Culture: Working with Deleuze in Film Theory* (Stanford, Calif.: Stanford University Press, 2003).

11. Ian Buchanan, "Is a Schizoanalysis of Cinema Possible?" *Journal of Film Studies* 16 (Spring 2006): 124.

12. Patricia Pisters, "Delirum Cinema or Machines of the Invisible?" in *Schizoanalysis and Cinema,* ed. Ian Buchanan and Patricia McCormack (London: Continuum, 2008).

13. See T. Elsaesser, "Mind Game Cinema" in *Puzzle Films: Complex Story Telling in World Cinema,* ed. Warren Buckland (New York: Blackwell, 2008).

14. Anna Powell, *Deleuze: Altered States and Film* (Edinburgh: Edinburgh University Press, 2007).

15. See Patricia Pisters, *The Matrix of Visual Culture: Working with Deleuze in Film Theory* (Stanford: Stanford University Press, 2003); David Martin-Jones, *Deleuze, Cinema and National Identity: Narrative Time in National Contexts* (Edinburgh: Edinburgh University Press, 2006).

16. Gregg Lambert and Gregory Flaxman, "Ten Propositions on the Brain," *Pli* 16 (2005): 114–28.

17. Deleuze, *Cinema 2*, 266.

18. Ibid., 209.

19. Ibid., 265.

20. Ibid., 267.

21. Ibid., 266.

22. Gilles Deleuze and Félix Guattari, *What Is Philosophy?* trans. Hugh Tomlinson and Graham Burchell (New York: Columbia University Press, 1994), 36.

23. See Powell, *Deleuze: Altered States and Film*; Steven Holtzman, *Digital Mosaics: The Aesthetics of Cyberspace* (New York: Touchstone, 1997).

24. G. N. Elston and B. Zietsch, "Fractal Analysis as a Tool for Studying Specialization in Neuronal Structure: The Study of the Evolution of the Primate Cerebral Cortex and Human Intellect," *Advances in Complex Systems* 8 (2005): 217–27.

25. Martin P. Paulus and David L. Braff, "Chaos and Schizophrenia: Does the Method Fit the Madness?" *Biological Psychiatry* 53 (2003): 3–11.

26. Darren Aronofsky, "The Writer/Director of *Pi* Discusses the Limits of Filmmaking and Human Knowledge," interview with Darren Aronofsky by Andrea Chase for ChitChatMagazine.Com, 1998, http://aronofsky.tripod.com/interview21.html (accessed October 2007).

27. Darren Aronofsky, "The Whiz Kid—Darren Aronofsky, writer/director of *Pi*," interview with Anthony Kaufman for Indiewire.com, 1998, http://aronofsky.tripod.com/interview3.html (accessed October 2007).

28. Darren Aronofsky, "An interview with Darren Aronofsky and Sean Gullette of *Pi*," interview with Anthony Kaufman, 1998, http://aronofsky.tripod.com/interview2.html (accessed October 2007).

29. Darren Aronofsky, "Feature Interview with Darren Aronofsky," interview for Netribution by Stephen Applebaum, 2001, http://aronofsky.tripod.com/interview12.html (accessed October 2007).

30. Aronofsky, "The Writer/Director of *Pi* Discusses the Limits of Filmmaking and Human Knowledge," 5.

31. Teun Koetsier and Luc Bergmans, *Mathematics and the Divine: A Historical Study* (Amsterdam: Elsevier Science Publishers, 2005).

32. Deleuze, *Cinema 2*, 277.

33. Darren Aronofsky, "Darren Aronofsky: *The Fountain*," interview for Suicide Girls.Com by Daniel Robert Epstein, 2005, http://suicidegirls.com/interviews/Darren+Aronofsky+-+The+Fountain (accessed October 2007).

34. Deleuze, *Cinema 2*, 293.

35. Ibid., 297.

36. Gilles Deleuze, *Difference and Repetition*, trans. Paul Patton (New York: Columbia University Press, 1994), 71–81, 101–3, 273–76.

37. Martin-Jones, *Deleuze, Cinema and National Identity*, 59–62.

38. Gilles Deleuze, *The Fold*, trans. Tom Conley (Minneapolis: University of Minnesota Press, 1993) 16–17.

39. Deleuze, *The Fold*, 17.

40. Aronofsky, "Darren Aronofsky: *The Fountain*," 2.

41. Peter Parks on Wikipage *The Fountain*, http://en.wikipedia.org/wiki/The_Fountain_(film), 5.

42. Deleuze, *The Fold*, 4.

43. Ibid., 4.

44. Patricia Pisters, "Touched by a Cardboard Sword: Aesthetic Creation and Non-Personal Subjectivity in *Dancer in the Dark* and *Moulin Rouge*" in *Discernments: Deleuzian Aesthetics*, ed. Joost de Bloois, Sjef Houpermans, and Frans-Willem Korsten (Amsterdam and New York: Rodopi, 2004).

45. Claire Colebrook, "Mathematics, Vitalism, Genesis," in *A/V Actual/Virtual Journal*, (September 2007) http://www.eri.mmu.ac.uk/deleuze (audio-visual lecture from The Deleuzian Event conference, September 2007).

10. The Image of Thought and the Sciences of the Brain after *What Is Philosophy?*

1. Gilles Deleuze and Félix Guattari, *What Is Philosophy?* trans. Hugh Tomlinson and Graham Burchell (New York: Columbia University Press, 1994), 37.

2. The concept owes its provenance primarily to physics. I have discussed it in "Chaosmologies: Chaos and Thought in Gilles Deleuze and Félix Guattari's *What Is Philosophy?* with Quantum Field Theory."

3. Although I can only mention this point (which I addressed in "Chaosmologies"), chaos as incomprehensible implies the irreducible nature of chance in our interactions with it, and thus the luck of any possible causality hidden behind the appearance of chance. I shall discuss some relationships between these conceptions of chaos later.

4. Of course, there are other creative endeavors, in principle conforming to Deleuze and Guattari's view of thought. It appears, however, that they see philosophy, art, and science as the primary ways in which thought and the brain itself confront chaos. Insofar as other such endeavors represent thought's confrontation with chaos, they could be seen in terms of philosophy, art, and science, or their combinations.

5. Cf. Georges Bataille, *Lascaux: Or, the Birth of Art*, trans. Austrin Wainhouse (Milan: Skira, 1955); and *The Tears of Eros*, trans. Peter Connor (San Francisco, Calif.: City Lights, 1989).

6. Deleuze and Guattari's analysis proceeds in part via Jakob von Uexkühl's work on the artistic-like "technique of nature" (*What Is Philosophy?* 183–86; 232,

note 25). This last phrase is Kant's, who gives it a more causal and teleological meaning (*The Critique of Judgment*, 409).

7. Marcel Proust, *The Guermantes Way, The Remembrance of Things Past*, 3 vols., trans. Scott K. Moncrieff and Terence Kilmartin (New York: Vintage, 1982), 435.

8. Tianming Yang and Mihael N. Shadlen, "Probabilistic Reasoning by Neurons," *Nature* 447 (2007): 1075–80.

9. (Jakob von) Uexkühl is generally spelled with the Estonian overlong "ll": Uexküll. We have chosen to keep the alternate spelling (Uexkühl) because it appears this way in passages cited from both *What Is Philosophy?* and Berthoz's *The Brain's Sense of Movement*.

10. As noted earlier, these connections extend beyond neuroscience or biological sciences, in particular, to Riemann's mathematical ideas concerning spatiality, which, as I shall explain, are also relevant in the present, neurological context. On the role of biology in Deleuze and Guattari, see John Marks's "Molecular Biology in the Work of Deleuze and Guattari," *Paragraph* 29, no. 2 (July 2006): 81–97, and references there.

11. Gilles Deleuze and Félix Guattari, *A Thousand Plateaus*, trans. Brian Massumi (Minneapolis: University of Minnesota Press, 1987), 51, 257, 314–15.

12. Rodolfo R. Llinás, *The I of the Vortex: From Neurons to Self* (Cambridge, Mass.: The MIT Press, 2002), as well as A. J. Pellionisz and R. Llinás, "Tensorial Approach to the Geometry of Brain Function. Cerabellar Coordination via Metric Tensor," *Neuroscience* 5 (1980): 1761–70; and Alain Berthoz, *The Brain's Sense of Movement* (originally published in French in 1997). Berthoz further develops his ideas in *La Décision* (2003).

13. Rodolfo R. Llinás, *The I of the Vortex: From Neurons to Self* (Cambridge, Mass.: MIT Press, 2002), 15.

14. Alain Berthoz, *The Brain's Sense of Movement*, trans. Giselle Weiss (Cambridge, Mass.: Harvard University Press, 2000), 10.

15. Berthoz's research also addresses our perception/conception of the (straight) line, and in particular, the relationships in our brain between the visual system, the motor system, and the vestibular system. These ideas extend insights by Poincaré and earlier work by A. J. Pellionisz and Llinás ("Tensorial Approach to the Geometry of Brain Function"), in turn linked to mathematical ideas of tensors (mathematical objects that transform vectors in space) (*The Brain's Sense of Movement* 37–38, 48–49). (The symmetry of animals' bodies, vis-à-vis plants, is related to these relationships.) See also a beautiful paper by Bernard Teissier, "Protomathematics, Perception and the Meaning of Mathematical Objects" (in Pierre Grialou, Giuseppe Longo, and Mitsushiro Okada, eds., *Images and Reasoning* [Tokyo: Keio University Press, 2005] 135–46). Teissier argues that it may be possible to claim that the evolution of our perceptual systems has created

an isomorphism between the visual line and the vestibular line, and shows the importance of this isomorphism for our concept of the mathematical line.

16. Berthoz, *The Brain's Sense of Movement*, 242.

17. Ibid., 11, 269n.10; quoted from Merleau-Ponty, *The Visible and the Invisible* (translation follows Berthoz's translator).

18. Berthoz, *The Brain's Sense of Movement*, 22, emphasis added.

19. Berthoz, *The Brain's Sense of Movement*, 175; Maurice Merleau-Ponty, *La Nature: notes-cours du Collège de France* (Paris: Seuil, 1995), 220.

20. Berthoz, *The Brain's Sense of Movement*, 175.

21. Llinás, *The I of the Vortex: From Neurons to Self*, ix, 21, 23.

22. Berthoz, *The Brain's Sense of Movement*, 115.

23. This aspect of the argument becomes more pronounced in Berthoz's work on the Bayesian probabilistic reasoning in cognition, on which I comment later (note 25).

24. Yang and Shadlen, "Probabilistic reasoning by neurons," 1075.

25. This thematic is related to the so-called Bayesian approach to probability, which deals with predictions concerning the outcome of individual events (including those we have never confronted before) on the basis of the available information and, hence, memory, rather than on statistical inferences based on frequencies of repeated events. The argument concerning chance and probability given below is essential Bayesian because, for example, every potential move from one point of a rhizome of events and connection to another is always a unique event. Berthoz's group is involved in the "Bayesian Approach to Cognitive Science" project, and his *La Décision* is linked to this problematic as well.

26. According to Berthoz, however, while "Bergson accepts the idea that the brain is used in the internal simulation of movement, . . . he limits its contribution to this mental structuring and refuses to grant that it plays any role whatsoever in the highest cognitive functions that are [according to Bergson] the exclusive domains of the mind" (*The Brain's Sense of Movement* 262). In other words, Bergson's "*movementalism*," as it may be called, which brings him close to Berthoz and Deleuze alike, is ultimately overtaken by his "mentalism," which drives him apart from both.

27. Deleuze and Guattari, *What Is Philosophy?* 65–66.

28. Motion is also an image or concept that we construct, and there have indeed been arguments from Plato on that this image is an illusion, as against either static or some other reality, for example, as in quantum theory, a possible inconceivable "reality" (chaos as the incomprehensible). The image and concept of motion work both in everyday life (consistently with Berthoz's or Llinás's argument) and in classical physics, whose models are (mathematized) idealizations of our phenomenal perception of everyday life. It does not appear possible to use such models in quantum theory: we do not seem to be able to form a useful "illusion" or "hallucination" or to "dream up" the behavior of quantum objects

or of how they give rise to the famously strange quantum phenomena. I have considered this epistemology in *The Knowable and the Unknowable* (29–108).

29. This ontology may be seen as (philosophically) closer to the topological and geometrical ideas of Riemann, which entail a more heterogeneous, yet interactive, and more transformative view of spatiality, and are often invoked by Deleuze and Guattari (e.g., *A Thousand Plateaus*, 482–85). I have discussed the subject in "Manifolds: On the Concept of Space in Riemann and Deleuze." According to Pellionisz and Llinás our brain may function as a tensor-like system that appropriately transforms vectors of our motion (also *The Brain's Sense of Movement*, 48–49). Tensor is a concept developed in Riemannian geometry, especially in relation to the curvature of space, although it can also apply to Euclidean (flat) spaces. As the preceding analysis suggests, the mediation between the workings of our neural machinery and the construction of any phenomenal ontology is immense, and it may never be fully available to a scientific investigation, which can only establish certain correlations and causalities between them. This mediation is essentially at stake in the current discussions and debates in the cognitive and neurosciences. It is not possible to consider the spectrum of positions taken in this regard, and the literature (scientific, philosophical, or popular) on the subject is by now immense. Both Llinás's and Berthoz's arguments are, however, major contributions to these debates.

30. Cf. Gilles Deleuze and Félix Guattari, *Anti-Oedipus: Capitalism and Schizophrenia*, trans. Mark Seem, Helen R. Lane, and Robert Hurley (Minneapolis: University of Minnesota, 1985), 9–16.

31. Sigmund Freud, *Beyond the Pleasure Principle*, trans. James Strachey (New York: W. W. Norton, 1961), 54.

11. Deleuze, Guattari, and Neuroscience

1. Gilles Deleuze and Félix Guattari, *What Is Philosophy?* trans. Hugh Tomlinson and Graham Burchell (New York: Columbia University Press, 1994), 216.

2. Jonathan Moreno, *Mind Wars: Brain Research and National Defense* (Washington, D.C.: Dana Press, 2006); Joseph Dumit, *Picturing Personhood: Brain Scans and Biomedical Identity* (Princeton, N.J.: Princeton University Press, 2003); Brian Massumi, "Fear (The Spectrum Said)" *Positions* 13, no. 1 (2005): 31–48; Elizabeth A. Wilson, "The Work of Antidepressants: Preliminary Notes on How to Build an Alliance between Feminism and Psychopharmacology," *Biosocieties* 1 (2006): 125–31; Steven Rose, *The Future of the Brain* (Oxford: Oxford University Press 2006).

3. Mark Amerika, *Meta/Data* (Cambridge, Mass.: The MIT Press, 2007), 53.

4. Ibid., 32–33.

5. For two examples, see Watson discussing Edelman (Sean Watson, "The Neurobiology of Sorcery: Deleuze and Guattari's Brain," *Body and Society* 4 [1998]:

23–45; Gerald Edelman, *Bright Air, Brilliant Fire* [New York: Basic Books, 1993]); or Guattari on his taking up of Varela's work (Félix Guattari, "On Machines," *Journal of Philosophy and the Visual Arts* VI [1995]: 8–12).

6. Gilles Deleuze, *Negotiations*, trans. Martin Joughin (New York: Columbia University Press, 1995), 176.

7. Brian Massumi notes that this materialism might be found in "the quantum waves crossing the brain's synaptic fissures" in *A User's Guide to* Capitalism and Schizophrenia (Cambridge, Mass.: The MIT Press, 1992), 157.

8. Brian Massumi, *A User's Guide to* Capitalism and Schizophrenia, 33.

9. Gilles Deleuze, *The Logic of Sense*, trans. Mark Lester and Charles Stivale (New York: Columbia University Press, 1990), 240.

10. Deleuze, *Negotiations*, 60.

11. Ibid., 149.

12. Gilles Deleuze, *Difference and Repetition*, trans. Paul Patton (New York: Columbia University Press, 1994), 147.

13. Ibid., 143.

14. Deleuze, *Negotiations*, 176.

15. As in the title of Dumit's book, *Picturing Personhood*.

16. Deleuze, *Negotiations*, 149, emphasis added.

17. Ibid., 61.

18. Watson, "The Neurobiology of Sorcery," 38, emphasis added.

19. Deleuze, *Negotiations*, 61.

20. Charles Stivale, "N–Z Summary of *L'Abécédaire de Gilles Deleuze*," 2000, http://www.langlab.wayne.edu/CStivale/D-G/ABC3.html, accessed October 2, 2004, my emphases.

21. Steven Rose, *The Conscious Brain* (New York: Vintage Books, 1976).

22. Deleuze and Guattari, *What Is Philosophy?* 122.

23. For a neurological account of neural networks, see Donald Hebb, *The Organization of Behavior* (New York: Wiley, 1949); for an account of their sensory-cultural form, see Friedrich Hayek, *The Sensory Order* (Chicago: University of Chicago Press, 1952).

24. Deleuze, *Difference and Repetition*, 73.

25. V. S. Ramachandran and Sandra Blakeslee, *Phantoms in the Brain: Probing the Mysteries of the Human Mind* (New York: Harper Perennial, 1999).

26. Charles T. Wolfe, "De-ontologizing the Brain: From Fictional Self to the Social Brain," *Ctheory* 30, 2007, http://www.ctheory.net/articles.aspx?id=572, accessed October 30, 2007.

27. V. S. Ramachandran, "Mirror Neurons and Imitation Learning as the Driving Force Behind 'the Great Leap Forward' in Human Evolution," *Edge*, 2000, http://www.edge.org/3rd_culture/ramachandran/ramachandran_index.html, accessed September 14, 2007.

28. Paolo Virno, *Multitude between Innovation and Negation* (New York: Semiotext(e), 2008).

29. Brian Massumi, *Parables for the Virtual* (Durham, N.C.: Duke University Press, 2002), 187.

30. Paul Bains, "Subjectless Subjectivities" in *A Shock to Thought,* ed. Brian Massumi (New York: Routledge, 2002), 101–16; also see Eric Alliez, *The Signature of the World* (London: Continuum, 2005), 62–63. The immanence of formation as individuation is also taken from Gilbert Simondon. Guattari also takes the concept of autopoiesis from Franciso Varela ("On Machines"), although he qualifies it with an "allopoiesis" that respects a simultaneous, if still immanent, relation to exteriority in the process of "active formation."

31. Bains, "Subjectless Subjectivities," 101–16.

32. Ibid., 103.

33. Cf. Jean-Pierre Dupuy, *The Mechanization of the Mind: On the Origins of Cognitive Science* (Princeton, N.J.: Princeton University Press, 2000).

34. Cf. Norbert Wiener, *Cybernetics or Control and Communication in the Animal and the Machine* (Cambridge, Mass./New York: MIT/Wiley, 1949); and Norbert Wiener, "The History and Prehistory of Cybernetics," *Kybernetes* 27 (1998): 29–37. Gary Genosko has pointed this out, in personal conversation.

35. Cf. Maurizio Lazzarato, "The Machine," transversal, 2006, http://eipcp. net/transversal/1106/lazzarato/en, accessed October 10, 2007; and Paolo Virno, *A Grammar of the Multitude: For an Analysis of Contemporary Forms of Life* (New York: Semiotext(e), 2004).

36. Samuel Weber, *Targets of Opportunity: On the Militarization of Thinking* (New York: Fordham University Press, 2005).

37. Gilles Deleuze, *Foucault* (London: Athlone Press, 2001), 131.

38. John Marks, "Molecular Biology in the Work of Deleuze and Guattari," *Paragraph* 29 (2006): 95.

39. Deleuze, *Foucault*, 131.

40. Stivale, "N–Z Summary of *L'Abécédaire de Gilles Deleuze.*"

41. Deleuze, *Difference and Repetition*, 119.

42. Ibid., 119.

43. Stivale, "N–Z Summary of *L'Abécédaire de Gilles Deleuze.*"

44. Watson, "The Neurobiology of Sorcery," 38.

45. See Alliez on Guattari and Whitehead, in *The Signature of the World* (London: Continuum, 2005), 55.

46. Félix Guattari, *Cartographies Schizoanalytiques* (Paris: Éditions Galilée, 1989), 82ff.

47. Félix Guattari, *Chaosmosis: An Ethico-Aesthetic Paradigm* (Sydney: Power, 1995), 113.

48. Guattari, *Cartographies Schizoanalytiques*, 82, quoting Alfred North Whitehead, *Process and Reality* (New York: Free Press, 1979), 224.

49. For example, it is speculated that the mirror neuron creates a "unity of feeling" across bodies and brains, and this might create new potentials for future action. This does not, however, mean that this action will resemble that which has been "mirrored."

50. Deleuze, *Difference and Repetition*, 118.

51. Gilles Deleuze, *The Fold: Leibniz and the Baroque* (Minneapolis: University of Minnesota Press, 1992), 86.

52. Deleuze and Guattari, *What Is Philosophy?* 209.

53. Ibid., 209.

54. Deleuze, *The Fold*, 86.

55. William James, *The Principles of Psychology* (New York: H. Holt and Company, 1893), 609.

56. Deleuze, *The Fold*, 88. A simple example might be all the movements—and microperceptions—involved in a moving car. We will tend to notice a *shift* in velocity, which is the differential of distance covered over time, precisely because it changes, when we accelerate, and for a moment the new speed is "notable or remarkable." An example that Deleuze gives here is that of the "sound of the sea: at least two waves must be minutely perceived as nascent and *heterogeneous* enough to become part of a relation that can allow the perception of a third, one that "excels" over the others and comes to consciousness" (88).

57. Ibid., 89.

58. Ibid., 90. See also Tim Van Gelder, 1999, for a cognitive philosopher's view of these differential conditions, and Varela, 1999, for a neurologist's perspective.

59. Ibid., 91.

60. Deleuze, *Difference and Repetition*, 92.

61. Deleuze, *The Fold*, 92.

62. John Mullarkey, "Deleuze and Materialism," *South Atlantic Quarterly* 96 (1997): 446.

63. Gilles Deleuze, *Cinema 2: The Time-Image* (Minneapolis: University of Minnesota Press, 1989), 158.

64. Ibid., 204.

65. John Sutton, "Porous Memory and the Cognitive Life of Things," in *Prefiguring Cyberculture: An Intellectual History*, ed. Darren Tofts, Annemarie Jonson, and Alessio Cavallaro (Cambridge, Mass.: MIT Press, 2002), 130–41.

66. Deleuze, *Cinema 2: The Time-Image*, 266.

67. Ibid., 265.

68. John Rajchman, *The Deleuze Connections* (Cambridge, Mass.: MIT Press, 2000), 136.

69. Deleuze, *Negotiations*, 60.

70. Rajchman, *The Deleuze Connections,* 135. Here we might include Ramachandran's "10 universal laws of art," for example, as useful as they might be. See V. S. Ramachandran, "The Artful Brain," *Reith Lectures 2003* BBC, 2003, http://www.bbc.co.uk/radio4/reith2003/lecture3.shtml (accessed August 31, 2007).

71. Rajchman, *The Deleuze Connections,* 136.

72. Gary Genosko, "Félix Guattari: Towards a Transdisciplinary Metamethodology," *Angelaki* 8 (2003): 129–40.

73. Félix Guattari, *Soft Subversions,* trans. Sylvère Lotringer (New York: Semiotext(e), 1996), 268–69.

74. All "models" are metamodels in a sense, as there is no outside to immanence.

75. Félix Guattari, *L'Inconscient Machinique* (Paris: Recherches, 1979), 7.

76. Ibid., 8. Translated by Taylor Atkins at http://fractalontology.wordpress.com/2007/10/01/introduction-to-felix-guattaris-machinic-unconscious.

77. Guattari, *Soft Subversions,* 194.

78. Ibid., 196.

79. Cf. Sutton, "Porous Memory and the Cognitive Life of Things."

80. Guattari, *Soft Subversions,* 196–97.

81. Guattari in Charles Stivale, *The Two-Fold Thought of Deleuze and Guattari: Intersections and Animations* (New York: Guilford Press, 1998), 224.

82. Guattari, *Cartographies Schizoanalytiques,* 200.

83. Simply put, not fitting into systems of signification yet still operating as a kind of intensity. It does not follow a logic of representation, but rather creates new territories—it "opens fields of the possible that aren't at all in bi-univocal relationship with the description presented" (Guattari in Stivale, 1998: 221).

84. Ibid., 199, emphasis added.

85. Deleuze and Guattari, *What Is Philosophy?* 213.

86. Ibid., 216.

87. Guattari, *Cartographies Schizoanalytiques,* 218. A fractal could be seen as an infinitely complex breakdown of dimension across levels (most famously represented in a snowflake but also commonly seen in fractal diagrams).

88. Massumi, *A User's Guide to Capitalism and Schizophrenia,* 150.

89. Guattari, *Cartographies Schizoanalytiques,* 218–19.

90. Gilles Deleuze and Félix Guattari, *A Thousand Plateaus,* trans. Brian Massumi (Minneapolis: University of Minnesota Press, 1987), 15.

91. Ibid., 16.

92. Deleuze and Guattari, *What Is Philosophy?* 216.

93. Deleuze, *Logic of Sense,* 355.

94. Of field in self-survey at infinite speed, of the shifts in the relations between probabilistic series of microevents.

95. Alliez, *The Signature of the World,* 63.

96. Deleuze and Guattari, *What Is Philosophy?* 210.

97. Ibid., 210.

98. Ibid., 208.

99. Cf. Rose, *The Future of the Brain*, 11ff.

100. Deleuze and Guattari, *A Thousand Plateaus*, 64.

101. Alliez, *The Signature of the World*, 62.

102. Arkady Plotnisky, "Chaosmologies: Quantum Field Theory, Chaos and Thought in Deleuze and Guattari's *What Is Philosophy?*" *Paragraph* 29 (2006): 53.

103. Ibid., 54.

104. Deleuze, *Difference and Repetition*, 137.

105. Deleuze, "What Children Say," in *Essays Critical and Clinical*, trans. Daniel W. Smith and Michael A. Greco (Minneapolis: University of Minnesota Press, 1997), 63–64.

12. Mammalian Mathematicians

1. As W. V. Quine proposes in "Epistemology Naturalized," *Ontological Relativity and Other Essays* (New York: Columbia University Press, 1969), 69–90.

2. For a more complete description of phase space as it applies specifically to Deleuze, consult DeLanda's *Intensive Science and Virtual Philosophy,* which does an admirable job of explaining these ideas to the nonspecialist. Steven Strogatz's *Non-linear Dynamics and Chaos* is a very well-written technical introduction to the mathematics involved and its scientific application.

3. Manuel DeLanda, *Intensive Science and Virtual Philosophy* (New York: Continuum, 2002), 66. Technically, though they still constitute topological singularities of the phase space, these dynamic attractors are not singular points within the space, but instead singular trajectories known as limit cycles or even chaotic attractors.

4. DeLanda, *Intensive Science,* 20.

5. Ibid., 32.

6. Ibid., 67.

7. Ibid., 12–13.

8. Gilles Deleuze and Félix Guattari, *What Is Philosophy?* (New York: Columbia University Press, 1994), 36.

9. Gilles Deleuze and Félix Guattari, *A Thousand Plateaus* (Minneapolis: University of Minnesota Press, 1987), 249. "It amounts to the same thing to say that each multiplicity is already composed of heterogeneous terms in symbiosis, and that a multiplicity is continually transforming itself into a string of other multiplicities, according to its thresholds and doors."

10. P. W. Anderson, "More Is Different," *Science* 177 (1972), 393.

11. Sensitive to Deleuze's love of Spinoza, perhaps it would be best to refer to it as the triadic structure of *expression*. See, for example, Gilles Deleuze, *Expressionism in Philosophy* (New York: Zone Books, 1992), 334.

12. Deleuze and Guattari, *What Is Philosophy?* 19.

13. While I only refer to Nozick's discussion of consciousness, *Invariances* as a whole centers around the proposition that what is truly *objective* in this world are things that remain as invariants of transformations, an idea closely related to the topological forms of the virtual we are discussing here.

14. Robert Nozick, *Invariances*, (Cambridge, Mass.: Belnap Press of Harvard University, 2001), 177, 190. Nozick calls these "evolutionary choice points." Similar observations are made by Peter Godfrey-Smith, "Environmental Complexity and the Evolution of Consciousness," in *The Evolution of Intelligence*, ed. R. Sternberg and J. Kaufmann (Mahwah, N.J.: Lawrence Erlbaum Associates, Inc., 2002), 233–49. Of course, in the context of an essay about Deleuze and consciousness, it is impossible to overlook the similarity of these views to Henri Bergson's conception of organisms as "zones of indescernibility" or "centers of indetermination" in *Matter and Memory*.

15. Nozick, *Invariances*, 191.

16. Ibid., 197. Nozick also gives several interesting examples of the way common knowledge changes a game situation.

17. Ibid., 201.

18. In his discussion of phenomenology, Nozick emphasizes that *only* the topological features of a phenomenal experience can be important: Nozick, *Invariances*, 208. DeLanda (*Intensive Science*, 147) makes the same point, arguing that the value of models lies in their sharing singularities with the system modeled.

19. Nozick, *Invariances*, 208.

20. Ibid., 202–3.

21. Steven Strogatz's *Sync* is an excellent popular introduction to some of this research.

22. It is impossible to claim that there is anything approaching consensus, either scientific or philosophical, in the study of consciousness. Another recent scientific account of the "neural correlates of consciousness" is summarized by Christof Koch and Francis Crick in "A Framework for Consciousness," *Nature Neuroscience 6* (2003): 119–26 and more fully explored in Koch's recent book *The Quest for Consciousness: A Neurobiological Approach* (Englewood, Colo.: Roberts and Company, 2004). Crick and Koch explicitly state that they consider Edelman and Tononi's ideas very similar to their own. Due to limitations of space, I will only discuss Edelman and Tononi.

23. Gerald M. Edelman and Giulio Tononi, *A Universe of Consciousness* (New York: Basic Books, 2000), 18.

24. Edelman has for years used the term "reentry" to describe reciprocal connections between brain regions. While I understand the reasoning behind this choice of terminology (see Edelman and Tononi, *A Universe of Consciousness,* 85), I will continue to use "feedback." I find that "reentry" obfuscates a real continuity between feedback as self-organizing synchronization, feedback as homeostatic self-maintenance (autopoesis), and feedback as (potentially) creative self-destruction (the body without organs and the war machine).

25. Edelman and Tononi, *A Universe of Consciousness,* 120.

26. Ibid., 29–30.

27. Ibid., 131.

28. Ibid., 130.

29. Gilles Deleuze, *The Logic of Sense* (New York: Continuum, 2004), 25.

30. Edelman and Tononi, *A Universe of Consciousness,* 17. Tononi puts this very precisely in "An Information Integration Theory of Consciousness," *BMC Neuroscience* 5, no. 42 (2004) http://www.biomedcentral.com/1471-2202/5/42: "Consciousness is characterized here as a disposition of *potentiality*—in this case as the potential differentiation of a system's responses to all possible perturbations, yet it is undeniably *actual.*"

31. Edelman and Tononi, *A Universe of Consciousness,* 138.

32. Ibid., 109.

33. DeLanda, *Intensive Science,* 55. "This property of stimulus-independence must be added to the mechanism-independence I discussed before as part of what defines the 'signature' of the virtual."

34. Ibid., 35. See also the discussion of "explanatory problems" in DeLanda, *Intensive Science,* 129.

35. Deleuze and Guattari, *What Is Philosophy?* 210.

36. Edelman and Tononi, *A Universe of Consciousness,* 83–85.

37. Ibid., 88.

38. Ibid., 88–89.

39. Ibid., 46.

40. Ibid., 97–98.

41. Deleuze and Guattari, *A Thousand Plateaus,* 263.

42. Ibid., 294.

43. Ibid., 256–57.

44. Ibid., 274.

45. Edelman and Tononi, *A Universe of Consciousness,* 210. "Examples of such structures range from the immune system to reflexes and, finally, to consciousness. In this view, there are as many memory systems as there are systems capable of autocorrelation with their previous states over time."

46. Deleuze and Guattari, *What Is Philosophy?* 164. The "self-positing" aspect of the territory is obviously another example.

47. The suggestion that *everything* is emergent, that all things are processes, and "it's turtles all the way down," often strikes people as falling somewhere between the mystical and the simply confused. In light of this, it is notable that precisely this idea forms the basis of Nobel Prize–winning physicist Robert Laughlin's *A Different Universe: Reinventing Physics from the Bottom Down* (2005), where he argues that organization *precedes* law, rather than vice versa.

48. Deleuze and Guattari, *A Thousand Plateaus,* 309.

49. Edelman and Tononi, *A Universe of Consciousness,* 109.

50. Ibid., 101.

51. Ibid., 107.

52. There are some interesting connections to be made in this context between Deleuze's and Daniel Dennett's work. The memory systems we are discussing here are constituted by semi-stable compositions between an organism and its environment. They are relations of feedback between perception and action that, if looked at from the point of view of the composite system, appear as relatively stable attractors populating an environment like vortices. As Edelman and Tononi suggest, there are any number of these memory systems at work in the environment, and even in the brain of a single animal, ranging from those operating at very rapid timescales like consciousness, to the glacially slow accumulation and dispersion of genetic singularities in DNA, all of which interact with one another by sharing the same earth (Edelman and Tononi, *A Universe of Consciousness,* 98–99). This collection of mechanisms bears analogy to the spectrum of information transmission systems Dennett outlines in his *Freedom Evolves* (see especially pp. 172–73) and the "tower of generate and test" he describes in *Darwin's Dangerous Idea*. These span the distance between genes and memes with goat trails and grooming behavior lodged somewhere in between. Dennett, however, imagines these memory systems as always transmitting discreet chunks of information that correspond to the reproduction of some "thing," and not primarily as feedback loops where there is some stabilized *flow* of information that may or may not take the form of discreet reproduction. Naturally, Deleuze's idea that there are singularities expressing these stabilities, self-consistent processes that tend to grow in intensity (being feedback loops), would relegate reproduction to special case of intensive growth, and the replicating organism to, "that which life sets against itself in order to limit itself" (Deleuze and Guattari, *A Thousand Plateaus,* 503). Finally, in this context, the expressive character of the territory and the refrain fit neatly between the singularities of DNA and those of human culture, making it possible to see the territory as a sort of proto-meme.

53. Deleuze and Guattari, *A Thousand Plateaus,* 509.

54. Ibid., 280.

55. DeLanda, *Intensive Science,* 111. The discussion of the "mechanisms of immanence" begins on p. 103.

56. It may objected that these two planes are not in fact that same thing, which is perhaps true. Perhaps *the* plane of immanence is woven from many planes of consistency, in the same way Deleuze and Guattari repeatedly describe *the* plane of consistency of *A Thousand Plateaus* as being woven from the intersection of many bodies without organs, or many strata (see Deleuze and Guattari, *A Thousand Plateaus*, 165–66). In addition, we have, "*The* plane of immanence is *interleaved*" (Deleuze and Guattari, *What Is Philosophy?* 50).

57. Deleuze and Guattari, *What Is Philosophy?* 38.

58. Ibid., 39.

59. Ibid., 152–53.

60. Ibid., 213. They say, "Not every organism has a brain, and not all life is organic, but everywhere there are forces that constitute microbrains, or an inorganic life of things. We can dispense with Fechner's or Conan Doyle's splendid hypothesis of a nervous system of the earth only because the force of contracting or of preserving, that is to say, of feeling appears only as a global brain in relation to the elements contracted directly and to the mode of contraction which differ depending on the domain and constitute precisely irreducible varieties."

61. Deleuze and Guattari, *A Thousand Plateaus*, 40.

Afterword

1. Interview with Arnaud Villani, reprinted in *La guêpe et l'orchidée: Essai sur Gilles Deleuze* (Paris: Éditions de Belin, 1999), 129–31. See commentary in the Introduction to the present volume.

2. Karl Sims's Web site can be found at http://www.karlsims.com (accessed September 26, 2008).

Contributors

MANOLA ANTONIOLI teaches at the Collège international de philosophie in Paris, where she teaches courses on Deleuze, Guattari, and Blanchot. Her books include *Géophilosophie de Deleuze et Guattari, Deleuze et l'histoire de la philosophie,* and *L'Écriture de Maurice Blanchot.*

CLARK BAILEY has degrees in physics and philosophy and studied neuroscience at Rockefeller University. He now works for a hedge fund.

ROSI BRAIDOTTI is Distinguished Professor in the Humanities in a Globalised World at Utrecht University, The Netherlands. Her work has been translated into nineteen languages, and her books include *Transpositions: On Nomadic Ethics; Metamorphoses: Towards a Materialist Theory of Becoming;* and *Nomadic Subjects: Embodiment and Sexual Difference in Contemporary Feminist Theory.*

MANUEL DELANDA is the author of *War in the Age of Intelligent Machines; A Thousand Years of Nonlinear History; Intensive Science and Virtual Philosophy; A New Philosophy of Society;* and *Philosophy, Emergence, and Simulation.* He teaches philosophy seminars in the architecture department at the University of Pennsylvania.

ADEN EVENS is assistant professor of English at Dartmouth College, where he teaches new media studies, poststructuralist theory, and postmodern literature. He is author of *Sound Ideas: Music, Machines, and Experience* (Minnesota, 2005).

GREGORY FLAXMAN is an assistant professor in the Department of English and Comparative Literature at the University of North Carolina. The editor of *The Brain Is the Screen: Deleuze and the Philosophy of Cinema* (Minnesota, 2000), he is also the author of *Gilles Deleuze and the Fabulation of Philosophy: Powers of the False, Volume I* (forthcoming from Minnesota). Currently, he is writing a monograph on *Chinatown* and the representation of history.

PETER GAFFNEY is visiting assistant professor at Haverford College and the Curtis Institute of Music, where he teaches film studies, philosophy, and literature. He is working on a book with Thomas Kelso that confronts the problematic relationship between artistic production and ontological realism. He is codirector of the Red Light Cabaret and Musical Theater Company in Philadelphia.

THOMAS KELSO is an independent scholar based in Turkey. His fields of specialization are Italian and French literature, film studies, rhetoric, and translation. He is author of *Italian Dreams: Neorealism and Deleuze* and has published several translations from Italian, including *Who Loves You Like This* by Edith Bruck and "Checchina's Virtue" by Matilde Serao.

JULIE-FRANÇOISE KRUIDENIER is visiting assistant professor in the French Department at Hamilton College.

ANDREW MURPHIE is at the School of English, Media, and Performing Arts at the University of New South Wales, Sydney. He is editor of the *Fibreculture Journal* and coauthor of *Culture and Technology*. He works on media philosophy and sociology, the social impact of models and practices of mind, electronic arts and design, open-access publishing, cultural theory, and the digital humanities.

PATRICIA PISTERS is professor of film studies in the Department of Media Studies at the University of Amsterdam. She has published on film-philosophical questions on the nature of perception, the ontology of the image, politics of contemporary screen culture, and the idea of the "brain as screen" in connection to neuroscience. Her books include *The Matrix of Visual Culture: Working with Deleuze in Film Theory*; *Shooting the Family: Transnational Media and Intercultural Values*; and *Mind the Screen*.

ARKADY PLOTNITSKY is professor of English, director of the Theory and Cultural Studies Program, and codirector of the Philosophy and Literature Program at Purdue University. He has published several books on British and European Romanticism, modernism, continental philosophy, philosophy of physics and mathematics, and the relationships among literature, philosophy, mathematics, and science. His most recent books are *Epistemology and Probability: Bohr, Heisenberg, Schrödinger, and the Nature of Quantum-Theoretical Thinking*; *Reading Bohr: Physics and Philosophy*; and a coedited collection of essays, *Idealism without Absolute: Philosophy and Romantic Culture*.

STEVEN SHAVIRO is DeRoy Professor of English at Wayne State University. His books include *The Cinematic Body* (Minnesota, 1993); *Doom Patrols: A Theoretical Fiction about Postmodernism*; *Connected, or What It Means to Live in the Network Society* (Minnesota, 2003); and *Without Criteria: Kant, Whitehead, Deleuze, and Aesthetics*. He blogs at The Pinocchio Theory (http://www.shaviro.com/Blog).

ARNAUD VILLANI teaches at the Première Supérieure au Lycée Masséna de Nice. In addition to numerous articles, poems, and translations of poems, he has published three books, *Précis de philosophie nue; La guêpe et l'orchidée: Essai sur Gilles Deleuze, Kafka: L'ouverture de l'existant, Petites méditations métaphysiques sur la vie et la mort,* and *Court Traité du rien.* He is coeditor of *Le Vocabulaire de Gilles Deleuze* (a special edition of *Revue Noésis*) and is the translator of philosopher and mathematician Alfred North Whitehead.

Index

Abécédaire de Gilles Deleuze, L' (Bou-
     tang), 280
abstraction, 147–48
abstract machine, 19, 25, 117, 122, 234
actual, 1–12, 21, 27–28, 35–36, 47, 49,
     54–55, 57–58, 60–61, 64, 72–73, 76,
     79–80, 82, 85, 89, 92–93, 100, 105,
     107–8, 113–15, 124, 128–30, 142,
     145, 177, 302–3, 306–7, 310, 316,
     318, 321–23, 326–27; and digital,
     148–53, 155, 157–58, 162, 166–67,
     172; and negation, 149, 357n15;
     and physical properties, 325–26;
     and plane of immanence, 322; as
     solution, 4–6, 153; and stratifica-
     tion, 63. *See also* actual and vir-
     tual; actualization
actual and virtual, 1–2, 21, 49, 63–64,
     82, 85, 100, 124, 149, 177, 235, 303,
     322. *See also* actual; actualization;
     virtual
actualization, 1–6, 10–12, 14–15, 18–22,
     24–28, 30–31, 34–37, 40–41, 46–47,
     57–61, 72, 124, 130, 172–75, 177, 204,
     214, 217–18, 222, 226–28, 273, 302,
     304–5, 307, 337–38n50, 343n127; in
     Bergson, 106–7; and brain, 314, 316,
     320–21, 323, 327, 332; counteractu-
     alization, 130, 138; and digital, 151,
     155, 158; and divergent actualities,

110; and event, 109; and evolution,
     138, 142; and life, 28; and scientific
     image, 89–94, 104–17; subject and
     object as result of, 45–46; and
     thought, 90, 91
affect, 191, 325
agriculture, 32
Aion, 221
Alliez, Eric, 15, 89–90
Amerika, VJ Mark, 277
anthropocentrism, 137–38, 215–17
Anti-Oedipus (Deleuze and
     Guattari), 39
Antonioli, Manola, 36
appetition, 3, 43, 142–44; of self-pres-
     ervation, 145
applied science, 33
architecture, 36–37, 169–88; architec-
     tural machines, 185; and duration,
     184; Event-Architecture, 178–79;
     Japanese, 184; monumental, 169,
     170, 184; Morphosis, 178; poly-
     phonic, 186–87; and urban space,
     175–76; virtual, 188
Aristotle, 125
Aronofsky, Darren, 53, 244, 251
art, 85, 331–32; Japanese, 184–85; and
     monument, 169–70; and philoso-
     phy, 191–93; and science, 191–93
Artaud, Antonin, 41, 232

93–95, 112–13. *See also* biology;
physics; scientific discourse; scientific method
scientific discourse: and technologies of control, 216; and teleology, 133–35, 137
scientific image (model), 13, 91, 93–101, 107; and difference, 98; and duration, 101; and historical adaptation, 102–3, 108; and representation, 25, 101
scientific method: and empirical observation, 88, 104, 111; knowledge produced by, 87, 89, 100; primary and secondary qualities (Locke) in, 12, 96; repeatability and reproducibility in, 87–90, 350n42; and scientific observer, 88. *See also* science
secondary qualities (Locke), 88
Selleri, Franco, 105
sensation, 18–19, 38–41, 45, 191, 195, 197. *See also* plane of sensation
sense, 193, 197. *See also* common sense; good sense
series, 204–9; and chaos, 205; numerical vs. ordinal, 326–27
Serres, Michel, 6
Shaviro, Steven, 28–31, 38
Signature of the World, The (Alliez), 15
Simondon, Gilbert, 172
Sims, Karl, 331–32
singularities, 3, 37, 88, 129, 325; and digital object, 152–53; and fold, 173; and phase space, 303; singularization, 32
Sloterdijk, Peter, 180–81
Sophists, 71
soul, 172
space: and art, 119–22, 170; Cartesian, 119, 124; and digital, 154; Euclidean (geometric), 326; and

instrumental knowledge, 99; and intensity, 119–21, 126, 127, 128; nonmetric, 326; phase, 302–4; and physical processes, 105; schematization of relations in, 95; and scientific observer, 88; and subjectivity, 183; topological, 193, 304–5; urban, 175–76; virtual, 121–22, 124, 154, 305
Spellbound (Hitchcock), 231
Spheres (Sloterdijk), 180
Spinoza, Baruch, 16, 215, 217, 222
Stengers, Isabelle, 17, 208
strata. *See* stratification
stratification, 31, 60–64, 115, 181, 341n4; and identity, 63
subjectivity, 38, 44; and body, 183; and brain, 194, 195, 278; and common sense, 200; as contingent unity, 173–74; fractured, 242–43, 251; Hume's concept of, 43; knowing, 211–12; larval, 19; machinic, 14; as monad, 172; and nomadic thought, 211–15; psychological determination of, 57; rational vs. empirical, 138–39; as residue of recording process, 57; rhizomatic, 212; and *Sainte famille*, 56; and series, 206; subjectivation, 37, 40, 46, 50, 54; and superject, 194–95; and three syntheses, 45, 283
superject, 194–95
superposition: as crystal-image, 100–101; not possible with intensities, 95–96; of quantitative multiplicities, 93–95, 107; and quantum theory, 106–7; of science and metaphysics, 100; and stratigraphic time, 100
surplus code, 31, 32
symbiosis, 141, 374n9